To Victor E.

in appreciation for his contribution to

Volume 4

March 1975

Klaus Florey

Analytical Profiles of Drug Substances

Volume 4

Academic Press Rapid Manuscript Reproduction

Analytical Profiles of Drug Substances

Volume 4

Edited by

Klaus Florey

The Squibb Institute for Medical Research
New Brunswick, New Jersey

Contributing Editors

Norman W. Atwater
Glenn A. Brewer, Jr.
Jack P. Comer
Salvatore A. Fusari
Boen T. Kho
Gerald J. Papariello
Frederick Tishler

Compiled under the auspices of the
Pharmaceutical Analysis and Control Section
Academy of Pharmaceutical Sciences

Academic Press New York • San Francisco • London 1975
A Subsidiary of Harcourt Brace Jovanovich, Publishers

ACADEMIC PRESS, INC.
111 Fifth Avenue, New York, New York 10003

United Kingdom Edition published by
ACADEMIC PRESS, INC. (LONDON) LTD.
24/28 Oval Road, London NW1

Library of Congress Cataloging in Publication Data
Main entry under title:

Analytical profiles of drug substances.

Includes bibliographical references.
Compiled under the auspices of the Pharmaceutical Analysis and Control Section, Academy of Pharmaceutical Sciences.
1. Drugs–Collected works. 2. Chemistry, Medical and pharmaceutical–Collected works. I. Florey, Klaus, ed. II. Brewer, Glenn A. III. Academy of Pharmaceutical Sciences. Pharmaceutical Analysis and Control Section. [DNLM: 1. Drugs–Analysis–Yearbooks. QV740 AA1 A55]
RM300.A56 615'.1 70-187259
ISBN 0-12-260804-6 (v.4)

PRINTED IN THE UNITED STATES OF AMERICA

CONTENTS

AFFILIATIONS OF EDITORS, CONTRIBUTORS, AND REVIEWERS

N. W. Atwater, Searle and Company, Chicago, Illinois

J. I. Bodin, Carter-Wallace Inc., Cranbury, New Jersey

G. A. Brewer, Jr., The Squibb Institute for Medical Research, New Brunswick, New Jersey

L. Chafetz, Warner-Lambert Research Institute, Morris Plains, New Jersey

E. M. Cohen, Merck, Sharp and Dohme, West Point, Pennsylvania

J. L. Cohen, School of Pharmacy, University of Southern California, Los Angeles, California

J. P. Comer, Eli Lilly and Company, Indianapolis, Indiana

L. F. Cullen, Wyeth Laboratories, Philadelphia, Pennsylvania

R. D. Daley, Ayerst Laboratories, Rouses Point, New York

N. J. DeAngelis, Wyeth Laboratories, Philadelphia, Pennsylvania

F. Eng, Parke, Davis and Company, Detroit, Michigan

K. Florey, The Squibb Institute for Medical Research, New Brunswick, New Jersey

S. A. Fusari, Parke, Davis and Company, Detroit, Michigan

D. E. Guttman, School of Pharmacy, University of Kentucky, Lexington, Kentucky

W. W. Holl, Smith, Kline and French Laboratories, Philadelphia, Pennsylvania

E. H. Jensen, The Upjohn Company, Kalamazoo, Michigan

H. Kadin, The Squibb Institute for Medical Research, New Brunswick, New Jersey

B. T. Kho, Ayerst Laboratories, Rouses Point, New York

J. Kress, Carter-Wallace Inc., Cranbury, New Jersey

E. P. K. Lau, Searle and Company, Chicago, Illinois

H. H. Lerner, The Squibb Institute for Medical Research, New Brunswick, New Jersey

L. P. Marrelli, Eli Lilly and Company, Indianapolis, Indiana

D. L. Mays, Bristol Laboratories, Syracuse, New York

A. F. Michaelis, Sandoz Pharmaceuticals, East Hanover, New Jersey

J. E. Moody, USV Pharmaceutical Corporation, Tuckahoe, New York

E. S. Moyer, Ortho Research Foundation, Raritan, New Jersey

N. G. Nash, Ayerst Laboratories, Rouses Point, New York

G. J. Papariello, Wyeth Laboratories, Philadelphia, Pennsylvania

V. E. Papendick, Abbott Laboratories, North Chicago, Illinois

D. M. Patel, William H. Rorer Inc., Fort Washington, Pennsylvania

R. B. Poet, The Squibb Institute for Medical Research, New Brunswick, New Jersey

A. Post, Smith, Kline and French Laboratories, Philadelphia, Pennsylvania

J. A. Raihle, Abbott Laboratories, North Chicago, Illinois

N. H. Reavey-Cantwell, William H. Rorer Inc., Fort Washington, Pennsylvania

P. Reisberg, Carter-Wallace Inc., Cranbury, New Jersey

R. E. Schirmer, Eli Lilly and Company, Indianapolis, Indiana

A. P. Schroff, Ortho Research Foundation, Raritan, New Jersey

B. Z. Senkowski, Hoffmann-LaRoche, Inc., Nutley, New Jersey

L. A. Silvieri, Wyeth Laboratories, Philadelphia, Pennsylvania

A. M. Sopirak, Wyeth Laboratories, Philadelphia, Pennsylvania

J. L. Sutter, Searle and Company, Chicago, Illinois

D. Szulczewski, Parke, Davis and Company, Detroit, Michigan

F. Tishler, Ciba-Geigy, Summit, New Jersey

A. J. Visalli, William H. Rorer Inc., Fort Washington, Pennsylvania

J. J. Zalipsky, William H. Rorer Inc., Fort Washington, Pennsylvania

A. F. Zappala, Smith, Kline and French Laboratories, Philadelphia, Pennsylvania

PREFACE

Although the official compendia list tests and limits for drug substances related to identity, purity, and strength, they normally do not provide other physical or chemical data, nor do they list methods of synthesis or pathways of physical or biological degradation and metabolism. For drug substances important enough to be accorded monographs in the official compendia such supplemental information should also be made readily available. To this end the Pharmaceutical Analysis and Control Section, Academy of Pharmaceutical Sciences, has undertaken a cooperative venture to compile and publish Analytical Profiles of Drug Substances in a series of volumes of which this is the fourth.

The concept of Analytical Profiles is taking hold not only for compendial drugs but, increasingly, in the industrial research laboratories. Analytical Profiles are being prepared and periodically updated to provide physico-chemical and analytical information of new drug substances during the consecutive stages of research and development. Hopefully then, in the not too distant future, the publication of an Analytical Profile will require a minimum of effort whenever a new drug substance is selected for compendial status.

The cooperative spirit of our contributors had made this venture possible. All those who have found the profiles useful are earnestly requested to contribute a monograph of their own. The editors stand ready to receive such contributions.

This volume of Analytical Profiles is dedicated to the memory of David E. Guttman, an enthusiastic member of the Editorial Board until his tragic and untimely death in 1974.

Klaus Florey

CEFAZOLIN

Alfred F. Zappala, Walter W. Holl, and Alex Post

Contents

1. Description

1.1 Name, Formula, Molecular Weight

Cefazolin is 3-[[(5-Methyl-1,3,4-thiadiazol-2-yl)thio]methyl]8-oxo-7-[2-(1H-tetrazol-1-yl)acetomido]-5-thia-1-azabicyclo[4.2.0]oct-2-ene-2-carboxylic acid. It also exists as the sodium salt. Parenteral products are known as Ancef and Kefzol.

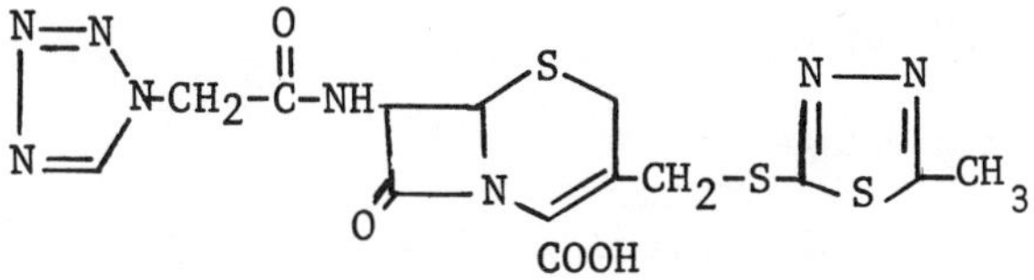

$C_{14}H_{14}N_8O_4S_3$ (Na, -H) Mol. wt. 454.512 (acid), 476.495 (salt)

1.2 Appearance, Color, Odor

White to slightly off white, odorless.

2. Physical Properties

2.1 Infrared Spectrum

The infrared spectrum of cefazolin is presented in Figure 1. The spectrum taken was that of a mineral oil dispersion of the standard using a Perkin-Elmer 457A Grating IR Spectrophotometer. A list of the assignments made for some of the characteristic bands is given in Table I (1).

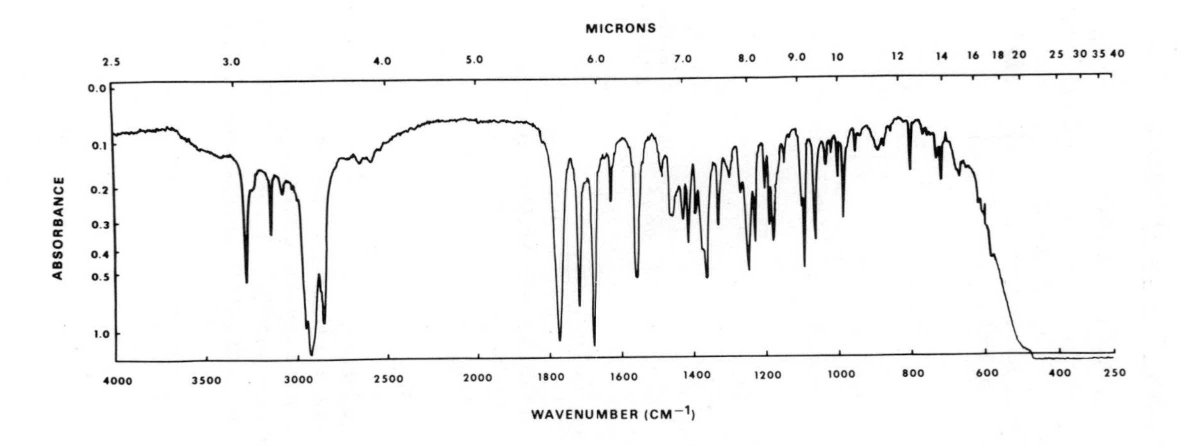

<u>Figure 1:</u> <u>Infrared Spectrum of Cefazolin</u>
Reference Standard, mineral oil dispersion.
Instrument: Perkin-Elmer 457A

Table I - IR Spectral Assignments for Cefazolin

Frequency (cm^{-1})	Characteristic of	
3280	-NH-	
3140	-N=N-	tetrazole ring
3075	-C=N-	tetrazole ring
2620	-OH, bonded, -COOH	
2580	-OH, bonded, -COOH	
1770	>C=O	lactam
1715	>C=O	acid
1670	>C=O	amide I
1555	>C=O	amide II

2.2 Nuclear Magnetic Resonance Spectrum

The 60 MHz NMR spectrum of cefazolin presented in Figure 2 was obtained in trifluoroacetic acid at a concentration of about 100 mg/ml and tetramethylsilane as internal standard. The spectral assignments are listed in Table II (1).

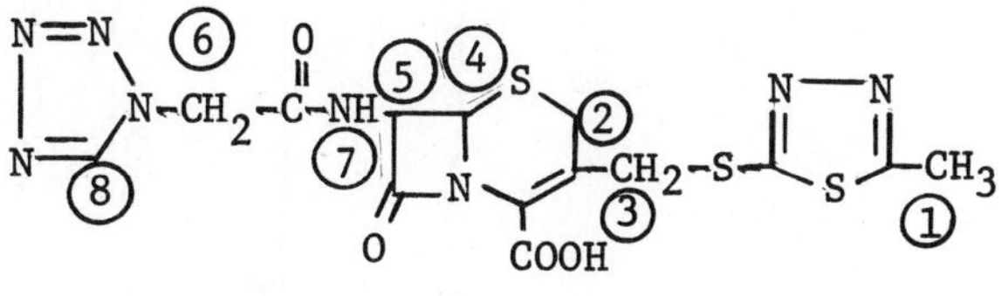

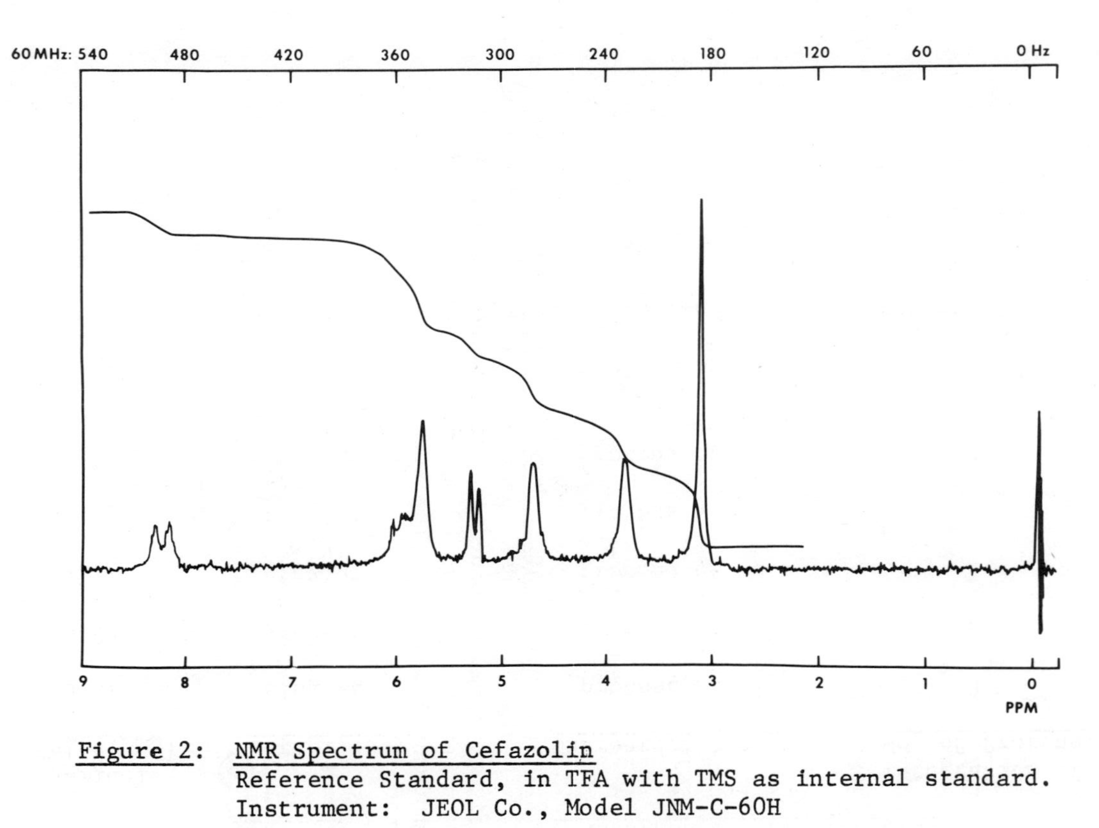

Figure 2: NMR Spectrum of Cefazolin
Reference Standard, in TFA with TMS as internal standard.
Instrument: JEOL Co., Model JNM-C-60H

Table II - NMR Spectral Assignments for Cefazolin

Chemical Shift (ppm)	Multiplicity	Characteristic of	Integration of No. of Protons
3.11	singlet	protons at 1	3
3.85	singlet	protons at 2	2
4.71	singlet	protons at 3	2
5.40	doublet	protons at 4	1
5.75	overlapping singlet & doublet	protons at 5 and 6	3
8.21	doublet	protons at 7	1

proton at 8 is beyond 9 ppm; however, it is masked by the solvent

2.3 Ultraviolet Spectrum

The ultraviolet absorption spectrum of cefazolin in 0.1M $NaHCO_3$ is shown in Figure 3. When scanned between 350 and 220 nm, cefazolin exhibits a single band with an absorption maximum at 270 - 272 nm (ε = 13,100).

2.4 Optical Rotation

The specific rotation of a 5% solution of cefazolin in 0.1M $NaHCO_3$ when measured at 25°C in a 1 decimeter tube is -17° $\pm$ 7°.

2.5 Melting Range

Cefazolin starts to decompose at about 190°C under USP conditions for Class I substances (2).

2.6 Differential Thermal Analysis

A differential thermal analysis was performed on cefazolin and the thermogram is presented in Figure 4. The typical melting endotherm is absent and only the decomposition exotherm at about 205°C is present.

2.7 Solubility

The approximate solubilities obtained for cefazolin at room temperature (25°C $\pm$ 1°C) are listed in Table III.

Table III - Approximate Solubilities of Cefazolin

Solvent	mg Cefazolin/ml
acetone	5.7
acetone:water (4:1 v:v)	21.2
chloroform	0.02
95% ethanol	1.1
ethyl acetate	0.24
isobutylacetate	0.05
isopropyl alcohol	0.21
methanol	1.7
methylene chloride	0.02
methylisobutylketone	0.25
sodium chloride, half-saturated	0.83
sodium chloride, saturated	0.44
water	1.1

2.71 pH Solubility Profile

The pH solubility profile of cefazolin is presented in Figure 5.

2.8 pKa

The pKa is 2.15 determined spectrophotometrically. A pKa of 2.05 determined titrimetrically has been reported (3).

2.9 Crystal Properties

Several crystal forms of sodium cefazolin have been reported (4). The pathways of conversion of one form to another are shown below.

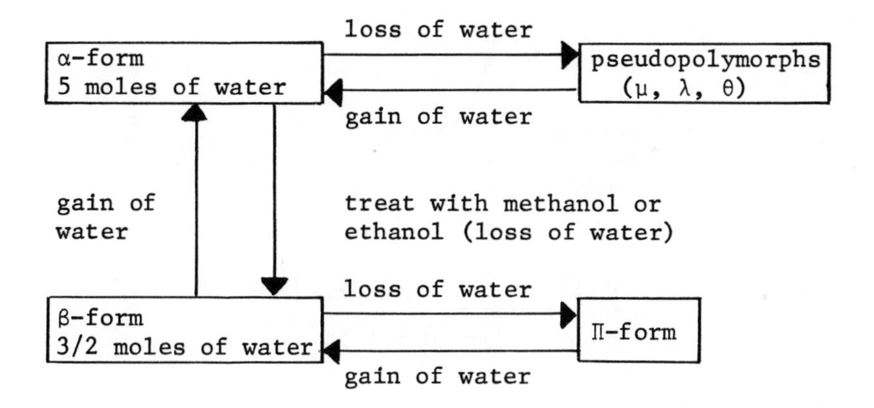

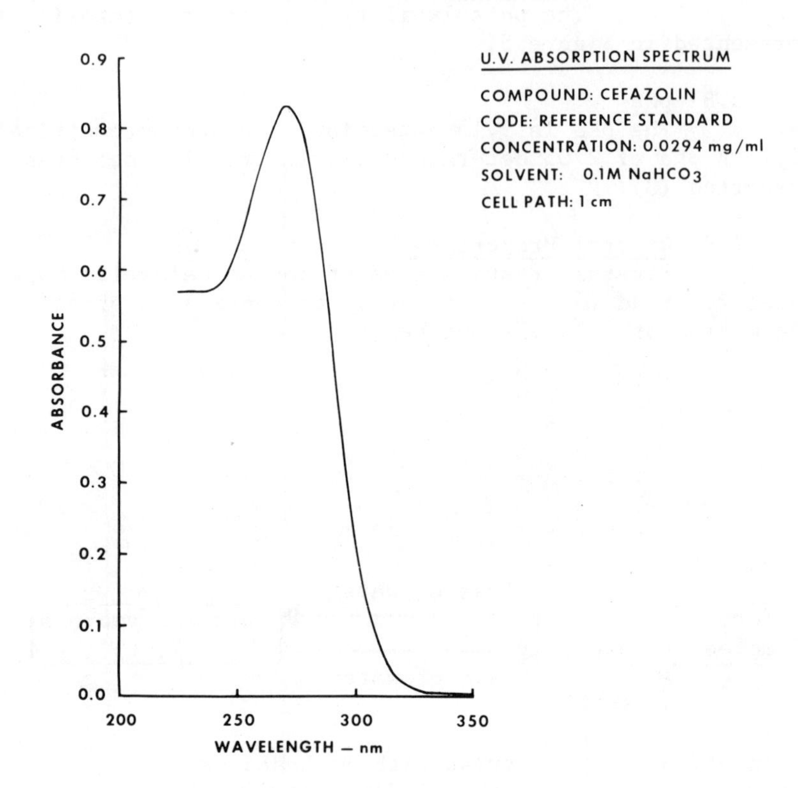

Figure 3: UV Absorption Spectrum of Cefazolin Reference Standard
Instrument: Cary Model 14

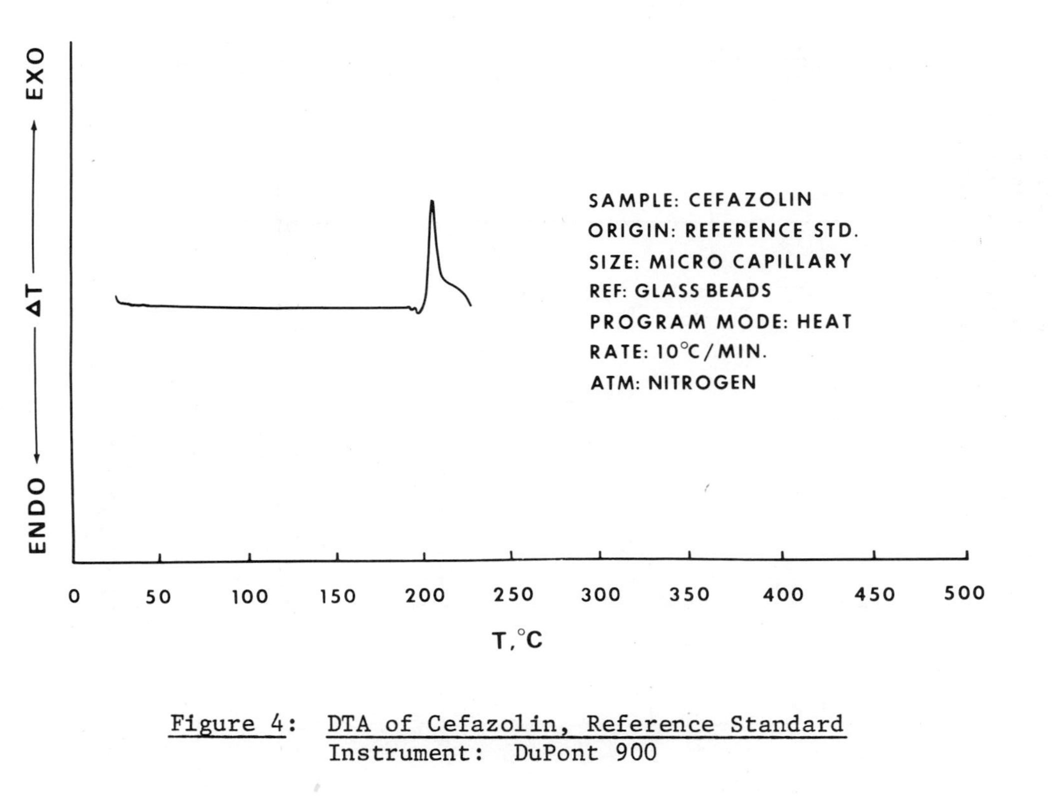

<u>Figure 4</u>: <u>DTA of Cefazolin, Reference Standard</u>
Instrument: DuPont 900

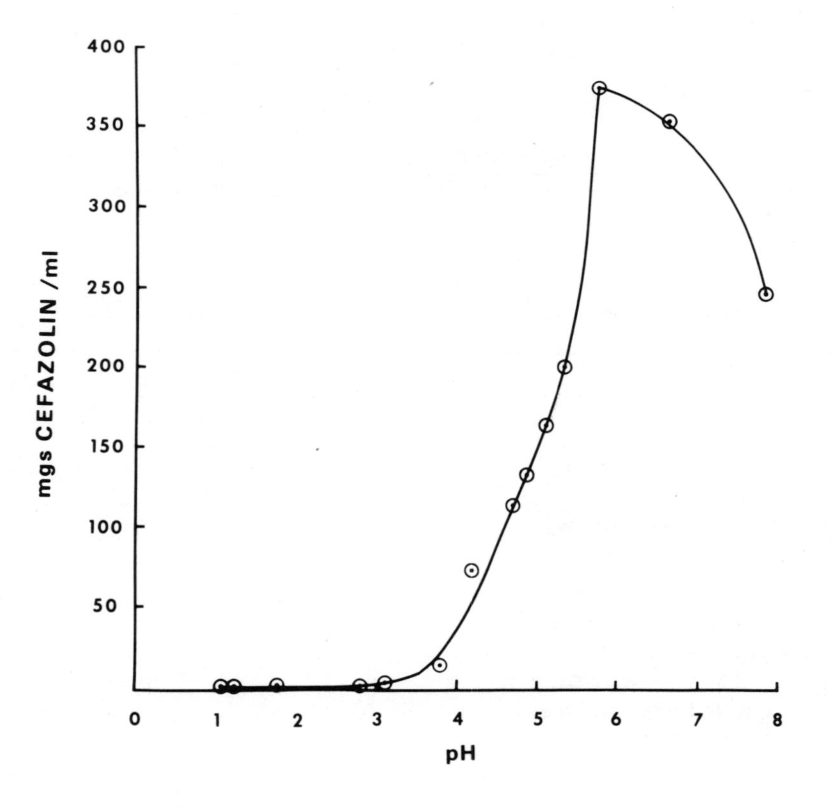

Figure 5: pH Solubility Profile of Cefazolin

3. Synthesis

Cefazolin may be prepared by the following route: (5)

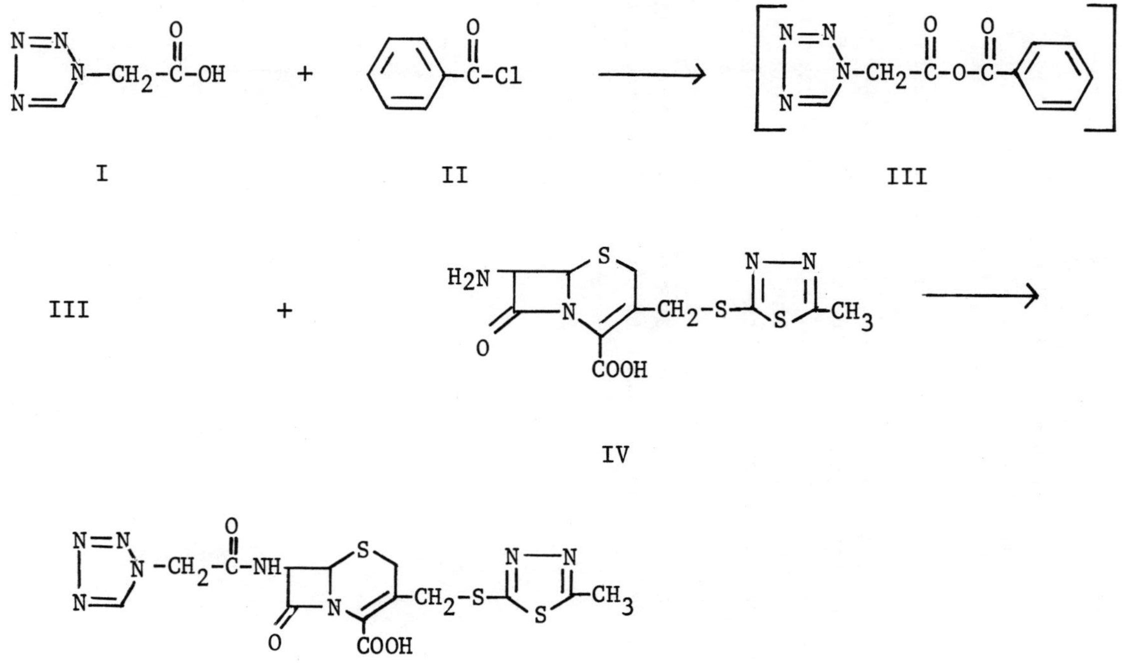

N=N, N= , N–CH_2–C(=O)–OH + (phenyl)–C(=O)–Cl ⟶ [N=N, N= , N–CH_2–C(=O)–O–C(=O)–(phenyl)]

I II III

III + H_2N, S, N, O, COOH, –CH_2–S–, N—N, S, –CH_3 ⟶

IV

N=N, N= , N –CH_2–C(=O)–NH, S, N, O, COOH, –CH_2–S–, N—N, S, –CH_3

V Cefazolin

4. Stability

The stability of Ancef vials after reconstitution with water for injection (a), bacteriostatics water (b), and normal saline solution (c) was determined (6). The resulting solutions were analyzed by the UV hydroxylamine method (Section 6.5) initially and again after storage at 5°C for 96 hours. No degradation was observed.

The physical and chemical stability of cefazolin sodium in several intravenous fluids (Table IV) was evaluated (7). The solutions were assayed by the UV hydroxylamine method (Section 6.5) initially and periodically up to 72 hours. Inspection indicated no apparent physical change. The chemical stability of the solutions can be seen from the data in Table IV.

Lyophilized cefazolin sodium is stable for at least two years in the dry state at room temperature. A comparison of the chemical (8) and microbiological (9) assay methods is illustrated in Table V.

Table IV - Stability of Cefazolin Sodium in Selected Intravenous Solutions*

	RT				5°C	
Time in hrs.	0	24	48	72	24	72
10% dextrose in H_2O	100	101	98.2	95.8	104	99.1
5% dextrose in H_2O	100	98.3	98.2	94.5	99.4	99.7
5% dextrose with lactated Ringer	100	98.3	96.6	94.4	98.6	97.8
5% dextrose in Ringer's injection	100	101	100	96.3	100	93.2
5% dextrose in 0.9% NaCl	100	100	99.4	98.5	102	97.4
5% dextrose in 0.45% NaCl	100	100	100	96.0	98.0	100
10% dextrose in 0.9% NaCl	100	100	97.7	97.7	99.4	99.7
0.9% NaCl	100	101	99.7	98.9	105	100
lactated Ringer injection	100	103	98.9	96.9	101	99.4
Ringer injection	100	102	101	96.9	100	99.7

*Expressed as a percent of initial concentration

Table V - Comparison of Chemical and Microbiological Methods*

Months	1	3	6	9	12	18	24
Chem. assay	100	101	97.3	99.0	96.0	99.0	100
Micro. assay	99.4	103	97.5	98.2	97.0	101	98.0

*Expressed as a percent of initial concentration.

5. Drug Metabolic Products

Studies conducted thus far indicate there is very little biotransformation of parenterally administered cefazolin sodium in the body. Between 94 - 98% is excreted in the urine unchanged. Only trace amounts of metabolites have been seen but not identified (10).

6. Methods of Analysis

6.1 Elemental Analysis

The results from an elemental analysis of cefazolin reference standard are listed in Table VI .

Table VI

Element	% Theory	% Found
C	37.00	36.75
H	3.10	3.29
N	24.65	24.44
S	21.16	21.31
O (by difference)	14.09	14.23

6.2 Non-Aqueous Titration of Cefazolin

Reagents

(1) Dimethylsulfoxide (DMSO)

(2) Tetrabutylammonium hydroxide (TBAH) - 0.05N in 9:1 benzene:methanol; this solution is standardized against benzoic acid (National Bureau of Standards).

Procedure

An accurately weighed sample of about 200 mgs of cefazolin is dissolved in 70 - 80 ml of DMSO. The resulting solution is titrated potentiometrically with standard 0.05N TBAH using a glass-calomel electrode pair or combination electrode. Each milliliter of 0.05N TBAH is equivalent to 0.02273 g of cefazolin.

6.3 Non-Aqueous Titration of Cefazolin Sodium

An accurately weighed sample of about 200 mgs of cefazolin sodium is dissolved in 70 - 80 ml of dimethylsulfoxide (DMSO). The resulting solution is titrated potentiometrically with standard 0.05N acetous perchloric acid using a glass-calomel electrode pair or combination electrode. Each milliliter of 0.05N perchloric acid is equivalent to 0.02383 g of cefazolin sodium.

6.4 Thin-Layer Chromatography

The following thin-layer method may be used for the qualitative purity evaluation of cefazolin and its sodium salt.

About 50 and 100 micrograms of cefazolin, dissolved in a 4:1 mixture of acetone:water, are spotted two cm from the edge of a Silica Gel GF plate. The plate is placed in a suitable chromatographic chamber lined with filter paper saturated with the developing solvent (ethylacetate:acetone:acetic acid:water, 5:2:1:1), and allowed to equilibrate for ten minutes. The solvent is then allowed to rise to a line drawn across the plate 10 cm from the origin. The plate is removed from the chamber and allowed to air dry in a fume hood until solvent vapors are no longer detectable. The developed chromatogram may be visualized under ultraviolet light (254 and 365 nm), exposure to iodine vapors, and spraying with potassium permanganate. Cefazolin has an Rf value of about 0.45.

6.5 Spectrophotometric-UV Hydroxylamine Method

Reagents

(1) 0.5 Molar sodium bicarbonate

(2) Acetate Buffer - Equal volumes of 0.1M acetic acid and 0.1M sodium acetate are mixed together and the resulting solution is adjusted to pH 4.0.

(3) Alkaline Sodium Acetate - 86.5 g of sodium hydroxide and 10.3 g of sodium acetate are dissolved in sufficient water to make 1000 ml.

(4) Hydroxylamine Solution - One volume of 5M hydroxylamine hydrochloride is mixed with two volumes of alkaline sodium acetate and three volumes of water.

Procedure

An accurately weighed sample of approximately 50 mg is dissolved in 5.0 ml of 0.5M sodium bicarbonate and diluted to 1000 ml with water. Five ml aliquots of this solution are transferred to each of two 100 ml volumetric flasks. To one flask is added 5.0 ml of hydroxylamine solution. The flask is swirled and allowed to stand for 45 minutes, after which both solutions are diluted to 100 ml with acetate buffer. Two aliquots of a standard solution of cefazolin are treated in the same manner. The ultraviolet absorption spectrum of the unreacted solution is recorded versus that of the reacted solution in the reference cell from 350 to 240 nm in 1 cm cells. The calculation of the purity of the sample is accomplished by comparison of the absorbance difference between 270 nm and 340 nm for the sample to that of the standard. This procedure has also been automated (8).

6.6 High Pressure Liquid Chromatographic Procedure

Reagents

(1) Mobile Phase: 0.02 Molar monobasic sodium phosphate adjusted to pH 6.2 $\pm$ 0.1 with 1N sodium hydroxide.

(2) Standard Solution: Approximately 20 mg of reference standard is accurately weighed into a 50 ml volumetric flask and dissolved in and diluted to volume with 0.05M sodium bicarbonate.

Instrumental Conditions
Column Packing: Strong anion exchange resin
Column Diameter: 2.1 mm I.D.
Column Length: 1 m
Column Temperature: Ambient
Column Pressure: 1000 psig
Flow Rate: 0.5 ml per minute
Detector: U.V., 254 nm

Procedure

An accurately weighed sample of approximately 20 mg is dissolved in and diluted to 50 ml with 0.05M sodium bicarbonate. Duplicate 20 µl aliquots of the standard and sample solutions are injected. The retention time for cefazolin is approximately 20 minutes. The calculation of the purity of the sample is accomplished by comparison of the average peak height of the sample solution to that of the standard solution. Instrumental conditions may require modifications with other HPLC units and different lots of column packing.

6.7 "Federal Register" 38:31505-31509, 1973

Additional methods listed in the Federal Register are (1) microbiological agar diffusion assay, (2) iodometric assay, and (3) hydroxylamine colorimetric assay.

7. References

1. R. J. Warren, SmithKline Corp., Personal Communication

2. U.S.P. XVIII, p. 935

3. Fujisawa Pharmaceutical Co., Ltd., Osaka, Japan, Personal Communication

4. J. of Antibiotics, Vol. XXIII, No. 3, 131 - 136 (1970)

5. J. Hill and H. Winicov, SmithKline Corp., Personal Communication

6. L. Ravin and E. Rattie, SmithKline Corp., Personal Communication

7. L. Ravin and E. Rattie, SmithKline Corp., Personal Communication

8. W. W. Holl et al, SmithKline Corp., Personal Communication, to be published

9. Antimicrobial Agents & Chemotherapy, Vol. 5, No. 3, 223 - 227 (1974)

10. J. of Antibiotics, Vol. XXV, No. 2, 86 - 93 (1972)

CEPHALEXIN

Louis P. Marrelli

TABLE OF CONTENTS

1. Description

1.1 Name: Cephalexin

Chemical Abstracts designates cephalexin as 5-thia-1-azabicyclo [4.2.0] oct-2-ene-2-carboxylic acid, 7-(2-amino-2-phenyl-acetamido)-3-methyl-8-oxo.

Cephalexin monohydrate is also known as 5-thia-1-azabicyclo [4.2.0] oct-2-ene-2-carboxylic acid, 7-[(amino-phenylacetyl) amino]-3-methyl-8-oxo, monohydrate[1], 7-(D-2-amino-2-phenylacetamido)-3-methyl-3-cephem-4-carboxylic acid monohydrate[2], and 7-(D-α-aminophenylacetamido)-3-methyl-3-cephem-4-carboxylic acid monohydrate[3].

1.2 Formula and Molecular Weight

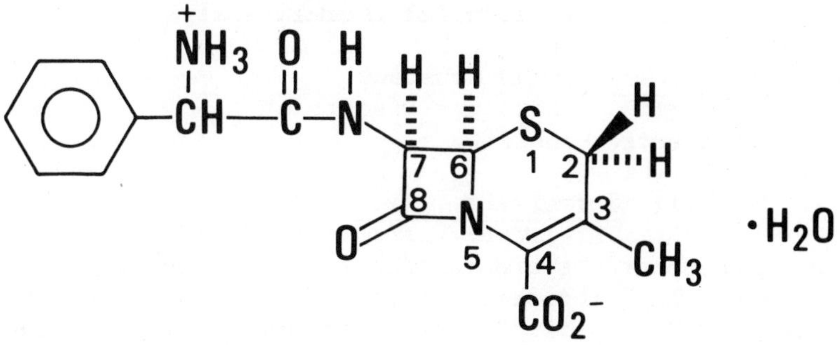

$C_{16}H_{17}N_3O_4S \cdot H_2O$ 365.41

1.3 Isomers

The synthesis of the L epimer of cephalexin has been reported[4]. The D-isomer exhibits considerably more biological activity than the L-isomer. Penicillins derived from D-α-amino acids also show more biological activity than their L-epimers[5,6].

1.4 Hydrates

Pfeiffer et. al.[7] provided x-ray powder diffraction data for the monohydrate and dihydrate of cephlexin. Cephalexin was found to crystallize from aqueous solutions at room temperature as the dihydrate but converted to the monohydrate when the relative humidity was below 70%. Refer to Section 2.21.

1.5 Appearance

Cephalexin is a white to cream-colored crystalline powder, having a characteristic odor.

2. Physical Properties

2.1 Spectra

2.11 Infrared Spectrum

The infrared spectrum of cephalexin monohydrate recorded as a potassium bromide disc is presented in Figure 1. Interpretation of the spectrum is given in Table I[8]. Changes in the β-lactam carbonyl stretching region (1760 cm^{-1}) can indicate opening of the β-lactam ring. Morin[9] and coworkers have shown a relationship between the β-lactam carbonyl stretching frequency and biological activity. The importance of this stretching frequency has been discussed in a recent review[10].

2.12 Nuclear Magnetic Resonance Spectrum

Figure 2 shows the proton magnetic resonance spectrum of cephalexin monohydrate. The solvent used was deuterium oxide containing a small amount of trifluoroacetic acid to enhance solubility. 3-(Trimethylsilyl)-propanesulfonic acid, sodium salt was added as the internal reference. The spectrum was recorded on a Varian T60-A instrument. The assignment of the spectrum is shown in Table II[8]. A most characteristic region of the NMR spectrum is that originating from the two β-lactam ring protons, H(6) and H(7).

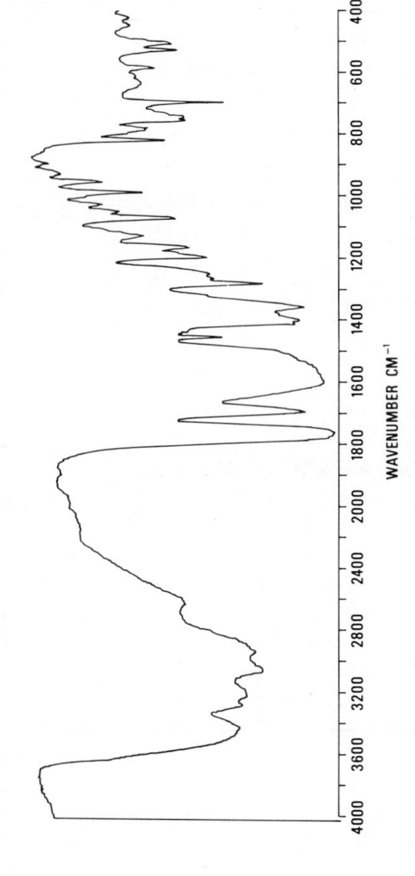

FIGURE 1. Infrared spectrum of cephalexin monohydrate (potassium bromide disc).

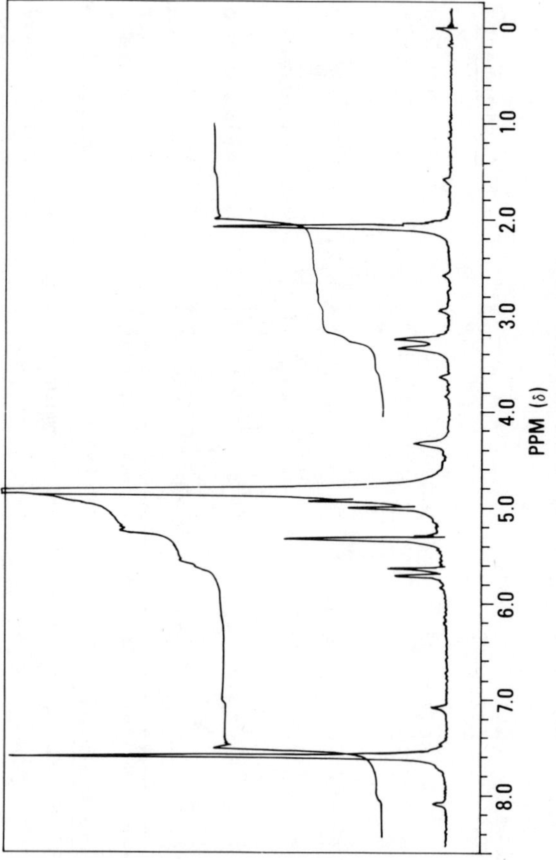

FIGURE 2. NMR spectrum of cephalexin monohydrate (D_2O + trifluoroacetic acid).

TABLE I

Infrared Spectrum of Cephalexin Monohydrate

Wavelength (cm^{-1})	Assignment
3500 - 3000 (series of broad bands)	OH from H_2O and amide NH stretch
2600 (broad)	$\overset{+}{N}H_3$
1760	β-lactam c = o stretch
1690	Amide c = o stretch
(1600 [very broad] (1400	$\overset{O}{\overset{\parallel}{C}}$-O - (carboxylate stretching)
1550 (unresolved)	Amide II band
820 - 690	Mainly skelectal vibrations including out-of-plane aromatic hydrogen bending, characteristic of mono-substituted aromatic ring

TABLE II

Proton Magnetic Resonance Spectrum

Peak Assignments

p.p.m. (δ)	Relative Intensity	Multiplicity	Assignment
2.07	3	singlet	CH_3 (3)
3.30	2	quartet (AB)	CH_2 (2)
4.85	-	singlet	HOD (solvent)
4.97	1	doublet ($J=4H_z$)	H (6)
5.34	1	singlet	H (benzyl)
5.67	1	doublet ($J=4H_z$)	H (7)
7.60	5	singlet	C_6H_5

2.13 Ultraviolet Absorbance

An aqueous solution of cephalexin exhibits a UV absorption maximum at 262 nm (Figure 3). The $E_{1\%}^{1\ cm}$ reported for cephalexin (on an anhydrous basis) was 236[11]. The ultraviolet absorbance of cephalexin as well as other cephalosporins has been attributed to the O = C-N-C = C- chromophore of the ring[12]. Chou[13] utilized the UV absorption at 262 nm to determine the cephalexin content of solution fractions isolated from human urine.

2.2 Crystal Properties

2.21 X-Ray Powder Diffraction

Cephalexin was found to occur in several solvated crystal forms, and often in widely varying mixtures of these forms[14]. Some of the solvated crystal forms prepared were the dihydrate, monohydrate, diacetonitrilate, formamidate, methanolate, and acetonitrile hydrate. X-ray powder diffraction data for cephalexin monohydrate is presented in Table III.

2.22 Differential Thermal Analysis

Differential thermal analysis[15] of cephalexin monohydrate was conducted on a DuPont Model 950 thermal analyzer in a nitrogen atmosphere. A heating rate of 20°C per minute was utilized. An endotherm was noted at 123°C indicating the loss of water, and an exotherm of 203°C indicating decomposition.

2.3 Solubility

The solubility of cephalexin monohydrate in the following solvents has been reported[16]:

Solvent	Mg. Cephalexin Monohydrate Per Ml. Solvent, 25°C.
Water	13.5
Methanol	3.4
N-octanol	0.03
Chloroform	<0.01
Ether	<0.01

Table IV relates the solubility of cephalexin monohydrate in water as a function of pH.

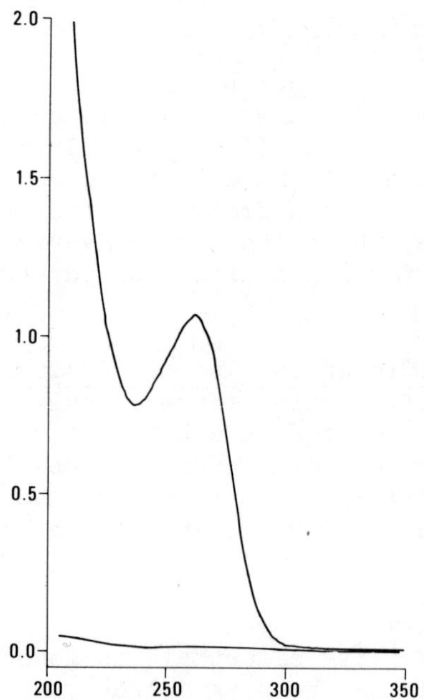

FIGURE 3. Ultraviolet absorption spectrum of cephalexin monohydrate, 49 mcg. per ml. in H_2O.

TABLE III

X-Ray Powder Diffraction Pattern of Cephalexin Monohydrate
Radiation; Cu/Ni. Norelco De Bye-Scherrer Camera

Cephalexin Monohydrate

d	I/I_1
15.15	0.40
11.85	1.00
11.00	0.30
9.36	0.20
8.55	0.50
7.86	0.50
6.89	0.20
5.98	0.40
5.39	1.00
4.97	0.50
4.76	0.40
4.57	0.40
4.39	0.60
4.22	0.60
4.00	0.70
3.86	0.70
3.60	0.80
3.46	0.30
3.24	0.60
3.10	0.60
2.98	0.40
2.90	0.60
2.81	0.40
2.73	0.20
2.68	0.40
2.63	0.10
2.47	0.30
2.41	0.15
2.31	0.30
2.25	0.30
2.12	0.10
2.09	0.05
2.01	0.02
1.93	0.05
1.87	0.05
1.85	0.05
1.82	0.10
1.72	0.05
1.66	0.02
1.62	0.02

TABLE IV

Solubility - pH Profile of Cephalexin in Water (37°C)

pH	Cephalexin Monohydrate mg./ml.	pH	Cephalexin Monohydrate mg./ml.
2.3	120	6.0	13
2.5	95	6.5	16
3.0	31	7.0	24
3.5	20	7.5	40
4.0	16	8.0	75
5.0	12	8.2	100

2.4 Dissociation Constant (pK_a)

The following dissociation constants were reported:

Solvent	pK_a Carboxyl	pK_a Amino	Reference
66% DMF	5.2	7.3	4
66% DMF	5.3	7.3	17
H_2O		7.1	18

2.5 Optical Rotatory Dispersion

Optical rotation has been used as an auxiliary method for the quantitation of cephalexin[3]. The specific rotation $[\alpha]_D$ reported for cephalexin, calculated on an anhydrous basis, was +153° (C = 1.0 in H_2O)[19].

3.0 Cephalexin Stability

The stability of cephalexin in solution is dependent on pH, degrading rapidly in basic media and remaining stable under mild acidic conditions. No loss in cephalexin activity occurred in 72 hours at 25°C in the pH range from 3 to 5. The rate of degradation found at pH 6 and pH 7 (25°C) was approximately 3% and 18% per day, respectively[20]. With refrigeration, no appreciable loss occurs between pH 3 and pH 7 after 72 hours. In U.S.P. hydrochloric acid buffer (pH 1.2), cephalexin lost 5% activity in 24 hours at 37°C as compared to a 45% loss in phosphate buffer at pH 6.5[21]. The antibiotic retains activity well in serum and urine as no loss in activity was noted after storage at -20°C for 14 days[22]. Cephalexin in serum was found to lose 10%, 50% and 75% activity, respectively, after storage at 5°C, 25°C, and 37°C for 48 hours[21,22]. Some organisms have been found to produce a β-lactamase (cephalosporinase) which can rapidly degrade cephalexin[23]. Degradation of cephalexin also results from heat, strong alkali, strong acids and ultraviolet light (260 nm).

4.0 Synthesis

Two synthetic routes of general applicability have been proposed for cephalexin[4].

The first method is based on the cleavage of the acetoxyl group from cephaloglycin[24] (I) by hydrogenation or, more satisfactorily, from N-t-butoxy-carbonylcephaloglycin (II) to produce the corresponding desacetoxy analogs V (cephalexin) and III as shown in Figure 4, Scheme I. The t-

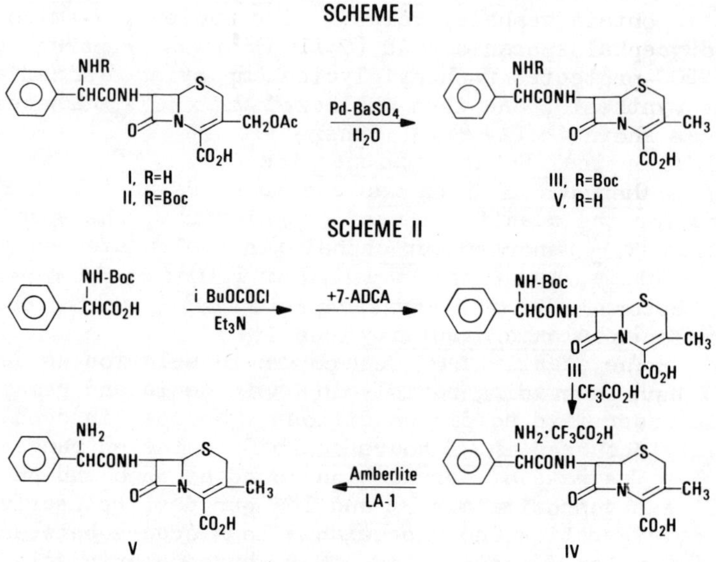

SCHEME I

SCHEME II

FIGURE 4. Synthesis routes for cephalexin.

butoxy-carbonyl (BOC) group was removed from III with trifluoroacetic acid and the reaction product was converted to desacetoxycephaloglycin (cephalexin, V) by treatment with Amberlite LA-1 resin.

The second method was similar to that previously used to obtain cephaloglycin[24]. The nucleus, 7-aminodesacetoxycephalosporanic acid (7-ADCA)[25] was prepared, acylated with BOC-protected D-phenylglycine employing a mixed anhydride synthesis, and then deblocked with trifluoroacetic acid as shown in Figure 4, Scheme II.

5.0 Methods of Analysis

5.1 Identification Tests

Cephalexin may be identified by infrared spectroscopy. British Pharmacopoeia[3] utilizes two characteristic color reactions for identity. Thin layer (Sec. 5.23), paper (Sec. 5.24), and column chromatography (Sec. 5.25) have been utilized for identity purposes.

5.2 Quantitative Methods

5.21 Titration

The iodometric titration procedure has been used for the determination of cephalexin[26,27]. The method is based on the fact that the intact cephalexin molecule does not consume iodine, whereas the alkali-hydrolysis product of cephalexin does. Alkaline hydrolysis of cephalexin results in cleavage of the β-lactam ring. Variations in hydrolysis time, temperature, pH of the iodine solution and concentration of cephalexin present influence the consumption of iodine by the test solution. The method compares favorably to the microbiological cylinder-plate method (Sec. 5.27) in accuracy, and is much more rapid. Possible intermediates used in the synthesis such as 7-ADCA will also respond to the test.

An automated iodometric assay has been used recently for the assay of cephalexin and formulations thereof[28]. The procedure incorporates a sample hydrolysis step (0.18 N NaOH) at 37°C for 10 minutes followed by a 5-minute iodine consumption step (pH 5.3-5.5, 37°C). Concentration of the sample is related to the decrease in iodine color measured at 350 nm. A reference standard is run concurrently through the analyzer for comparative purposes. The automated system gives excellent linearity of response for the recommended concentration range of cephalexin (0 - 1.5 mg. per ml. sample solution), with all standard curve plots passing through the origin. The reproducibility of the method on the same sample or standard solution on a given day is generally better than ±1% relative standard deviation (RSD). Cephalexin can be titrated with perchloric acid in a glacial acetic acid medium[29]. Crystal violet indicator (2% in glacial acetic acid) may be used to determine the endpoint.

Moll and Döker[30] have reported using a formol titration procedure for the determination of cephalexin, ampicillin and related compounds. In this procedure 4 ml. of dilute formaldehyde solution (neutralized to the phenolphthalein end point) is added to 10 ml. of an aqueous solution containing 15.0 mg. of cephalexin. After 2 minutes the solution is titrated with 0.02N sodium hydroxide. A precision of ±0.5% RSD could be achieved in the titration of cephalexin monohydrate raw material samples. Acidic compounds as well as amino acids must not be present as impurities in the sample. The formol titration takes advantage of the reaction between an amino acid and formaldehyde as a means of suppressing the basicity of the amino group and thus making possible the titration of the acid.

5.22 <u>Colorimetric Determination</u>

Reaction with hydroxylamine has been utilized for the colorimetric determination of cephalexin[31,32]. The method is based on the fact that hydroxylamine cleaves the β-lactam ring (pH 7.0) to form a hydroxamic acid which forms a colored complex with ferric ion. Degradation products or intermediates having an intact β-lactam ring react as well.

Kirschbaum[33] has described a procedure for the colorimetric determination of the antibiotic cephradine[1] and related cephalosporins. An aqueous solution of the compound (1 to 30 mcg. per ml.) is reacted with sodium hydroxide, partially neutralized, and then reacted with 5,5'-dithiobis-(2-nitrobenzoic acid) resulting in the formation of a yellow chromophore (412 nm.). The molar absorptivity E x 10^{-4} reported for cephalexin when carried through this procedure was 1.29. The formation of the yellow chromophore was attributed to the presence of the R - $CHNH_2$ -CO-cephalo-

[1] Cephradine is the generic name for 7-[D-2-amino-2-(1,4-cyclohexadienyl)acetamido] desacetoxycephalosporanic acid

sporin nucleus, in which R is a mono-, di- or tri- enyl cyclohexyl ring.

A specific colorimetric test was developed for the determination of cephalosporin derivatives having the following intact side chain in the 7- position: R - $CHNH_2$-CO-cephalosporin nucleus, R being a heterocyclic or aromatic ring[3,4]. The D-phenylglycine derivatives of both 7-ADCA (cephalexin) and 7-ACA (cephaloglycin)[2] respond well. These compounds (0.5 - 1.0 mg. per ml. in H_2O) react with acetone and sodium hydroxide at 100°C to form characteristic red chromophores (520 nm.). At the 1 mg. per ml. level, this test will visually differentiate cephalexin from cephradine.

[2] Cephaloglycin is the generic name for 7-(D-α-aminophenylacetamido) cephalosporanic acid

5.23 Thin Layer Chromatography

The following thin layer chromatographic systems have been reported:

TABLE V

Adsorbent	Solvent System	R_f	Ref.
Silica Gel	Acetonitrile/Water (3:1)	0.67	13
Silica Gel	Ethyl Acetate/Acetone/Acetic Acid/Water (5:2:2:1)	0.22	35
Cellulose Chromatogram Sheet	Butanol/Acetic Acid/Water (3:1:1)	-	36
Sheet	Ethyl Acetate/Acetic Acid/Water (3:1:1)	-	36
Sheet	Acetonitrile/Ethyl Acetate/Water (3:1:1)	-	36
Cellulose	Acetonitrile/Water (3:1)	0.50	37
Cellulose	Butanol/Acetic Acid/Water (3:1:1)	0.70	37

Cellulose is the preferred sorbent since it is inert toward cephalexin. Additional thin layer chromatography systems used for cephalexin and other cephalosporins have been tabulated[37]. Cephalexin may be detected by ultraviolet absorbance and quenching, ninhydrin, iodoplatinate, alkaline permanganate, and phosphomolybdic acid sprays. Iodine detection and vanillin-phosphoric acid spray have also been utilized. Of the microorganisms used, *Sarcina lutea* is preferred over *Bacillus subtilis* or *Staphylococcus aureus*.

5.24 Paper Chromatography

The following paper chromatographic systems have been reported:

Solvent Systems	R_f	Ref.
Butanol/Acetic Acid/Water (3:1:1)	0.60	13, 36, 38
Ethyl Acetate/Acetic Acid/ Water (3:1:1)	-	36

Whatman No. 1*, untreated, was used with both solvent systems and the equilibrating solvent was the same as the developing solvent. Additional paper chromatographic systems for cephalexin and other cephalosporins have been tabulated[38]. The butanol/acetic acid/water (3:1:1) system will separate cephaloglycin from cephalexin, cephaloglycin being less mobile. An acetonitrile/water (9:1) system[39] using Whatman No. 3 paper buffered at pH 5.0 (16 hour development) has been utilized to separate cephradine from cephalexin, cephradine being less mobile. The developed chromatogram can be examined under ultraviolet light, dipped in ninhydrin, or bioautographed, using Sarcina lutea[40].

5.25 Column Chromatography

The Moore-Stein amino acid analyzer has been used for the determination of cephalexin in urine samples[41]. Beckman Custom Research Resin Type PA-35, packed to a height of 9.0 cm. in a water-jacketed column (0.9-cm. i.d. x 23-cm. length) was used for the separation. The urine sample was diluted with an equal volume of sodium citrate buffer (pH 2.2) and 100 λ applied to the column. The elution time for cephalexin was approximately 61 minutes. Excellent agreement was found between the analyzer method and the microbiological method (Section 5.27) on a series of urine samples tested. In addition, this technique has been useful in determining low levels of 7-ADCA and phenylglycine in cephalexin[41] (Section 5.3). Determination of the less active (biologically) L-isomer in cephalexin by the Moore-Stein amino acid analyzer has been reported[4]. Chou[13] used the anionic resin, Bio-Rad AG2 - x8 (acetate form), to isolate cephalexin from human urine.

*Whatman chromatography paper, Reeve Angel, 9 Bridewell Place, Clifton, New Jersey

5.26 Electrophoresis

Paper electrophoresis has been utilized by the British Pharmacopoeia[3] as a test for impurities present in cephalexin (Section 5.3).

5.27 Microbiological Assays

Microbiological assays for cephalexin have been discussed by Marrelli[42], Wick[43], Mann[44], and Simmons[45] and are listed in the Federal Register[46]. Briefly, the two plate systems well suited for the determination of cephalexin in pharmaceutical formulations are the cylinder plate methods utilizing *Staphylococcus aureus* (ATCC 6535) and *Bacillus subtilis* (ATCC 6633). The assay range for both is approximately 2.5 to 20 mcg. of cephalexin per ml.[42,44] The *B. subtilis* plate test has an advantage over the *S. aureus* plate test in that better defined zones of inhibition are obtained, thereby increasing the assay precision[42]. Since degradation products of cephalexin possess practically no antimicrobial activity[42], rapid and precise photometric microbiological assays are possible with cephalexin. The test organism for the photometric assay is *Staphylococcus aureus* 9144, 3 to 3.5 hours being required for incubation. In Antibiotic No. 3 Broth, the concentration range is 0.2 to 2.0 mcg. per ml. of broth. If the automated AUTOTURB® System is used, precision in the order of 1-2% is possible[47]. Cephalexin in biological fluids may be assayed by a *Sarcina lutea* plate system. The concentration range for assay is 0.2 to 3.5 mcg./ml.[44]

5.3 Methods for Intermediates and Impurities

5.31 7-Aminodesacetoxycephalosporanic Acid (7-ADCA)

Cole[41] utilized the Moore-Stein amino acid analyzer for the determination of 7-ADCA in cephalexin. Column specifications were outlined in Section 5.25. Twenty-five milligrams of cephalexin sample were dissolved with sodium citrate buffer (pH 2.2) to a total volume of 10.0 ml. and 1.0 ml. applied to the column. The elution time for 7-ADCA was approximately 45 minutes. The sensitivity of the assay was 0.1% 7-ADCA.

A colorimetric procedure was developed by Marrelli[48] which permitted the direct determination of low levels of 7-ADCA in cephalexin (0.4 - 1.5%). The procedure is based on the interaction of 7-ADCA with ninhydrin under controlled conditions to produce a specific chromophore (λ max. at 480 nm.). Compounds having an α-amino group adjacent to a β-lactam ring respond in general to the test. British Pharmacopoeia[3] utilized a paper electrophoresis technique which estimated the 7-ADCA content in cephalexin at the 1% level. The buffer solution specified in the test consisted of a mixture of 5 ml. of formic acid, 25 ml. of glacial acetic acid, 30 ml. of acetone and water to a total volume of 1000 ml.

A 2.0 μl. aliquot of the cephalexin sample solution (5.0% w/v in 0.5N HCl) along with the specified amounts of reference standards and markers were applied to the paper. The voltage was adjusted to about 20 volts per cm. of paper and electrophoresis was allowed to proceed until the crystal violet spot moved 9 cm. from the base line. Both ninhydrin spray and UV were used for detection purposes.

5.32 Phenylglycine

The chromatographic procedures outlined in Section 5.31 for the determination of 7-ADCA in cephalexin have been concurrently used for the determination of phenylglycine in cephalexin. The elution time for phenylglycine in the amino acid analyzer assay was approximately 36 minutes. The sensitivity of the assay was 0.01%. The paper electrophoresis technique permitted estimation of phenylglycine at the 1% level. Hussey[48] has utilized a chromatographic procedure for phenylglycine similar to that of the amino acid analyzer assay but incorporating a Fluram™ detection system+.

+Hoffman-LaRoche Inc., Nutley, New Jersey

6. Protein Binding

Various values have been reported for the percentage of cephalexin bound to serum protein. Wick[22] concluded that serum inactivation or protein binding of cephalexin is low. Addition of serum to broth medium did not affect *in vitro* minimal inhibitory concentration determinations. Cephalexin assays in pH 7 buffer and human serum resulted in identical standard curves when 6-mm discs were saturated with solutions and tested by a *Sarcina lutea* microbiological assay. Utilizing a similar method, Griffith and Black[49] found that protein binding of cephalexin in human serum was 9% at concentrations above 1.0 μg./ml. and 41% at 0.2 μg./ml. Naumann and Fedder[50] also found that the amount of cephalexin bound to serum proteins varied with the cephalexin concentration. Using an ultrafiltration method, Kind *et. al.*[51] estimated the serum binding as being 15%. O'Callaghan and Muggleton[52] obtained a value of 43% by utilization of the ultrafiltration technique.

7. Pharmacokinetics

Oral doses of cephalexin are rapidly absorbed by animals and man, result in rather high blood serum levels, and are excreted unchanged in the urine. Wick[22] found that a 20-mg./kg. oral dose of cephalexin in mice gave a blood serum level of 18 μg./ml. Wells *et. al.*[53] reported a similar value (17 μg./ml.) with a 10 mg./kg. oral dose in dog. The rapidity of oral absorption was demonstrated by the fact that antibacterial activity in the serum was the same within 1.5 hours after oral or intramuscular administration; thereafter, the levels were higher for the oral dose. In man, after a 500-mg. dose the mean of the peak antibiotic activities found by several investigators was 15 μg./ml.[49,50,54-58] and usually occurred at 1-1.2 hours after treatment. The compound is almost completely absorbed from the upper small intestine in both man and animals[56,36]. In addition, the antibiotic is excreted unchanged in urine with almost 100% recovery[55,13,59]. A meal just prior to treatment resulted in lower blood levels and increased the time required for peak titer[59,60].

Kirby *et. al.*[61] calculated the serum half-life of intravenously administered cephalexin as 36 minutes. Kabins *et. al.*[62] and Naumann and Fedder[50] calculated the serum half-life after oral dosage as 54 minutes. Thornhill *et. al.*[58] found that probenecid increased the peak serum concentration by 50%, but Meyers *et. al.*[55] found a lesser effect. Linquist *et. al.*[63] examined the disappearance of cephalexin from the blood sera of anephric patients. Half-life values ranged from 23.5 - 41.0 hours with a mean of 31 hours, clearly demonstrating the dependence upon the kidney for excretion.

8. References

1. The United States Pharmacopoeia, XIX, Proof p. 2122.
2. Federal Register, 21CFR148w.6.
3. British Pharmacopoeia, 1973, p. 87.
4. Ryan, C.W., Simon, R.L., and Van Heyningen, E.M., J. Med. Chem., 12, 310 (1969).
5. Doyle, F.P., Fosker, G.R., Naylor, J.H.C., and Smith, H., J. Chem. Soc., 1440 (1962).
6. Analytical Profiles of Drug Substances, Vol. 2 (K. Florey, ed.) p.4, Academic Press, New York and London, 1973.
7. Pfeiffer, R.R., Yang, K.S. and Tucker, M.A., J. Pharm. Sci., 59, 1809 (1970).
8. Underbrink, C.D., Eli Lilly Analytical Development, Unpublished Data.
9. Morin, R.B., Jackson, B.G., Mueller, R.A., Lavagnino, E.R., Scanlon, W.B., and Andrews, S.L., J. Amer. Chem. Soc., 91, 1401 (1969).
10. Flynn, E.H., ed., Cephalosporins and Penicillins. Chemistry and Biology, Academic Press, New York and London (1972), p. 315.
11. Flynn, op. cit., p. 631.
12. Chauvette, R.R., et. al., J. Am. Chem. Soc., 84, 3401 (1962).
13. Chou, T.S., J. Med. Chem., 12, 925 (1969).
14. Pfeiffer, R.R., Eli Lilly Analytical Development, Personal Communication, (1969).
15. Cole, T.E., Eli Lilly Analytical Development, Personal Communication, (1968).
16. Pfeiffer, R.R., Eli Lilly Analytical Development, Personal Communication, (1970).
17. Flynn, op. cit., p. 310.
18. Hargrove, W.W., Eli Lilly and Company, Personal Communication, (1967).
19. Flynn, op. cit., p. 633.
20. Winely, C.L., Eli Lilly Analytical Development Laboratories, Unpublished Data.
21. Simmons, R. J., Anal. Microbiol., II., 193 (1972).
22. Wick, W.E., Appl. Microbiol., 15, 765 (1967).
23. Ott, J.L., and Godzeski, C.W., Antimicrob. Ag. Chemother. 1966, 75 (1967).
24. Spencer, J.L., Flynn, E.H., Roeske, R.W., Siu, F.Y., and Chauvette, R.R., J. Med. Chem., 9, 746 (1966).
25. Stedman, R.J., Swered, K., Hoover, J.R.E., J. Med. Chem., 7, 117 (1964).
26. Federal Register, 21CFR141.506.

27. British Pharmacopoeia, p. 88 (1973).
28. Stevenson, C.E., and Bechtel, L.D. (1971) Publication submitted for review in J. Pharm. Sci.
29. Marrelli, L.P., Eli Lilly and Company, Personal Communication (1967).
30. Moll, F., and Döker, H., Arch. Pharm., Berl., 305 (7), 548 (1972).
31. Federal Register, 21CFR141.507.
32. Flynn, op. cit., p. 615.
33. Kirschbaum, J., J. Pharm. Sci., 63, 923 (1974).
34. Marrelli, L.P., J. Pharm. Sci., 61, 1647 (1972).
35. Thomas, P.N., Eli Lilly and Company, Personal Communication (1972).
36. Sullivan, H.R., Billings, R.E., and McMahon, R.E., J. Antibio., 22, 195 (1969).
37. Flynn, op. cit., p. 621.
38. Flynn, op. cit., p. 620.
39. Marrelli, L.P., Eli Lilly and Company, Personal Communication (1972).
40. Miller, R.P., Antibiot. and Chemother., 12, 689 (1962).
41. Flynn, op. cit., pp. 629, 680.
42. Flynn, op. cit., p. 610.
43. Flynn, op. cit., p. 497.
44. Mann, J.M., Anal. Microbiol., II., 207 (1972).
45. Simmons, R.J., Anal. Microbiol., II., 193 (1972)
46. Federal Register, CFR148w6.
47. Kuzel, N.R. and Kavanagh, F.W., J. Pharm. Sci., 60, 767 (1971).
48. Hussey, R.L., Eli Lilly Analytical Development, Personal Communication (1974).
49. Griffith, R.S. and Black, H.R., Postgrad. Med. J., 47, February Suppl., 32 (1971).
50. Naumann, P. and Fedder, J., Int. J. Clin. Pharmacol., Suppl., 2, 6 (1970).
51. Kind, A.C., Kestle, D.G., Standiford, H.C. and Kirby, W.M.M., Antimicrob. Ag. Chemother., 405 (1968).
52. Flynn, op. cit., p. 438.
53. Wells, J.S., Froman, R.O., Gibson, W.R., Owen, N.V., and Anderson, R.C., Antimicrob. Ag. Chemother., 489 (1968).
54. Kunin, C.M., and Finkelberg, Z., Ann. Inst. Med., 72, 349 (1970).
55. Meyers, B.R., Kaplan, K., and Weinstein, L., Clin. Pharmacol. Ther., 10, 810 (1969).
56. Muggleton, P.W., O'Callaghan, C.H., Foord, R.O., Kirby, S.M., and Ryan, D.M., Antimicrob. Ag.

Chemother., 353 (1968).
57. Perkins, R.L., Apicella, M.A., Lee, I., Cuppage, F.E., and Saslaw, S., J. Lab. Clin. Med., 71, 75 (1968).
58. Thornhill, T.S., Levison, M.E., Johnson, W.D., and Kaye, D., Appl. Microbiol., 17, 457 (1969).
59. Gower, P.E. and Dash, C.H., Br. J. Pharmac., 37, 738 (1969).
60. O'Callaghan, C.H., Footill, J.P.R., and Robinson, W.D., J. Pharm. Pharmac., 23, 50 (1971).
61. Kirby, W.M.M., de Maine, J.B., and Serrill, W.S., Postgrad. Med. J., 47, February Suppl., 46 (1971).
62. Kabins, S.A., Kelner, B., Walton, E., and Goldstein, E., Amer. J. Med. Sci., 259, 133 (1970).
63. Linquist, J.A., Siddiqui, J.Y., and Smith, I.M., New Engl. J. Med., 283, 720 (1970).

ACKNOWLEDGEMENTS

The author wishes to express his indebtedness to Dr. C. L. Winely for his contribution of the sections on microbiological assays, protein binding and pharmacokinetics.

CHLORAMPHENICOL

Dale Szulczewski and Fred Eng

CONTENTS

CONTENTS (Cont'd)

1. Description

1.1 Nomenclature

1.11 Chemical Names

a. D_g-(-)-threo-2,2-dichloro-N-[β-hydroxy-α-(hydroxymethyl)-p-nitrophenethyl]acetamide.

b. D_g-(-)-threo-1-(p-nitrophenyl)-2-(2,2-dichloroacetamido)-1,3 propanediol

1.12 Generic Name

Chloramphenicol

1.13 Trade Names

Chloromycetin (The Merck Index[1] lists 45 other trade names.)

1.2 Formulae

1.21 Empirical

$C_{11}H_{12}Cl_2N_2O_5$

1.22 Structural and Stereochemical

CH_2OH — $Cl_2CHCONH$ ► C ◄ H — H ► C ◄ OH — p-nitrophenyl (NO_2)

D_g-(or L_s)-threo-1-(p-nitrophenyl)-2-(2,2-dichloroacetamido)-1,3 propanediol[2]

or

(1R, 2R) 1-(p-nitrophenyl)-2-(2,2-dichloroacetamido)-1,3-propanediol

1.3 Molecular Weight 323.14

1.4 Elemental Composition

C-40.88; H-3.74; Cl-21.95; N-8.67; O-24.76

1.5 General[3]

Fine, white to grayish white or yellowish white, needle-like crystals or elongated plates. Its solutions are practically neutral to litmus.

2. Physical Properties

2.1 Crystal Properties

2.11 Crystallinity

Chloramphenicol is a crystalline solid. A typical photomicrograph[4] of chloramphenicol is shown in Figure 1.

2.12 X-ray Diffraction

Analytical X-ray diffraction powder data[5] indexed using single crystal diffraction data[6] for chloramphenicol follows:

Radiation. CuKα(λ1.5418); Filter. Ni.

I/I_1 Diffractometer.

System. Orthorhombic. Space Group. $C222_1(D_2^5)$

a_o 17.6A; b_o 7.35A; C_o 22.3A.

Z. 8.

$\varepsilon\alpha$. 1.519; $n\omega\beta$. 1.601; $\varepsilon\gamma$. 1.668.

Sign. Negative. 2V. 78°.

Measured Density. 1.49.

Fig. 1 Photomicrograph of Chloramphenicol.

2.12 X-ray Diffraction (continued)

Powder data for sample of chloramphenicol. (Chloromycetin sample from Parke, Davis & Company, Detroit, Lot # 573326.)

dA	2θ, deg. CuKα	I/I_1	h k ℓ
11.08	7.98°	12	002
8.76	10.10	1	200
8.15	10.85	32	201
6.87	12.89	64	202
5.64	15.72	57	203
4.98	17.81	32	113
4.68	18.98	61	204
4.47	19.85	57	311
4.37	20.30	79	400
4.28	20.74	100	401,114
3.95	22.50	11	205
3.88	22.91	11	313
3.70	24.05	18	115,006
3.66	24.31	64	020
3.61	24.63	4	021
3.44	25.93	54	404
3.38	26.37	7	220
3.34	26.68	7	221
3.28	27.17	7	023
3.23	27.61	18	222
3.16	28.25	4 br*	510,315
3.13	28.55	10 br*	511,405
3.05	29.25	7	024
2.97	30.05	7	207
2.92	30.64	79	600,513
2.89	30.90	50	601
2.87	31.19	14	316,117
2.82	31.69	89	406,602,025
	plus other lines		

* broad

2.12 X-ray Diffraction (continued)

Further information regarding confirmation of the chemical structure of chloramphenicol through analysis of x-ray diffraction patterns obtained on the antibiotic and its bromo analog are given by Dunitz[6].

2.13 Melting

2.131 Range

Bartz[7] determined the melting range of chloramphenicol to be 149.7 - 150.7°C (corrected).

2.132 As a criteria of acceptability

The United States Food and Drug Administration[8] and the U.S.P.[3] both specify a melting range of 151±2°C.

2.133 In relation to purity determination

Information pertaining to the purity of chloramphenicol can be obtained through interpretation of the thermograms obtained via Differential Scanning Colorimetry. A typical thermogram[9] obtained on chloramphenicol follows as Figure 2.

2.2 Solubility

2.21 Single Solvents

The following data are abstracted from the Merck Index[1]:

Solubility at 25° in water = 2.5 mg./ml.; in propylene glycol = 150.8 mg./ml. Very soluble in methanol, ethanol, butanol, ethyl acetate, acetone. Fairly soluble in ether; insoluble in benzene, petr. ether, vegetable oils. Solubility in 50% acetamide soln. about 5%. Solubilities determined by Weiss et al., Antibiotics & Chemotherapy 7,374 (1957) in mg./ml. at about 28°: Water 4.4; methanol >20; ethanol >20; isopropanol >20; isoamyl alcohol 17.3; cyclohexane 0.13; benzene 0.26; toluene 0.145; petr. ether 0.085; isooctane 0.022; carbon tetrachloride 0.295; ethyl acetate >20; isoamyl acetate >20; acetone >20; methyl ethyl ketone >20; ether >20; ethylene chloride 2.3; dioxane >20; chloroform 1.95; carbon disulfide 0.35; pyridine >20; formamide >20; ethylene glycol >20; benzyl alcohol 14.6.

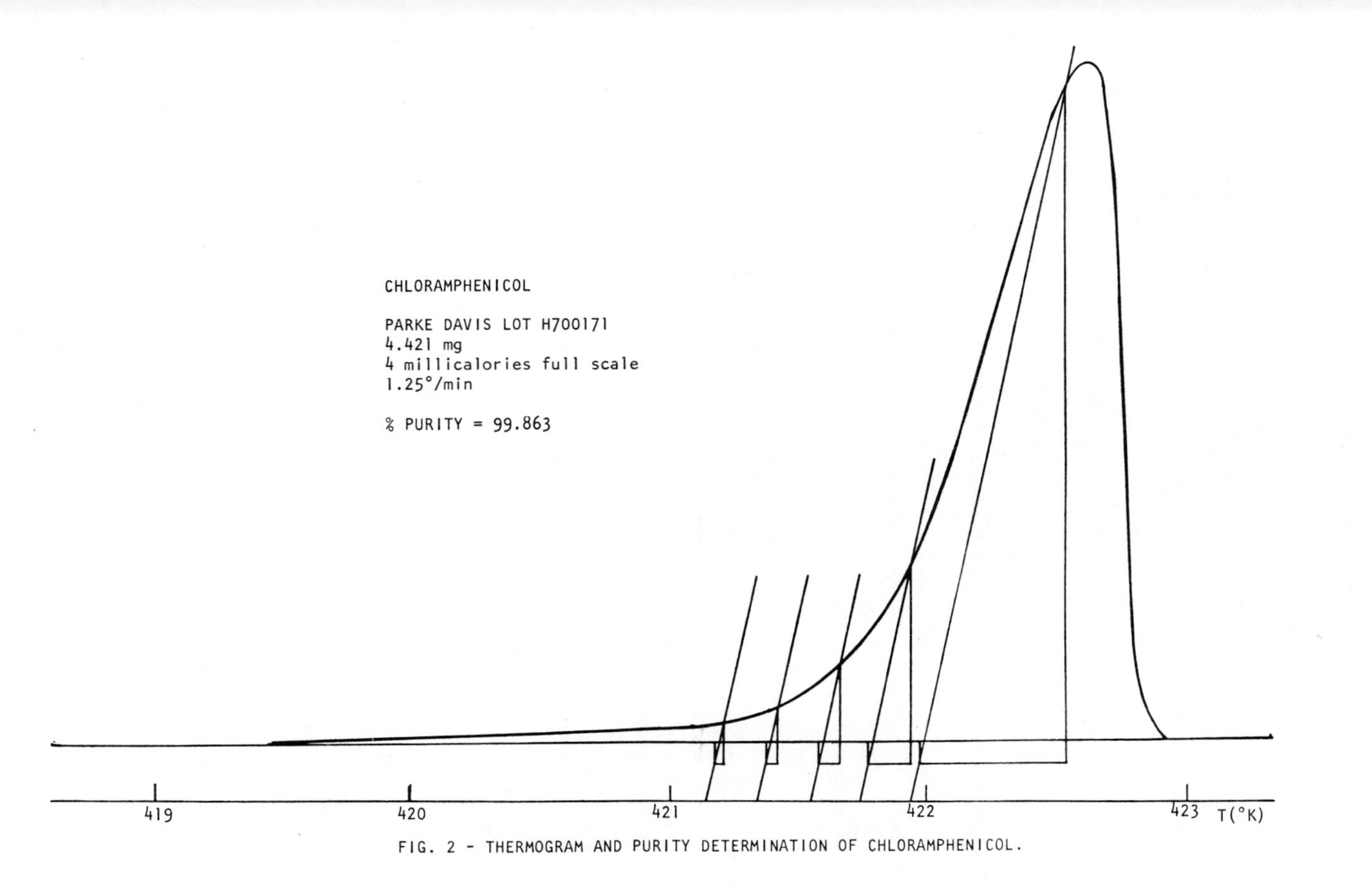

FIG. 2 - THERMOGRAM AND PURITY DETERMINATION OF CHLORAMPHENICOL.

2.22 In mixed solvents or as a result of complexation

The solubility profiles for chloramphenicol in several aqueous solvent mixtures were determined by Negoro and Associates[10]. His results are summarized in Figure 3.

Kostenbauder[11] determined the solubility of this antibiotic in aqueous solutions of N, N, N', N'-tetramethylphthalamides as part of a study on the complexing properties of these amides. Results obtained indicated a moderate influence on solubility as shown in Figure 4.

Aqueous solvents containing 5% Tween 20-80 increase the water solubility of chloramphenicol approximately 3 fold[12]. The solubility of chloramphenicol in serum and urine is approximately the same as it is in water[13].

The solubility of chloramphenicol in water is increased by addition of borax[14]. This solubility increase is explained on the basis of the formation of a 1:2 complex between the borate ion and the antibiotic. Results are summarized in Table 1.

Table 1[14]

Solubility of Chloramphenicol in Borax Solutions

Molar Borax Concentration	Solubility %	pH of Solution
0	0.375	4.70
0.0001	0.391	7.15
0.001	0.438	8.65
0.005	0.614	8.65
0.01	0.732	8.65
0.02	1.23	8.65
0.05	2.14	8.70
0.11	3.46	8.90
0.125	3.67	8.90
0.15	3.87	9.00

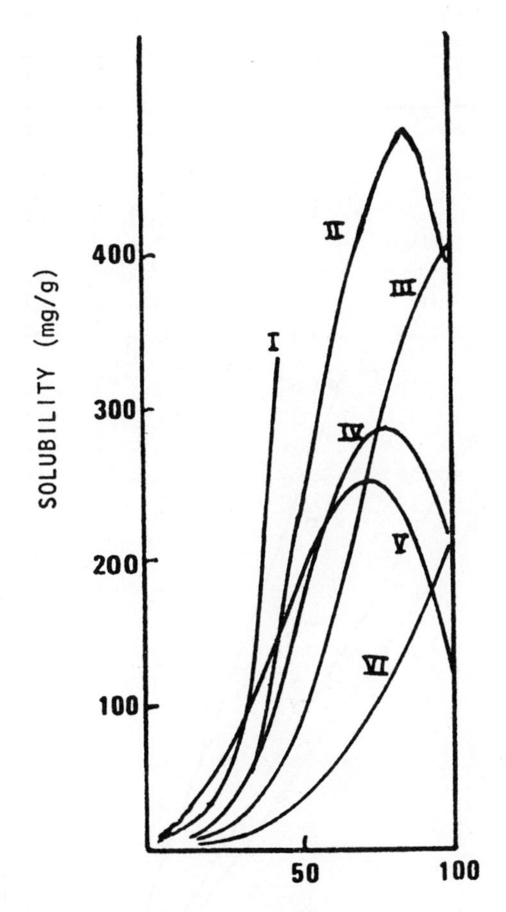

I. Dimethylacetamide, 31°C
II. Acetone, 0°C
III. Methanol
IV. Ethanol
V. N-Propanol
VI. Propylene Glycol

FIG. 3[10] - SOLUBILITY CURVES OF CHLORAMPHENICOL IN MIXED AQUEOUS SOLVENTS

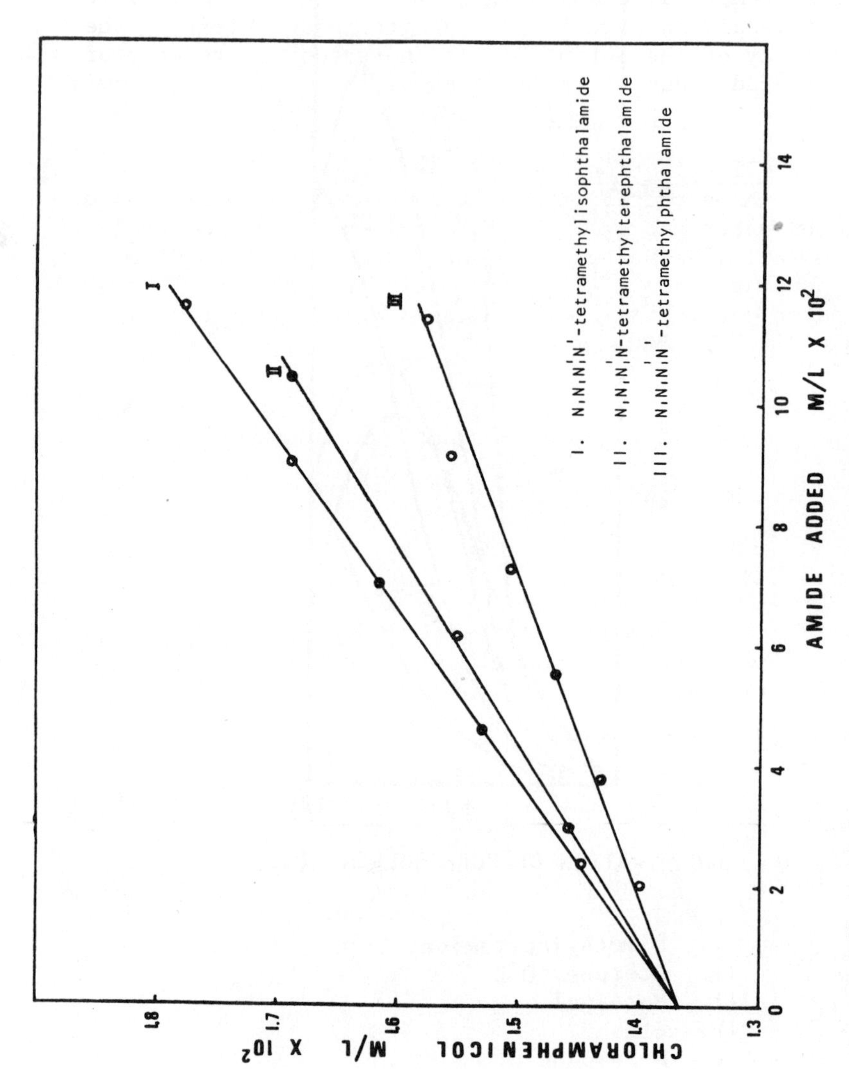

FIG. 4 [11] SOLUBILITY OF CHLORAMPHENICOL IN WATER IN PRESENCE OF CERTAIN AMIDES.

2.23 pH Effect

Since chloramphenicol is an essentially neutral compound, changes in pH (over the pH region 3 to 9) do not result in significant changes in solubility. The solubility of the antibiotic is increased in presence of strong acid[15] due to protonation of the weakly basic amido nitrogen.

2.3 Distribution

As expected, the distribution of chloramphenicol between water and an immiscible solvent is not markedly pH dependent. The per cent of total chloramphenicol found in the aqueous phase after distribution between equal volumes of water and immiscible organic solvents is tabulated in Table 2[7,16].

Table 2
% Chloramphenicol in Aqueous Phase

Immiscible Solvent	%	Ref.
Cyclohexanone	8	7
n-Butanol	8	7
Ethyl acetate	3	16
Methyl isobutyl ketone	8	7
Nitrobenzene	38.0	7
Nitromethane	17.0	7
Ethyl ether	20	16
Chloroform	82	16
Benzene	93	16
Petroleum ether	96	16
Ethylene dichloride	50.0	7

Brunzell[16] discusses the utility of simple extraction techniques in the analysis of formulations containing this antibiotic. The same author provides extraction methods for analysis of the drug in the presence of hydrolytic decomposition products.

2.4 Spectral Properties

2.41 Ultraviolet

Chloramphenicol in solution absorbs ultraviolet radiation over a broad range to produce a spectrum with a maximum near 278 nm and a minimum near 240 nm (see Fig. 5 for a typical spectrum).

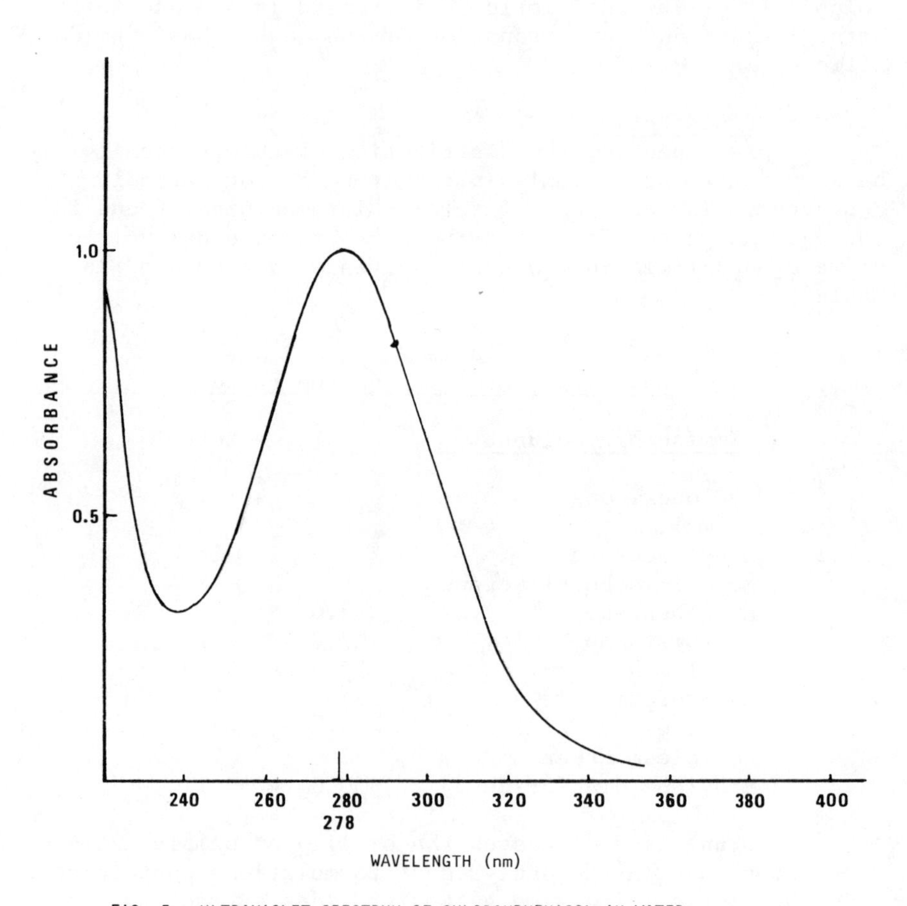

FIG. 5 - ULTRAVIOLET SPECTRUM OF CHLORAMPHENICOL IN WATER.

As was established by Vandenbelt[7,17], this absorption is due to the p-nitrophenyl chromophore and provides both a distinguishing characteristic of the antibiotic and a useful method for analysis.

An ultraviolet specification is included in both the Federal Register[8] and the U.S.P.[3] as a criteria of acceptability.

The ultraviolet spectrum of chloramphenicol in aqueous solvents is not significantly influenced by changes in pH. Aqueous Borate buffers (pH 9.0) perturb the spectrum, shifting the maximum from 278 nm to 284 nm as the result of complex formation.

2.42 Infrared

The infrared spectrum of chloramphenicol (KBr dispersion) is shown in Figure 6. Further information with regard to correlation of functional groups to infrared absorption maxima is given by Suzuki and Shindo[18] who determined infrared spectra of racemic erythro and threo isomers of the antibiotic as well as the spectra of related compounds. These authors obtained evidence for intramolecular hydrogen bonding from infrared spectra determined on dilute solutions.

Assignments for the accompanying spectrum follow:

Functional Group	Wavenumber cm^{-1}
bonded OH, NH	3340, 3260
amide portion of 2,2-dichlor-acetamide moiety	
amide I	1697
amide II	1568
nitro group (nitro-phenyl)	1530, 1358
hydroxyl	1068

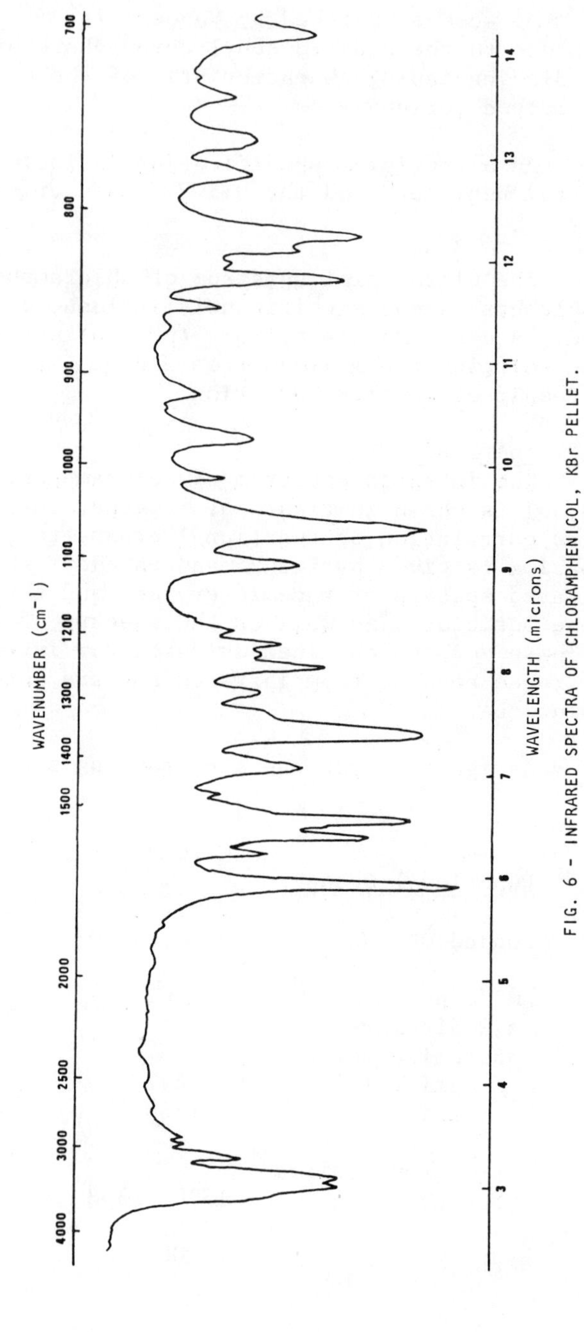

FIG. 6 - INFRARED SPECTRA OF CHLORAMPHENICOL, KBr PELLET.

2.43 Nuclear Magnetic Resonance

A typical nuclear magnetic resonance spectrum of chloramphenicol in deuterated acetone is given in Fig. 7. Further information concerning the interpretation of the NMR spectra of chloramphenicol and its diastereoisomer, the L(+)-erythro isomer, can be obtained from Jardetsky's careful study[19] concerning the conformation of chloramphenicol in solution. Jardetsky's chemical shift assignments of the chloramphenicol protons were determined from spectra in deuterated acetone of chloramphenicol samples lyophilized from D_2O and H_2O.

Jardetsky's assignments are as follows (with respect to benzene as external standard with 60-megacycle NMR).

Proton(s) Associated with	cps
$O_2N-C_6H_4-R$	-46.5 (center of H_2B_2 quartet)
$R'-CHCl_2$	+32.6
$C_1(H)$	+97.2
$C_2(H)$	+156.2
$C_3(H_2)$	+177.5 +187.7
$HN(R'')-R'''$	-29.3
C_1OH	+99.6
C_3OH	+151.0

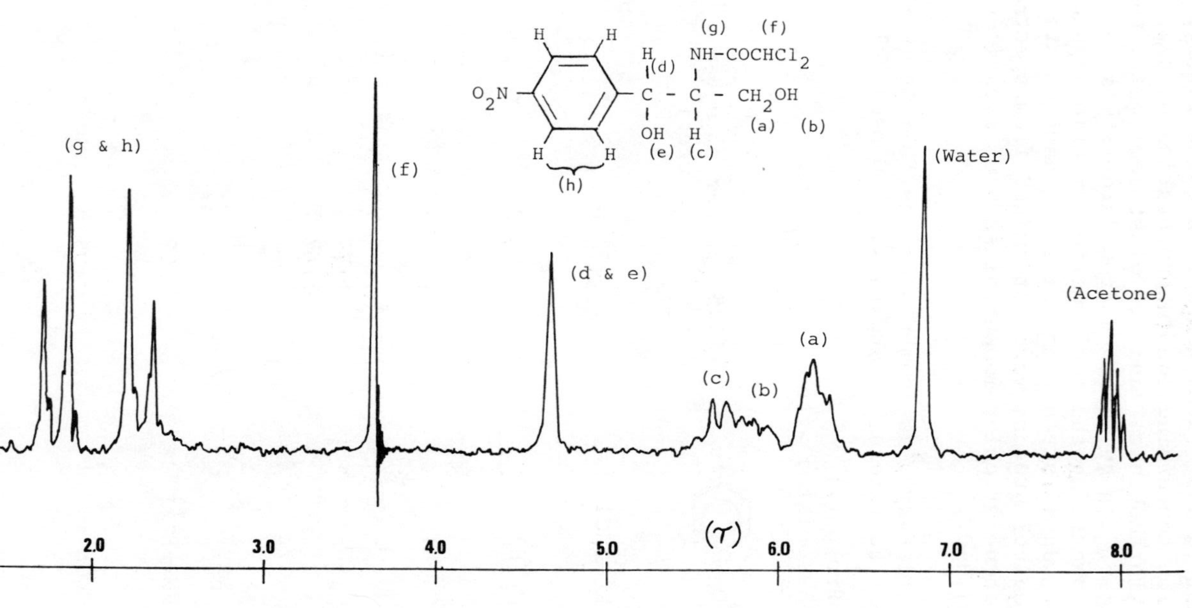

FIG. 7 - NMR SPECTRUM OF CHLORAMPHENICOL, PARKE DAVIS LOT H700171 IN DEUTERATED ACETONE CONTAINING TRIMETHYLSILANE AS INTERNAL STANDARD.

2.44 Optical Rotation

In the chloramphenicol series consisting of four diastereoisomers, therapeutic antibiotic activity resides in the Dg-threo (or IR, 2R) isomer. Specifications on specific rotation as a criteria of acceptability are contained in the Federal Register[8] as cited by the U.S.P.[3].

The Food and Drug Administration[8] has established the following specification:

"Its (chloramphenicol's) specific rotation in absolute ethanol at 20°C is +20 ± 1.5° and at 25° is +18.5 ± 1.5°." (Na_D or 589 nm)

The specific rotation[20] (C=5%, ethanol, 25°C) of chloramphenicol at some other wave lengths follows:

λ nm	$[\alpha]^{25°C}$
578	+19.8°
546	+23.8°
436	+59.7°

Both magnitude and sign of optical rotation are solvent dependent, i.e., in ethanol the antibiotic is dextrorotatory while in ethyl acetate it is levorotatory.

Circular dichroism measurements on the four chloramphenicol isomers have been made and are recorded in the literature[21]. Analysis of the antibiotic in combination with sulfonamides was accomplished by polarimetric means[22].

2.45 Mass Spectra

The mass spectra of chloramphenicol obtained by conventional electron impact ionization does not exhibit a parent peak. Brunnée and associates[23] obtained the mass spectrum of chloramphenicol using a combined field ionization/electron impact ion source. A parent peak of only 5% height is obtained (see Fig. 8). The base peak is at m/e 152 and is attributed to fragment I.

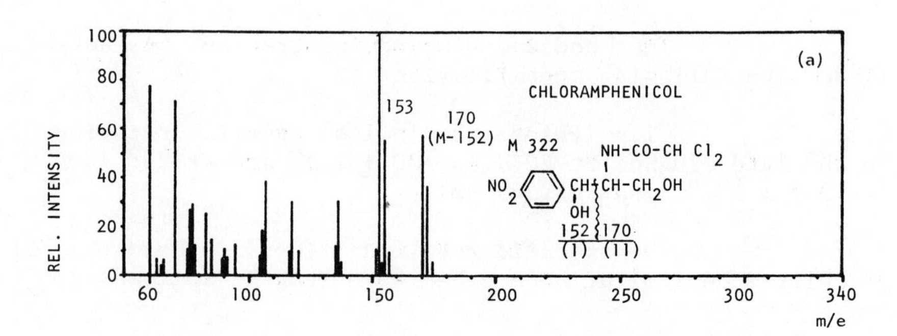

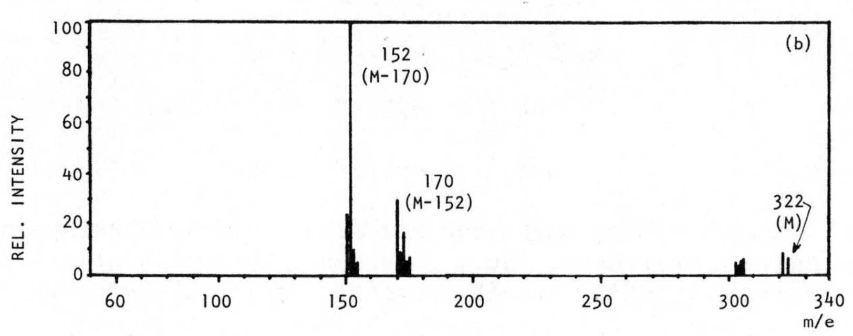

FIG. 8[23] - MASS SPECTRA OF CHLORAMPHENICOL - (a) ELECTRON IMPACT IONIZATION
(b) FIELD IONIZATION

$$O_2N-C_6H_4-\underset{OH}{CH} \;\big|\; \overset{NH-COCHCl_2}{CH}-CH_2OH$$

152	170
I	II

3. Production and Synthesis

Chloramphenicol was originally produced by isolating the antibiotic from cultures of Streptomyces Venezualae [24,25]. After it was demonstrated that synthesis of the antibiotic was possible[26], several synthetic processes were developed for manufacture. A flow diagram for one such process follows:

$$O_2N-C_6H_4-COCH_2Br \xrightarrow{\text{hexamethylene-tetramine}} O_2N-C_6H_4-COCH_2NH_2$$

$$O_2N-C_6H_4-COCH_2NH_2 \xrightarrow{Ac_2O} O_2N-C_6H_4-COCH_2NHCOCH_3$$

$$O_2N-C_6H_4-COCH_2NHCOCH_3 \xrightarrow{CH_2O} O_2N-C_6H_4-CO\overset{CH_2OH}{CH}NHCOCH_3$$

$$\xrightarrow{\text{Pondorf red}} O_2N-C_6H_4-\underset{OH}{\overset{H}{C}}-\underset{H}{\overset{HNCOCH_3}{C}}-CH_2OH$$

DL-threo

$$\xrightarrow{HCl} O_2N-C_6H_4-\underset{OH}{\overset{H}{C}}-\underset{H}{\overset{NH_2}{C}}-CH_2OH$$

$$\longrightarrow O_2N-C_6H_4-\underset{OH}{\overset{H}{C}}-\underset{H}{\overset{NH-COCHCl_2}{C}}-CH_2OH$$

Detailed information on various steps in this synthesis are found in references 27-40. Other synthetic schemes are contained in references 41-47. Processes for conversion of erythro-p-nitrophenyl-2-amino-1,3-propanediol, produced as a by-product, to the desired threo isomer generally proceed via oxazoline formation and are detailed in references 48-54. Resolution of the threo-(p-nitrophenyl)-2-amino-1,3-propanediol produced can be accomplished by conventional means[55,56] or by fractional crystallization[57]. A rather complete review of the synthetic chemistry of chloramphenicol is available[58]. This review also documents structure-activity relationships in the chloramphenicol series.

4. Stability and Decomposition Products

4.1 Crystalline solid and solid dosage forms

Chloramphenicol in the solid state as a bulk drug or present in solid dosage forms is a very stable antibiotic. Reasonable precautions taken to prevent excessive exposure to light or moisture are adequate to prevent significant decomposition over an extended period.

4.2 In Solution

The stability of chloramphenicol in aqueous solution is governed by the rate at which hydrolytic processes occur. The two primary routes of decomposition have been determined[15,59,60,61] to be (a) amide hydrolysis with the formation of 1-(p-nitrophenyl)-2-amino-1,3-propanediol

$$O_2N\text{-}C_6H_4\text{-}CH(OH)\text{-}CH(NH\text{-}CO\text{-}CHCl_2)\text{-}CH_2OH \xrightarrow{H_2O} O_2N\text{-}C_6H_4\text{-}CH(OH)\text{-}CH(NH_2)\text{-}CH_2OH + CHCl_2CO_2H$$

and (b) hydrolysis of covalent chlorine of the dichloroacetamide moiety

$$O_2N\text{-}C_6H_4\text{-}CH(OH)\text{-}CH(NH\text{-}CO\text{-}CHCl_2)\text{-}CH_2OH \longrightarrow O_2N\text{-}C_6H_4\text{-}CH(OH)\text{-}CH(CH_2OH)\text{-}NH\text{-}CO\text{-}CH(OH)_2 + 2HCl$$

The hydrolytic cleavage of the amide linkage[59] is the major cause of chloramphenicol breakdown and is the only significant route of degradation in solutions below pH 7. The rate of amide hydrolysis is independent of pH over the pH region 2 to 6 and independent of the ionic strength of the medium. Studies involving phosphate, acetate, and citrate buffers indicate that the amide hydrolysis is general acid-base catalyzed[60].

The hydrolysis of covalent-bound chlorine is insignificant at pH values below 6 but increase dramatically as pH increases. This increase is attributed to hydroxyl ion catalysis[59].

Numerous secondary reactions can occur which give rise to a variety of decomposition products. Among these secondary reactions are those associated with subsequent hydrolysis of dichloroacetic acid[61] and oxidation-reduction reactions which involve the nitro group as oxidant and the side chain (particularly the aminodiol side chain of the primary hydrolysis product) as reductant. Products isolated from partially or completely decomposed chloramphenicol solutions exposed to a variety of conditions are given in Table 3.

The presence of borate buffers has been shown to increase the aqueous stability of chloramphenicol. Brunzell[67] studied the stability of 0.5% chloramphenicol ophthalmic solutions in pH 7.4 borate buffer and in unbuffered aqueous solutions. The results indicate that the antibiotic is more stable in the presence of this buffer than in its absence. Heward and associates[68] studied the stability of the borate-buffered chloramphenicol eye drops BPC[69]. The results obtained are listed below.

Temperature ^{o}C	Rate Constants k hrs^{-1}	Calculated $t_{10\%}$
115	0.2188	29 minutes
110	0.7413×10^{-1}	85 minutes
30	0.1153×10^{-3}	38 days
20	0.3589×10^{-4}	4 months
4	0.4592×10^{-5}	31 months

Table 3

Decomposition Products of Chloramphenicol

No.	Compound	Environmental Conditions	Ref.
1.	$O_2N-C_6H_4-CH(OH)-CH(NH_2)-CH_2OH$	Acidic or basic aqueous solution	17, 60
2.	Cl_2CHCO_2H	"	17
3.	$O_2N-C_6H_4-CHO$	Aqueous solution, ambient temperature.	62, 63, 64
4.	$H_2N-C_6H_4-CH(OH)-CH(NH-C(=O)-CH_3)-CH_2OH$	Aqueous solution, ambient temperature.	62
	$R_1-C_6H_4-N=N-C_6H_4-R_2$		65
5.	$R_1=R_2=CH_2OH$	Aqueous alkaline solution, high temperature.	65
6.	$R_1=R_2=CO_2H$		"
7.	$R_1 = R_2 = CHO$		"
8.	$R_1 = R_2 = OH$		"

Decomposition Products ofChloramphenicol (Continued)

9.	$R_1 = CO_2H$; $R_2 = OH$		65
10.	$R_1 = CO_2H$; $R_2 = CH_2OH$		"
11.	$R_1 = OH$ $R_2 = CH_2OH$		"
12.	$R_1 = OH$ $R_2 = CHO$		"
13.	$O_2N-C_6H_4-CO_2H$	Aqueous solution after exposure to light.	66
14.	$HO_2C-C_6H_4-N{=}\overset{O}{N}-C_6H_4-CO_2H$	Aqueous solution after exposure to light.	"
15.	HCl	Aqueous solution; high temperature.	"

James and Leach[70,14] suggested that complexation between the antibiotic and borate ion is responsible for the increased stability of chloramphenicol in this buffer system.

4.3 In the presence of microorganisms

Smith[71] studied the decomposition of chloramphenicol in the presence of various microorganisms. He defined the five routes by which chloramphenicol could undergo degradation. These possibilities are summarized in scheme 1.

5. Metabolism and Pharmacokinetics

Biochemical changes which occur during metabolism of chloramphenicol were determined by Glazko and associates[72]. Isolation and identification of metabolites found in various body fluids after administration of this antibiotic indicates that its metabolism can occur by routes shown in scheme 2. Other pertinent information in this regard is to be found in references 73-76.

The absorption characteristics of chloramphenicol from oral dosage forms were determined by means of blood level measurements and urinary excretion measurements of chloramphenicol and its metabolites[77]. Investigators concluded that absorption occurs mainly by simple diffusion mechanisms in the intestinal tract with minimal absorption from the stomach[75,77] and that the degree of absorption was influenced by pharmaceutical factors involved in capsule formulation[78].

6. Methods of Analysis

6.1 Colorimetry

6.11 Qualitative

The following color reaction is published in the U.S.P. XVI[79] and the B.P. 1968[80] as part of an identification scheme.

Dissolve 10 mg. of chloramphenicol in a mixture of 1 ml of diluted alcohol and 3 ml of dilute calcium chloride T.S. (1 in 10). Add 50 mg of zinc dust, and heat on a steam bath for 10 minutes. Decant the clear, supernatant liquid into a test tube, and add 100 mg of anhydrous sodium acetate and 2 drops of benzoyl chloride. Shake the

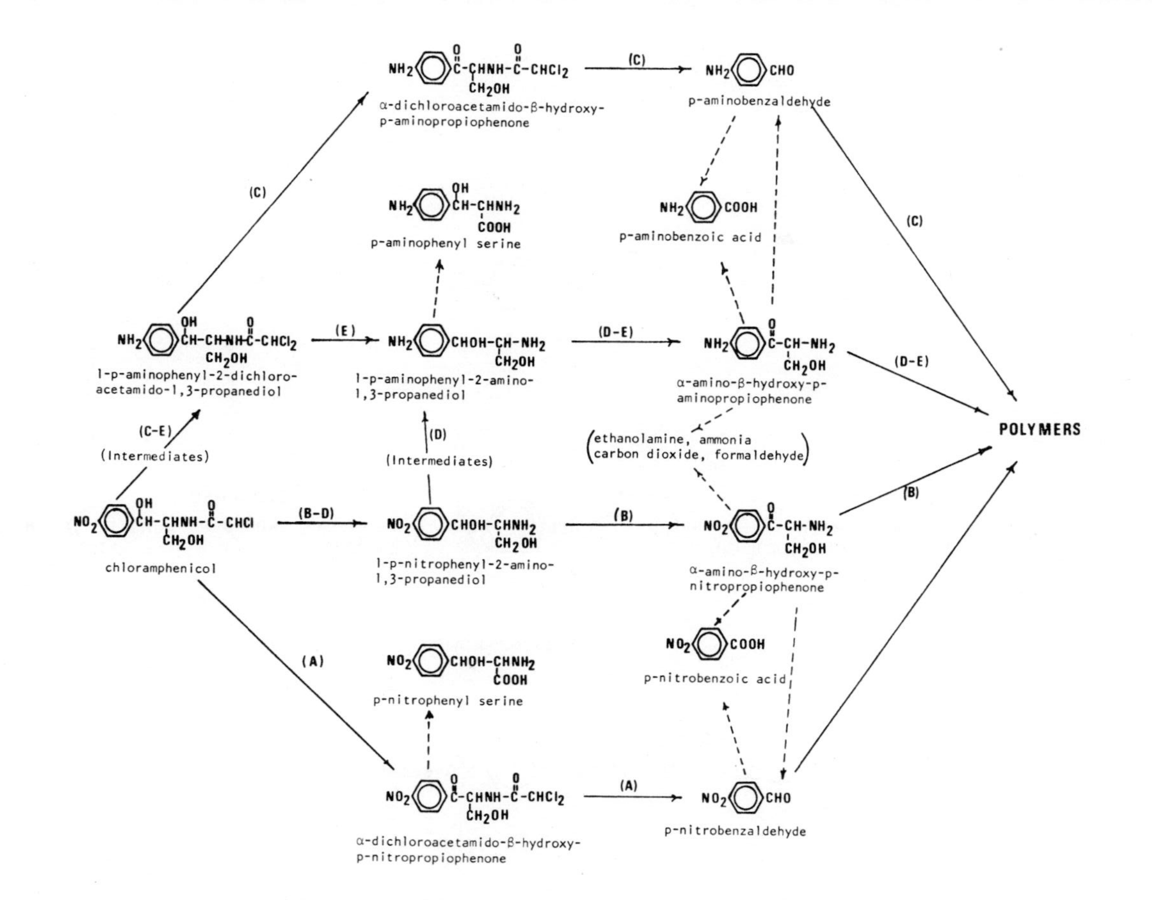

SCHEME 1 - MICROBIAL DECOMPOSITION PATHWAYS OF CHLORAMPHENICOL.

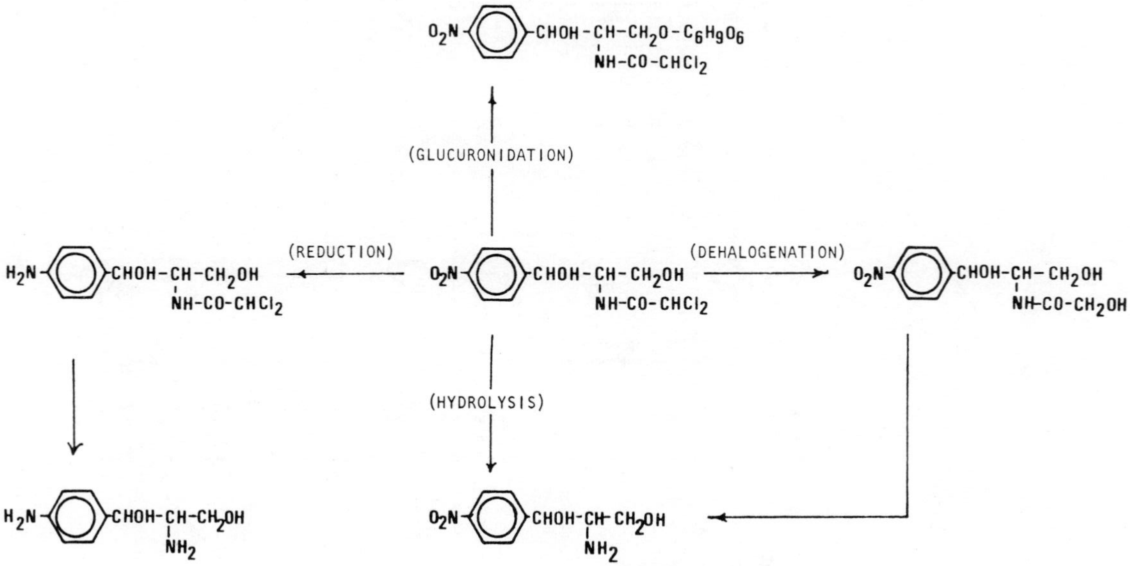

$O_2N-C_6H_4-CHOH-CH(NH-CO-CHCl_2)-CH_2O-C_6H_9O_6$

(GLUCURONIDATION)

$H_2N-C_6H_4-CHOH-CH(NH-CO-CHCl_2)-CH_2OH$ ← (REDUCTION) ← $O_2N-C_6H_4-CHOH-CH(NH-CO-CHCl_2)-CH_2OH$ → (DEHALOGENATION) → $O_2N-C_6H_4-CHOH-CH(NH-CO-CH_2OH)-CH_2OH$

(HYDROLYSIS)

$H_2N-C_6H_4-CHOH-CH(NH_2)-CH_2OH$

$O_2N-C_6H_4-CHOH-CH(NH_2)-CH_2OH$

SCHEME 2 - PATHWAYS IN THE METABOLIC DISPOSITION OF CHLORAMPHENICOL

mixture for 1 minute, add 0.5 ml of ferric chloride T.S. and, if necessary, diluted hydrochloric acid to produce a clear solution: a red violet to purple color is produced.

The formation of a yellow color by heating the antibiotic with concentrated sodium hydroxide served in this capacity in the third supplement of DAB6[81,82].

6.12 Quantitative Analysis

6.121 Reduction of Nitro Group followed by Diazotization and Coupling

This approach to analysis of chloramphenicol was among the first used for determination of the antibiotic in body fluids[83]. Glazko's original procedure was modified several times[84,85] with regard to the reducing agent used for reduction to the corresponding arylamine. Since the nitro group, the reactive moiety, exists in decomposition products and metabolites as well as in chloramphenicol, this procedure is not specific when directly applied. Specificity is imparted by including a separation step prior to analysis. The procedure involves reducing the nitro group to an amine followed by diazotization with acidic nitrous acid and coupling with N-(1-naphthyl)-ethylenediamine dihydrochloride.

6.122 Hydroxamic Acid Method

A higher degree of specificity is imparted to colorimetric analysis by using the hydroxamic acid reaction. In this method chloramphenicol is heated with hydroxylamine under basic conditions to form a hydroxamic acid which is then complexed with ferric ion to produce a red color[86]. This approach has also been used for analysis of chloramphenicol esters[87]. It should be noted that the principal hydrolysis product of chloramphenicol does not yield significant color under the conditions of this test.

6.123 Isonicotinic Acid Hydrazide Procedure

This procedure, as developed by Kakemi[88], has been studied in some depth[89,90,91]. The basis of the assay is the development of a yellow color which results from mixing chloramphenicol with isonicotinic acid hydrazide and sodium hydroxide in aqueous solution. The procedure is simple and can be performed relatively rapidly. Other antibiotics and the succinate and palmitate esters of chloramphenicol have been found not to interfere with the assay procedure.

6.124 Miscellaneous

Other colorimetric procedures have evolved for analysis of chloramphenicol which depend on reaction of acetone[92] or 1-naphthol[93] with the antibiotic under basic conditions. The chemistry of these methods is not defined and, although convenient, they offer little advantage over those previously discussed.

6.2 Polarographic Analysis

The presence of the nitrophenyl group makes it possible to utilize different instrumental techniques for detection and analysis of this antibiotic.

Methods which depend on the p-nitrophenyl group are not selective unless preceded by a separation step or accompanied by independent analysis to give assurance that the sample being analyzed contains only chloramphenicol and is not a mixture of the antibiotic and degradation or metabolic products.

Among the more convenient instrumental approaches to analysis of this antibiotic is polarography. It is less susceptible to interference from other materials than is, for example, ultraviolet spectroscopy, but, as discussed previously, could not be considered a specific procedure without modification.

A detailed study of the polarographic behavior of chloramphenicol was reported by Fossdal[94]. The antibiotic undergoes a 4-electron reduction at the dropping mercury electrode producing a well-defined diffusion-controlled polarographic wave of analytical utility. Results obtained indicated that chloramphenicol could be determined over the range 0.3 to 60 mcg/ml. Previous studies[95-99] reported application of polarography to analysis of chloramphenicol in pharmaceutical preparations.

The nature of polarography imparts a degree of selectivity to the assay of chloramphenicol. 2-(2,2-dichloracetamido)-3-hydroxy-4'-nitropropiophenone, a possible toxic contaminant in synthetic chloramphenicol was determined by direct polarographic measurement[100]. In this case the reduction potential of the impurity is sufficiently different from that of the drug's to permit direct instrumental analysis

6.3 Spectrophotometric

Because both spectrophotometric and polarographic methods depend on the existence of the p-nitrophenyl group, they are both subject to the same specificity considerations. Quantitative determination via direct ultraviolet measurement is not a specific analytical method since likely decomposition products absorb over the same region [101,102]. As previously mentioned, an ultraviolet procedure is official as a method to determine the potency of chloramphenicol[8].

Ultraviolet spectroscopy has been extensively applied to chloramphenicol determination in methods involving separation prior to quantitation[43,44,47,63]. It has also been employed to determine the antibiotic in pharmaceutical preparations[80].

6.4 Titrimetric Methods

Titrimetric methods have been developed for analysis of chloramphenicol. Such procedures are dependent on only a limited portion of the molecule, i.e., the nitro group in the p-nitro-benzene portion or the covalent chloride contained in the dichloracetamido moiety and hence would not be selective unless metabolic or degradative processes yielded products not containing these functional groups.

Procedures utilizing covalent chloride involve converting the covalently-bound chloride to its ionic form. The ionic chloride is then determined by argentometric titration[103,104].

Titrimetry of chloramphenicol using the nitro group has two variations both of which require reduction to an arylamine. Reduction of the nitro group with excess titanium chloride followed by determination of the excess reagent by back titration with ferric ammonium sulfate constitutes the titanometric method[105].

The bromatometric method[105] consists of a Zn-HCl reduction to form the arylamine. The arylamine is then determined by bromination in the presence of excess bromine followed by iodometric determination of the excess bromine.

6.5 Microbiological

Microbiological procedures have been developed for assay of chloramphenicol and applied to analysis of dosage forms, body fluids, and bulk drug[106]. Although time consuming, these methods are as accurate as physicochemical tests, provide analytical sensitivity equal to or surpassing many, and have the advantage of being directly related to use.

Since chloramphenicol's decomposition products or metabolites do not possess significant antibiotic activity, only intact chloramphenicol is measured providing that no other antibiotics or chemotherapeutic agents (i.e., a fixed combination dosage form, supplemental therapy) exist. Although the basic microbiological assay procedures may be subject to this kind of interference, the problem may be obviated by selective inactivation of the interfering antibiotics by using a microorganism sensitive for chloramphenicol but resistent to the interfering antibiotic, by including a separation scheme as part of the assay sequence, or by compensating for the presence of the interfering antibiotic by adding it to each solution of chloramphenicol used for the standard response curve[107].

In the case of chloramphenicol, two basic microbiological methods are in general use viz Cylinder Plate and Turbidimetric.

The Cylinder Plate Method for assay of chloramphenicol is an agar diffusion procedure using <u>Sarcina lutea</u> ATOC 9341 as the test organism. The response of the assay is produced by solutions of chloramphenicol in 1% phosphate buffer pH 6 diffusing through an agar layer uniformly inoculated with the test organism. To accomplish this response, stainless steel cylinders (8 mm o.d., 6 mm i.d., 10 mm long) are placed on the seeded agar surface and filled with the chloramphenicol solutions, and then inclubated overnight at 32°C. The responses that are produced are clear circular zones of inhibited growth around the cylinder on the agar surface otherwise totally covered with heavy growth. The dose-response relationship is a linear one in restricted limits of concentration when the dose is expressed logrithmically and the response arithmetically. In order to determine the concentration of an unknown, a reference standard must be used for comparison on each petri dish[108].

Two assay designs are commonly employed using the cylinder plate technique: the single dilution-standard curve design and a three by three (or 2 x 2) factorial design[109]. The first assay design is the official method of the F.D.A.[110]. The second has the advantage of being able to compare the parallelism of the dose response line of the standard and the unknown.

The turbidimetric method determines the concentration of chloramphenicol by measuring the turbidity that is produced by the actively growing test organism in a series of test tubes containing chloramphenicol and inoculated liquid culture medium. The test is incubated in a 37°C water bath for 2 to 5 hours. After the desired incubation, the growth is stopped by the addition of formaldehyde or other appropriate means, and the responses are read in terms of absorbance on a suitable photoelectric colorimeter or spectrophotometer. By comparing the turbidity of the unknown to that of the reference standard, the potency of chloramphenicol is found[111].

Like the cylinder plate assay, the assay design may vary. The single dilution-standard curve design and 3 x 3 (or 2 x 2) factorial assay are commonly used[109].

Several different organisms have been used for the turbidimetric assay of chloramphenicol. Escherichia coli ATCC 10536 is the organism used in the official method of the F.D.A. The dose-response line, log of concentration vs. response, produced by the organism is linear with a limited range of chloramphenicol concentration[111]. Shigella sonnei ATCC 11060 has been used for the turbidimetric assay of chloramphenicol. Because it is sensitive to lower concentrations of chloramphenicol than the plate assay, it is useful in determining chloramphenicol levels in blood serum and other clinical specimens. The dose-response line obtained with this organism is not linear over the wide range of concentration for which it can be used. Agrobacterium tumefaciens (Parke Davis culture No. 05057) has also been used in place of S. sonnei when a nonpathogenic organism is necessary. However, it is not as sensitive as S. sonnei to low levels of chloramphenicol[108].

In general turbidimetric techniques are faster and more easily adapted to automated techniques.

6.6 Chromatographic

6.61 Paper

The chromatographic behavior of chloramphenicol and related compounds likely to be encountered in metabolic studies or involved in enzymatic and chemical degradation work was established by Smith[112]. Whatman No. 1 paper was used together with a mobile phase consisting of water saturated n-butanol containing 2.5% acetic acid. Several reagents were used to detect various compounds after chromatography. These included p-dimethylaminobenzaldehyde (arylamines), reduction with stannous chloride, followed by p-dimethylaminobenzaldehyde (aromatic nitro compounds), Ninhydrin (aliphatic amino compounds, ammoniacal silver nitrate [Formyl or Carbonyl groups]). Table 4 lists R_f values and the response to various detection reagents.

6.62 Thin Layer

Several thin layer chromatographic systems have been developed to study the various aspects of chloramphenicol chemistry. The procedures described here have been used for the separation and identification of chloramphenicol derivatives, decomposition products, and synthetic intermediates.

Lin[113] achieved separation of chloramphenicol, chloramphenicol palmitate, and chloramphenicol succinate in two solvent systems using polyamide thin layer plates. The R_f values reported are:

	Solvent A	Solvent B
Chloramphenicol	0.35	0.80
Chloramphenicol Palmitate	0.95	0.90
Chloramphenicol Succinate	0.25	0.72

TABLE 4[112]

NAME	R_F	NITRO TEST	ARYL AMINE TEST	NINHYDRIN TEST	BENZIDINE TEST	$AgNO_3NH_3$ TEST	NaOH TEST
3-(p-Aminophenyl)serine	0.03	+	+	+	-	-	-
1-(p-Aminophenyl)-2-amino-1,3-propanediol	0.12	+	+	+	-	-	-
Ethanolamine	0.25	-	-	+	-	-	-
2-Amino-3-hydroxy-4'-nitropropiophenone—HCl	0.36	+	-	-	-	+	+
1-(p-Nitrophenyl)-2-amino-1,3-propanediol	0.45	+	-	+	-	-	-
1-(p-Aminophenyl)-2-(2,2-dichloroacetamido)-1,3-propanediol	0.69	+	+	-	-	-	-
p-Aminobenzoic acid	0.78	+	+	-	-	-	-
p-Nitrobenzoic acid	0.82	+	-	-	-	-	-
p-Aminobenzaldehyde	0.84	+	+	-	-	+	-
Formaldehyde	0.85	-	-	-	-	+	-
2-Acetamido-3-hydroxy-4'-nitropropiophenone	0.86	+	-	-	-	+	+
2-Acetamido-4'-nitroacetophenone	0.87	+	-	-	+	+	+
Chloramphenicol	0.89	+	-	-	-	-	-
2-(2,2-Dichloroacetamido)-3-hydroxy-4'-nitropropiophenone	0.95	+	-	-	-	-	-
p-Nitrobenzaldehyde	0.95	+	-	-	+	+	-

Solvent A - n-Butanol-$CHCl_3$-acetic acid (10:90:0.5)
Solvent B - n-Butanol-water-acetic acid (82:18:0.5)

Schlederer[114] accomplished a comparable separation using silica gel G plates and $CHCl_3$:MeOH (9:1) as developer.

The separation and identification of decomposition products of chloramphenicol by thin layer has been done mostly on silica gel plates using a variety of solvent systems. Tacharme[64] used silica gel G plates and two-dimensional chromatography to achieve separation. The R_f values he obtained are listed as follows:

	1st Solvent	2nd Solvent
p-nitrobenzaldehyde	0.98 (streak)	0.91
p-nitrobenzoic acid	0.10-0.15	0.80-0.81
Chloramphenicol	0.72	0.89-0.91
1-(p-nitrophenyl)-2-amino-1,3-propanediol	0.08-0.10	0.35

1st Solvent - ethyl acetate (saturated with water)
2nd Solvent - n-Butanol (saturated with 2.5% acetic acid)

Shih[63,66] used five binary developers with silica gel plates to detect and identify several secondary decomposition products as well as the azoxy compound formed from photolysis (see Table 3).

Aromatic decomposition products (see stability-decomposition) arising from chloramphenicol after heating in alkaline solution were detected by Knabe[65] using silica gel G plates. The results he obtained are as follows:

Compound	R_f	Developer
4,4'-azodiphenol	0.25	Methylene Chloride-Ether 9:1
p-nitrophenol	0.55	"
4,4'-azoxydiphenol	0.21	"
p-nitrosophenol	0.30	"
p-[(p-hydroxyphenyl)-azo]benzyl alcohol	0.15	"
p-[(p-hydroxyphenyl)-azo]benzaldehyde	0.45	"
4,4'-azodibenzaldehyde	0.92	Chloroform-Ether 9:1
p-[(α-hydroxy-p-tolyl)-azo]benzaldehyde	0.43	"
4,4'-azodibenzylalcohol	0.03	"

Thin layer chromatography has been used as part of a quantitative scheme of analysis. Schwarm[101] used silica gel HF_{254} thin layer plates to separate chloramphenicol from decomposition products. Once separation was accomplished, the intact antibiotic was desorbed and photometrically determined. $CHCl_3$-Isopropanol 4:1 was used as developer.

The same approach was used by Kassem[115] for analysis of intact drug. In this case, silica gel plates were used with $CHCl_3$-MeOH, 85:15, as mobile phase.

Libsovar[116] developed thin layer chromatographic systems for use in monitoring the classical synthesis of chloramphenicol. For this purpose, Alumina plates were used with binary developers of benzene-ethanol (2.5 - 20% ethanol).

6.63 Partition (Column)

Intact chloramphenicol can be determined in the presence of decomposition products by using a partition column[117]. The column is prepared with a silicic acid-water as internal phase. Chloroform followed by 10% ethylacetate in chloroform serve as eluents, the eluent being monitored spectrophotometrically at 278 nm.

This analytical procedure was validated by direct comparison with microbiological assay before being used to study the kinetics of degradation of chloramphenicol.

6.64 Gas Liquid

Shaw[118] developed a gas chromatographic procedure for analysis of chloramphenicol in the presence of structurally related compounds likely to be present in the growth medium or cell free extract of cultures of *S. venezuelae*. The procedure involved chromatography of these compounds after conversion to their corresponding trimethylsilyl derivatives. The column temperature was programmed over the region 110-260°C and hydrogen flame ionization was used for detection. A chromatogram of a synthetic mixture of these compounds is included as Fig. 9.

Resnick adapted Shaw's procedure to analysis of chloramphenicol in serum[90]. In brief, this procedure involves extraction of chloramphenicol from serum with amyl acetate, derivitization using Tri-Sil reagent followed by chromatography using Qf-1 on silanized Gas Chrom P. The sensitivity of the method was adequate for determination of chloramphenicol in the region of 0.6 mcg/ml. Yamamoto[119] reported that chloramphenicol can be determined in body fluids and pharmaceutical preparations by gas chromatography without prior derivatization. Two per cent diethyleneglycol succinate on PVP modified Anakrom was used with a column temperature of 195°C.

Davies[120] developed a gas chromatographic method for the determination of small amounts of the ortho and meta nitro isomers of chloramphenicol as possible impurities in chloramphenicol.

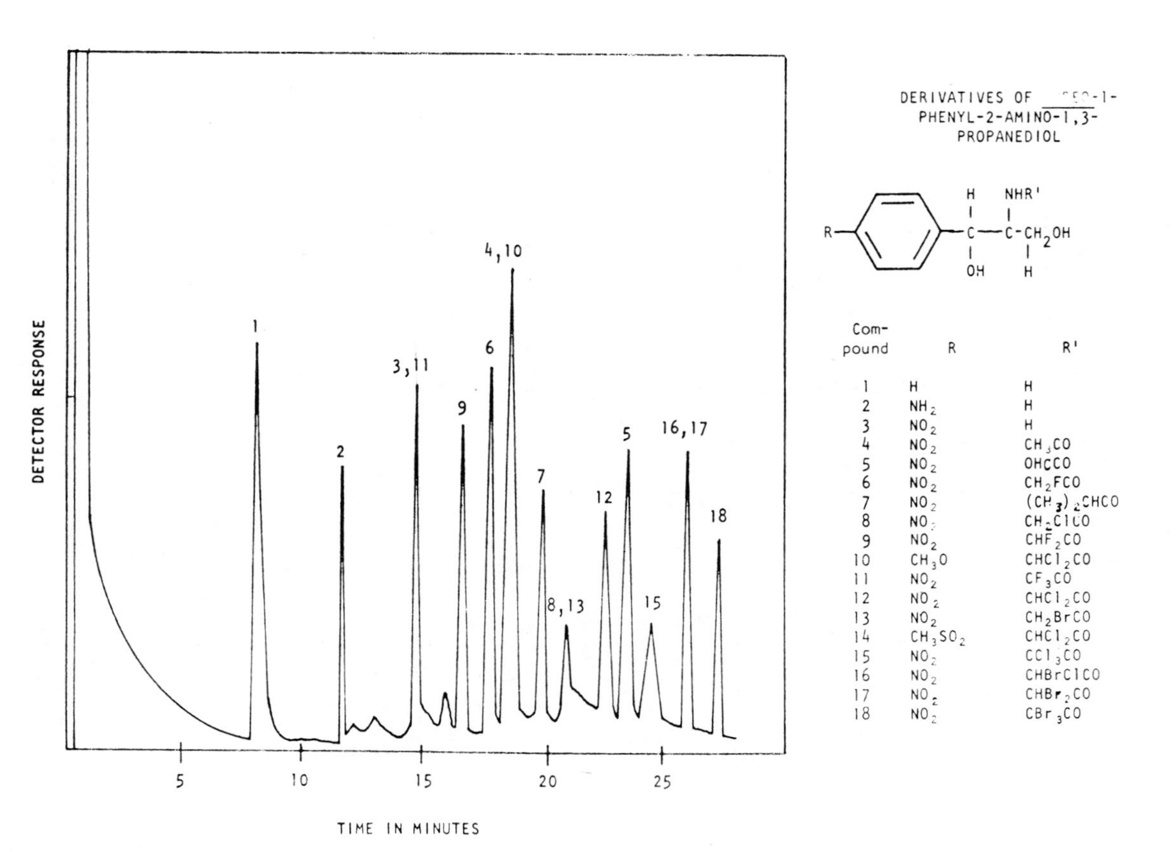

Compound	R	R'
1	H	H
2	NH_2	H
3	NO_2	H
4	NO_2	CH_3CO
5	NO_2	OHCCO
6	NO_2	CH_2FCO
7	NO_2	$(CH_3)_2CHCO$
8	NO_2	CH_2ClCO
9	NO_2	CHF_2CO
10	CH_3O	$CHCl_2CO$
11	NO_2	CF_3CO
12	NO_2	$CHCl_2CO$
13	NO_2	CH_2BrCO
14	CH_3SO_2	$CHCl_2CO$
15	NO_2	CCl_3CO
16	NO_2	CHBrClCO
17	NO_2	$CHBr_2CO$
18	NO_2	CBr_3CO

FIG. 9[118] - GAS CHROMATOGRAM OF CHLORAMPHENICOL AND STRUCTURALLY RELATED COMPOUNDS.

References

1. The Merck Index, 8th ed., Merck and Co., Inc., Rahway, N.J. (1968), p. 233.
2. W. H. Hartung and J. Andrako, J. Pharm. Sci., 50, 805 (1961).
3. The United States Pharmacopeia, 18th revision, Mack Publishing Co., Easton, Pa. (1970).
4. J. Krc, Parke Davis, personal communication.
5. J. Krc, Parke Davis, personal communication.
6. J. D. Dunitz, J. Am. Chem. Soc., 74, 955 (1952).
7. Q. R. Bartz, J. Biol. Chem., 172, 445 (1948).
8. Federal Register, 146 d. 301, July (1961).
9. C. Johnston, Parke Davis, personal communication.
10. H. Negoro, et al., Annual Report, Takanime Research Institute, 9, 77 (1957).
11. H. Kostenbauder and T. Higuchi, J. Am. Pharm. Assoc., Sci. Ed., 45, 518 (1956).
12. E. Regdon-Kiss, Pharmazie, 18, 755 (1963).
13. M. Suzuki, J. Antibiot., Ser. B., 15 323 (1961).
14. K. James and R. Leach, J. Pharm. Pharmacol., 22, 612 (1970).
15. T. Higuchi and A. Marcus, J. Amer. Pharm. Ass., Sci. Ed., 43, 530 (1954).
16. A. Brunzell, J. Pharm. Pharmacol., 8, 329 (1956).
17. M. Rebstock, H. Crooks, J. Controulis, and Q. Bartz, J. Amer. Chem. Soc., 71, 2458 (1949).
18. M. Suzuki and H. Shindo, Yakugaku Zasshi, 76, 927 (1956).
19. O. Jardetsky, J. Biol. Chem., 238, 2498 (1963).
20. D. Szulczewski, unpublished information.
21. L. A. Mitscher, F. Kautz, and J. Lapidus, Can. J. Chem., 47, 1957 (1969).
22. S. C. Ray, Ann. Biochem. Exp. Med., 23, 411 (1963).
23. C. Brunnee, G. Kappus, and K. H. Maurer, Z. Anal. Chem., 232 17 (1967).
24. U. S. Pat. 2,438,871
25. U. S. Pat. 2,483,892
26. J. Controulis, M. Rebstock, H. Crooks, J. Amer. Chem. Soc., 71, 2463 (1949).
27. L. M. Long, H. D. Troutman, J. Amer. Chem. Soc., 71, 2469 (1949).
28. ibid 71, 2473, (1949).
29. ibid 73, 481 (1951).

30. ibid 73, 542 (1951).
31. U. S. Pat. 2,681,364
32. U. S. Pat. 2,687,434
33. U. S. Pat. 2,562,107
34. U. S. Pat. 2,515,239
35. U. S. Pat. 2,515,240
36. U. S. Pat. 2,515,241
37. U. S. Pat. 2,546,762
38. U. S. Pat. 2,483,885
39. U. S. Pat. 2,692,897
40. U. S. Pat. 2,677,704
41. U. S. Pat. 2,538,763
42. U. S. Pat. 2,515,377
43. U. S. Pat. 2,686,788
44. U. S. Pat. 2,751,413
45. U. S. Pat. 2,543,957
46. U. S. Pat. 2,483,884
47. U. S. Pat. 2,699,451
48. G. Moersch, D. Hylander, J. Amer. Chem. Soc., 76, 1703 (1954).
49. S. Igiguma, Yakugaku Zasshi, 75, 673 (1952)
50. M. Myamoto, ibid, 72, 673 (1952)
51. U. S. Pat. 2,562,113
52. U. S. Pat. 2.718,520
53. U. S. Pat. 2,807,645
54. U. S. Pat. 2,562,114
55. U. S. Pat. 2,734,919
56. U. S. Pat. 2,727,063
57. U. S. Pat. 2,586,661
58. M. Suzuki, J. Antibiot., Ser. B., 14, 323 (1961).
59. T. Higuchi and C. Bias, J. Amer. Pharm. Ass., Sci. Ed., 42, 707 (1953).
60. T. Higuchi, A. Marcus, and C. Bias, ibid, 43, 129 (1954).
61. C. Trolle-Lassen, Arch. Pharm. Chemi., 60, 689 (1953).
62. R. Saba, D. Monnier, and F. R. Khalil, Pharm. Acta. Helv., 42, 335.
63. K. K. Shih, J. Pharm. Sci., 60, 786 (1971).
64. J. Lacharme and D. Netien, Bull. Trav. Soc. Pharm. Lyon. 8, 122 (1964).
65. J. Knabe and R. Krauter, Arch. Pharm., (1962) 190.
66. I. K. Shih, J. Pharm. Sci., 60, 1889 (1971).
67. A. Brunzell, Sv. Farm. Tidskr., 61, 129 (1957).

68. M. Heward, D. A. Norton, and S. M. Rivers, Pharm. J., 204, 386 (1970).
69. British Pharmaceutical Codex 1968 London, The Pharmaceutical Press, 1968.
70. K. C. James and R. H. Leach, Pharm. J., 204, 472 (1970)
71. G. Smith and C. Worrel, Arch. Biochem., 28, 232 (1950)
72. A. Glazko, Antimicrob. Ag. Chemother., (1966) 655.
73. W. A. Dill, E. M. Thompson, R. A. Fiskin, and A. J. Glazko, Nature, 185, 535 (1960).
74. A. Glazko, W. A. Dill, A. Kazenko, L. M. Wolf, and H. E. Carnes, Antibiot. Chermother., 8, 516 (1968).
75. A. J. Glazko, W. A. Dill, and L. M. Wolf, J. Pharmacol. Exp. Ther., 104, 452 (1952).
76. A. J. Glazko, L. M. Wolf, W. A. Dill, and A. C. Bratton, Jr., ibid, 96, 445 (1949).
77. A. J. Glazko, A. W. Kinkel, W. C. Alegnani, and E. L. Holmes, Clin. Pharmacol. Ther., 9, 472-483(1968)
78. A. J. Aguiar, L. M. Wheeler, S. Fusari, and J. Zelmer, J. Pharm. Sci., 57, 1555 (1968).
79. United States Pharmacopeia XVI. 1960
80. British Pharmacopeia 1968.
81. Deutsche Apotheker. 6
82. Dōll, Arzeim. Forsch., 5, 97 (1955).
83. A. Glazko, L. Wolf, and W. Dill, Arch. Biochem., 23, 411 (1949)
84. S. P. Bessman and S. Stevens, J. Lab. Clin. Med., 35, 127 (1950).
85. J. Levine and H. Fishback, Antibiot. Chemother., 1, 59 (1951).
86. T. H. Aihara, H. Machida, and Y. Yoneda, J. Pharm. Soc. Jap., 77, 1318 (1957).
87. M. S. Karawya and M. G. Ghourab, J. Pharm. Sci., 59, 1331 (1970).
88. K. Kakemi, T. Arito, and S. Ohasaki, Yakugaku Zasshi, 82, 342 (1962).
89. D. Hughes and L. K. Diamond, Science, 144, 296 (1964)
90. G. L. Resnick, D. Corbin, and D. H. Sandberg, Anal. Chem., 38, 582 (1966)
91. R. C. Shah, P. V. Raman, and P. V. Sheth, Indian J. Pharm., 30, 68 (1968).
92. F. M. Freeman, Analyst, 80, 299 (1956).
93. D. Masterson, J. Pharm. Sci., 57, 306 (1968).
94. K. Fossdal and E. Jacobson, Anal. Chim. Acta., 56, 105 (1971)

95. G. B. Hess, Anal. Chem., 22, 649 (1950).
96. C. G. Macros, Chem. Chron. A., 32 104 (1967).
97. C. Russo, I. Cruceanu, D. Monciu, and V. Barcaru, Il. Farmaco, 20, 22 (1965).
98. C. Russu, I. Cruceanu, and V. Barcaru, Pharmazie 18, 799 (1963).
99. A. F. Summa, J. Pharm. Sci., 54, 442 (1963).
100. P. Zuman, "Organic Polarographic Analysis", MacMillan Co., New York, 1964, p. 186.
101. E. Schwarm, C. Dabner, J. Wilson, and M. Boghosian, J. Pharm. Sci. 55, 744 (1966).
102. T. Higuchi, A. D. Marcus, and C. D. Bias, J. Amer. Pharm. Ass., Sci. Ed., 43, 135 (1959).
103. W. Awe and H. Stohlman, Arch. Pharm., 289, 61, 276 (1956).
104. M. Hadicke and G. Schmid, Pharm. Zentralh., 95, 387 (1956).
105. W. Awe and H. Stohlman, Arzeim. Forsch., 7 (8), 495 (1959).
106. D. C. Grove and W. A. Randall, "Assay Methods of Antibiotics: A Laboratory Manual", 238 pp Medical Encyclopedia, Inc., New York, 1955
107. B. Arret, M. R. Woodard, D. M. Wintermere, and A. Kirschbaum, Antibiot. Chemother., 7 545 (1957).
108. R. Hans, M. Galbraith, and W. C. Alegnani, "Analytical Microbiology," F. Kavanagh, Ed., Academic Press, 1963, p 271-281.
109. United States Pharmacopeia XVIII, 1970, pp 857-864, Antibiotics-Microbial Assays.
110. F.D.A. Regulations Title 21 Sec. 141.110.
111. F.D.A. Regulations Title 21 Sec. 141.111.
112. G. N. Smith and C. S. Worrel, Arch. Biochem., 28, 1 (1950).
113. Y. T. Lin and K. T. Wang, J. Chromatogr., 21, 158 (1966).
114. E. Schlederer, Cosm. Pharma., 3, 17 (1966).
115. M. A. Kassem and A. A. Kassem, Pharm. Ztg., 48, 1972 (1966).
116. J. Lebsovar, Cesk. Farm. 11, 73 (1962).
117. T. Higuchi, C. Bias, and A. Marcus, J. Amer. Pharm. Ass., Sci. Ed., 43, 135 (1954).
118. P. D. Shaw, Anal. Chem., 35, 1580 (1963).
119. M. Yamamoto, S. Iguchi, and T. Aoyama, Chem. Pharm. Bull., 15, 123 (1967).

120. V. Davies, Parke Davis, personal communication.

ACKNOWLEDGMENT

The authors express appreciation to Mrs. Pat Greenwood of the Microbiology Department at Parke, Davis & Company for assistance in preparing a portion of this profile.

CLORAZEPATE DIPOTASSIUM

James A. Raihle and Victor E. Papendick

Contents

Analytical Profile - Clorazepate Dipotassium

1. Description

1.1 Name, Formula, Molecular Weight

Clorazepate dipotassium is 7-chloro-1,3-dihydro-2-oxo-5-phenyl-1H-1,4-benzodiazepine-3-carboxylic acid, monopotassium salt, monopotassium hydroxide.

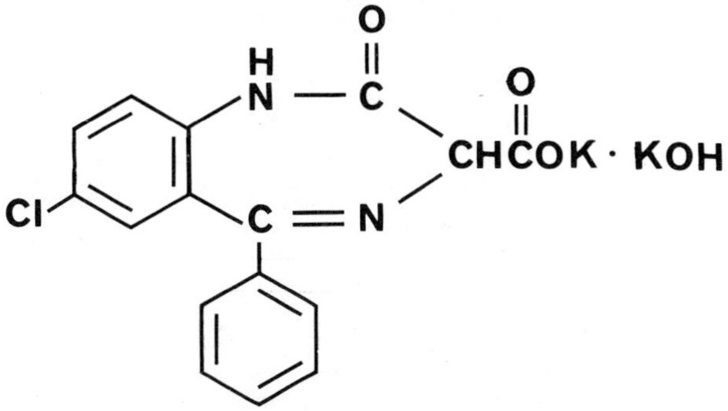

Clorazepate Dipotassium

$C_{16}H_{11}O_4N_2ClK_2$ Molecular Weight 408.93

1.2 Appearance, Color, Odor

Off white to pale yellow, fine crystalline powder which is practically odorless.

2. Physical Properties

2.1 Infrared Spectrum

The infrared spectrum of clorazepate dipotassium is presented in Figure 1. The spectrum was measured in the solid state as a mull in mineral oil. The following bands (cm^{-1}) have been assigned for Figure 1.(1)

a. 3530 cm^{-1} characteristic for hydroxyl

b. 1610 cm^{-1} characteristic skeletal stretching modes of the aromatic ring

c. 1560 cm^{-1} characteristic C=O stretching mode of the carboxyl salt

2.2 Nuclear Magnetic Resonance Spectrum (NMR)

The NMR spectrum shown in Figure 2 was obtained by dissolving 50 mg of clorazepate dipotassium in 0.5 ml of D_2O containing tetramethylsilane as an internal reference. Only the aromatic protons between 7.0 and 7.6 ppm are visible.(2)

Figure 1 INFRARED SPECTRUM OF CLORAZEPATE DIPOTASSIUM

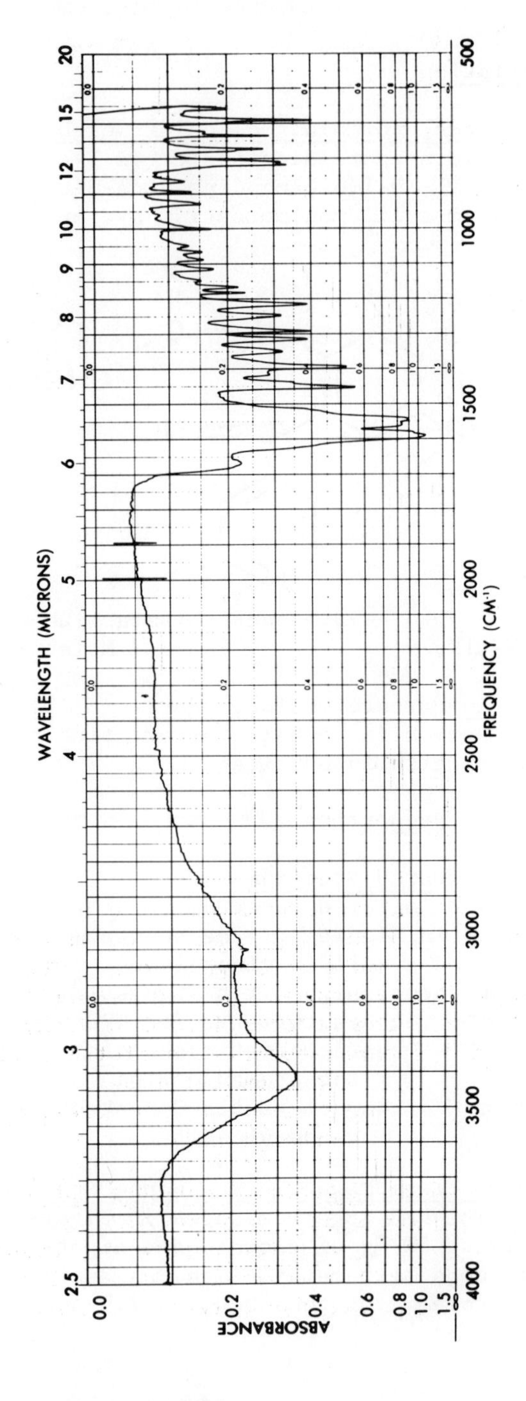

Figure 2 NUCLEAR MAGNETIC RESONANCE SPECTRUM OF CLORAZEPATE DIPOTASSIUM

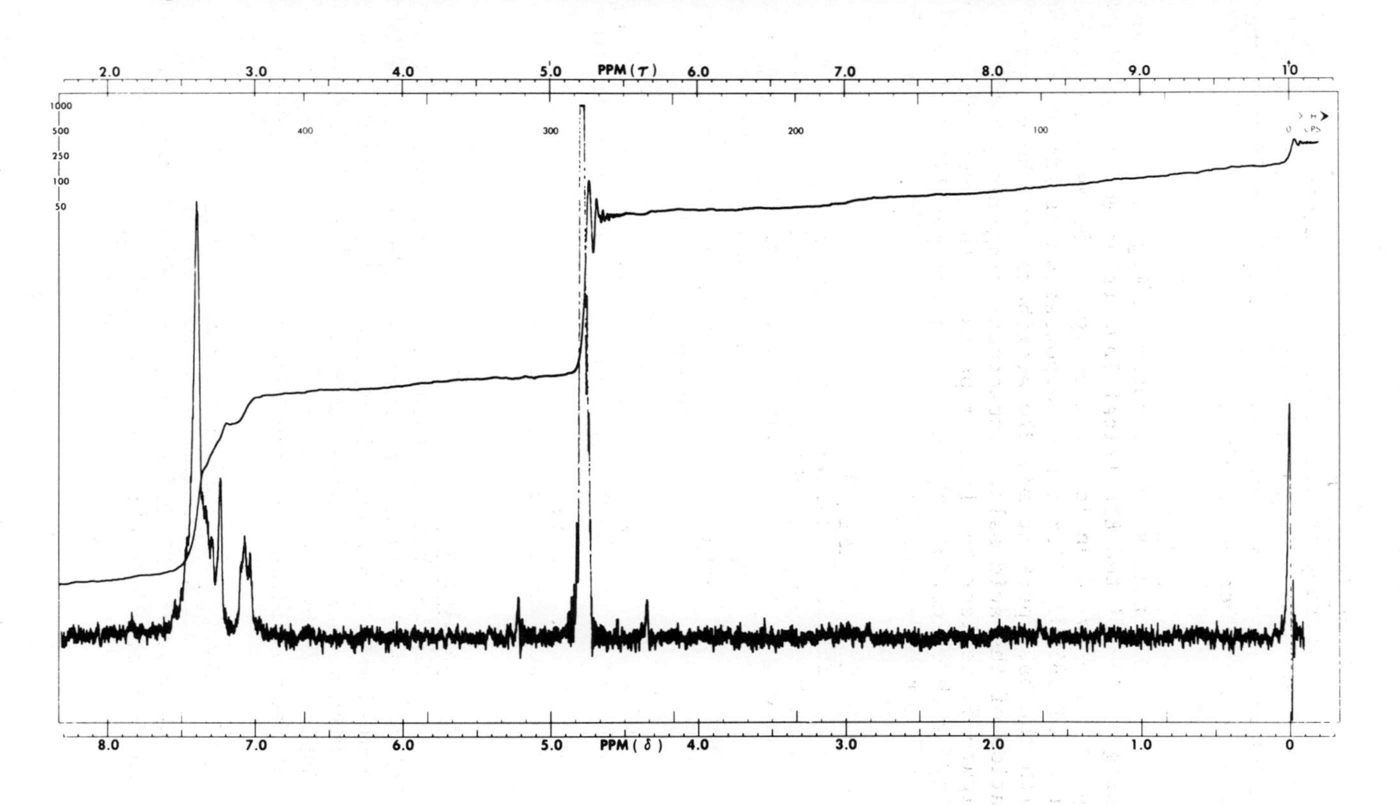

2.3 Ultraviolet Spectrum

Clorazepate dipotassium when scanned between 400 and 200 nm in 0.03% aqueous potassium carbonate exhibits a maximum at 230 nm as shown in Figure 3, (ϵ = 35,000) characteristic of benzodiazepines.

2.4 Mass Spectrum

The mass spectrum shown in Figure 4 was obtained using an Associated Electrical Industries Model MS-902 Mass Spectrometer with an ionizing energy of 50 eV and a temperature of 185°C. Clorazepate dipotassium yields a spectrum with the base peak at m/e 270 attributed to the decarboxylation of the acid salt. Subsequent fragments, Table I and Figure 5, reflect the loss of parts of the seven membered ring or chlorine.(3) The mass spectrum parallels that reported for diazepam.(4)

2.5 Raman Spectrum

The Raman spectrum of clorazepate dipotassium, as shown in Figure 6, was obtained in the solid state on a Ramalog Spectrophotometer at Spex Industries. The following bands (cm^{-1}) have been assigned for Figure 6.(1)

a. 1595 cm^{-1} skeletal stretching mode of aromatic ring
b. 1565 cm^{-1} C=N stretching vibration of the heterocyclic ring
c. 1495 cm^{-1} skeletal stretching mode of aromatic ring

2.6 Optical Activity

Clorazepate dipotassium exhibits no optical activity.

2.7 Melting Range

Clorazepate dipotassium does not have a definite melting range. Typical behavior when the material is slowly heated in a glass capillary tube may be described in the following manner: discoloration begins at about 215°C, shrinking is observed to begin between 225°C to 235°C with total decomposition occuring between 235°C and 295°C.

2.8 Differential Thermal Analysis (DTA)

The DTA curve obtained on a DuPont Model 900 Analyzer as shown in Figure 7 confirms the observed melting characteristics described in section 2.7.

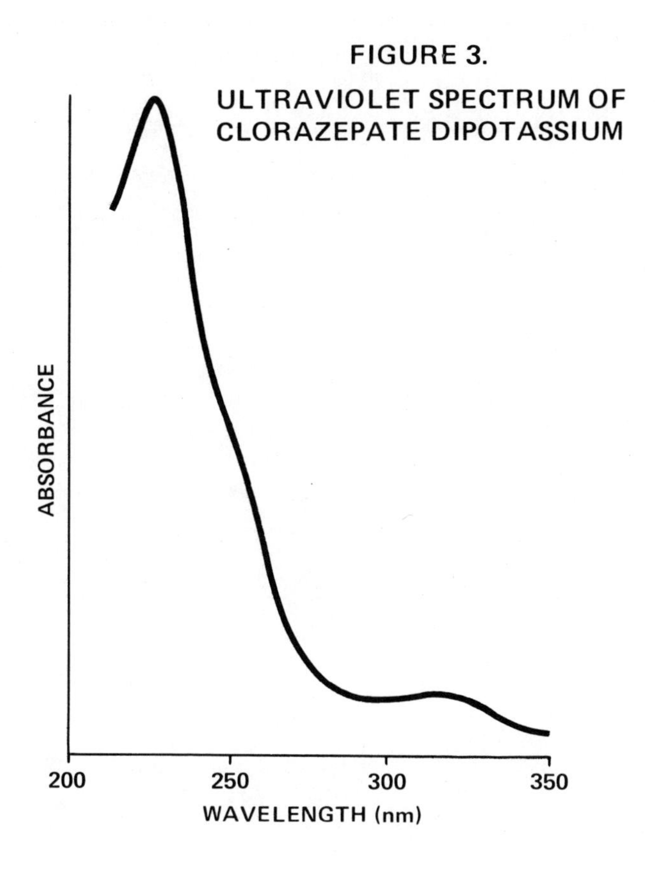

FIGURE 3.
ULTRAVIOLET SPECTRUM OF CLORAZEPATE DIPOTASSIUM

Figure 4 MASS SPECTRUM OF CLORAZEPATE DIPOTASSIUM

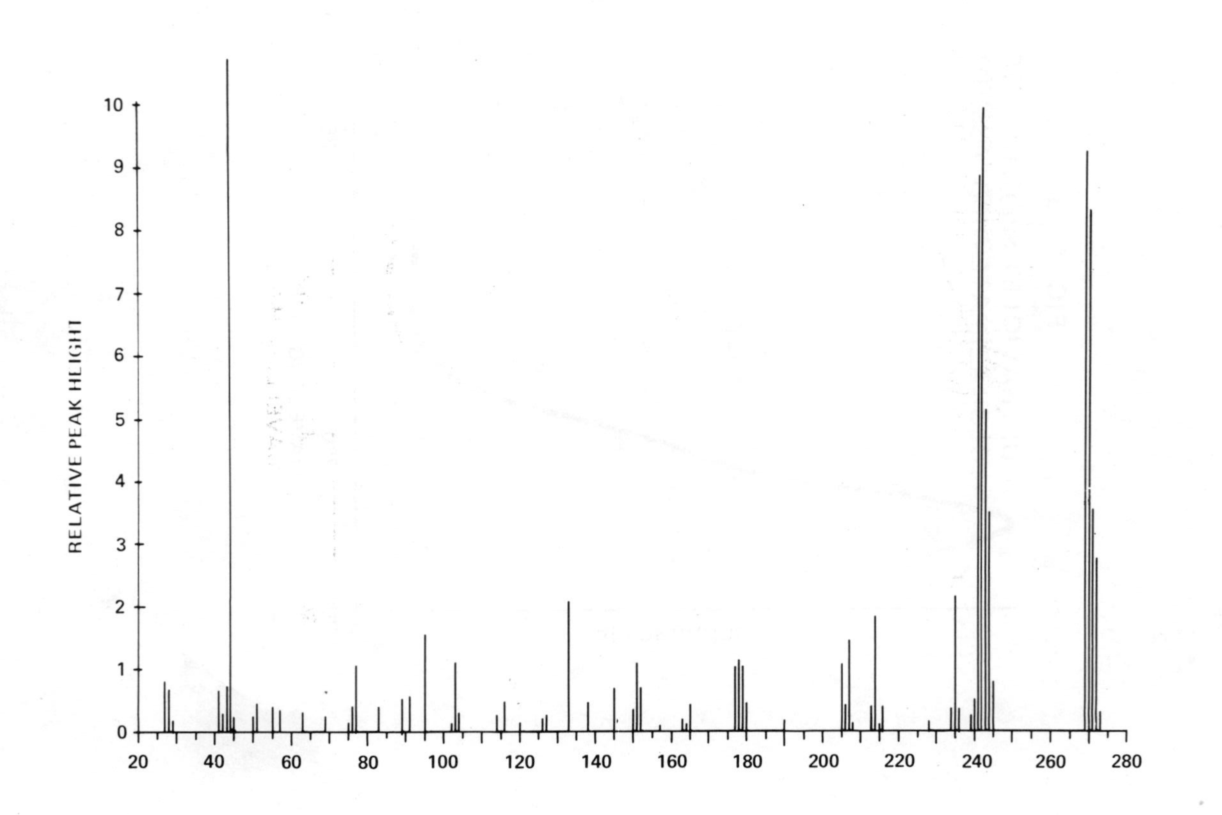

TABLE I

High Resolution Mass Spectrum of Clorazepate Dipotassium

Found Mass	Calculated Mass	C	H	N	O	Cl_{35}
270.0539	270.0559	15	11	2	1	1
269.0475	269.0481	15	10	2	1	1
242.0392	242.0372	14	9	1	1	1
241.0517	241.0532	14	10	2	0	1
235.0862	235.0871	15	11	2	1	0
214.0427	214.0423	13	9	1	0	1
179.0794	179.0735	13	9	1	0	0
178.0669	178.0656	13	8	1	0	0
177.0586	177.0578	13	7	1	0	0
77.0392	77.0391	6	5	0	0	0
43.9900	43.9898	1	0	0	2	0

FIGURE 5.
FRAGMENTATION PATHWAYS OF CLORAZEPATE DIPOTASSIUM

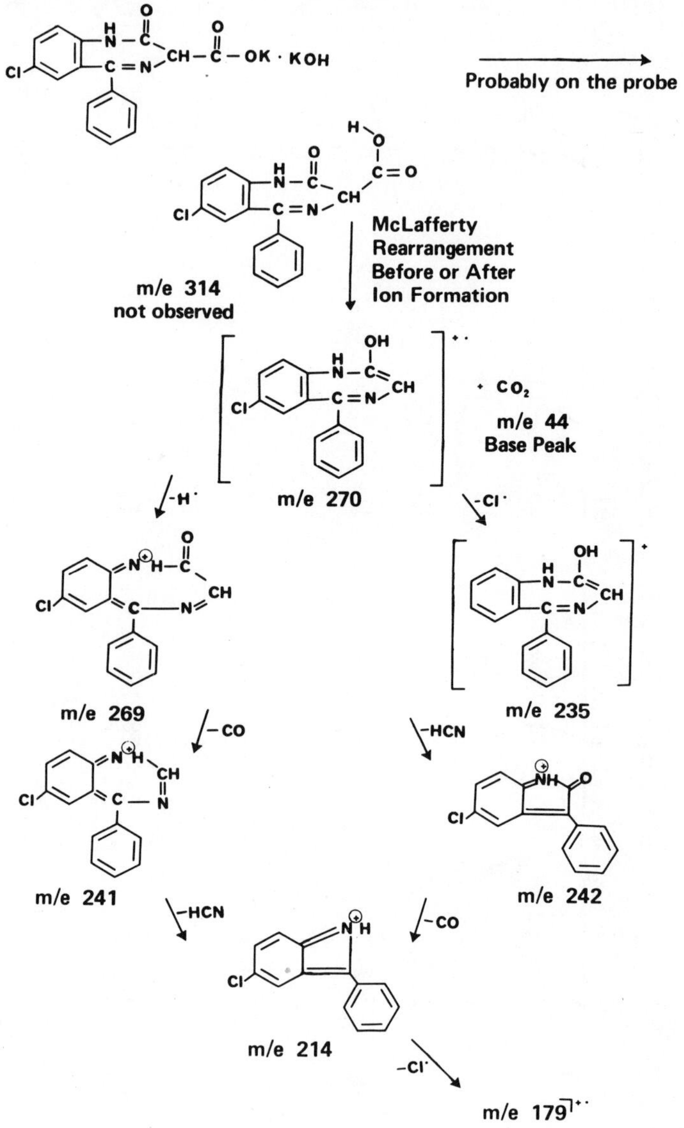

Probably on the probe

m/e 314
not observed

McLafferty
Rearrangement
Before or After
Ion Formation

m/e 270

+ CO_2
m/e 44
Base Peak

-H·

-Cl·

m/e 269

m/e 235

-CO

-HCN

m/e 241

m/e 242

-HCN

-CO

m/e 214

-Cl·

m/e 179

Figure 6 RAMAN SPECTRUM OF CLORAZEPATE DIPOTASSIUM

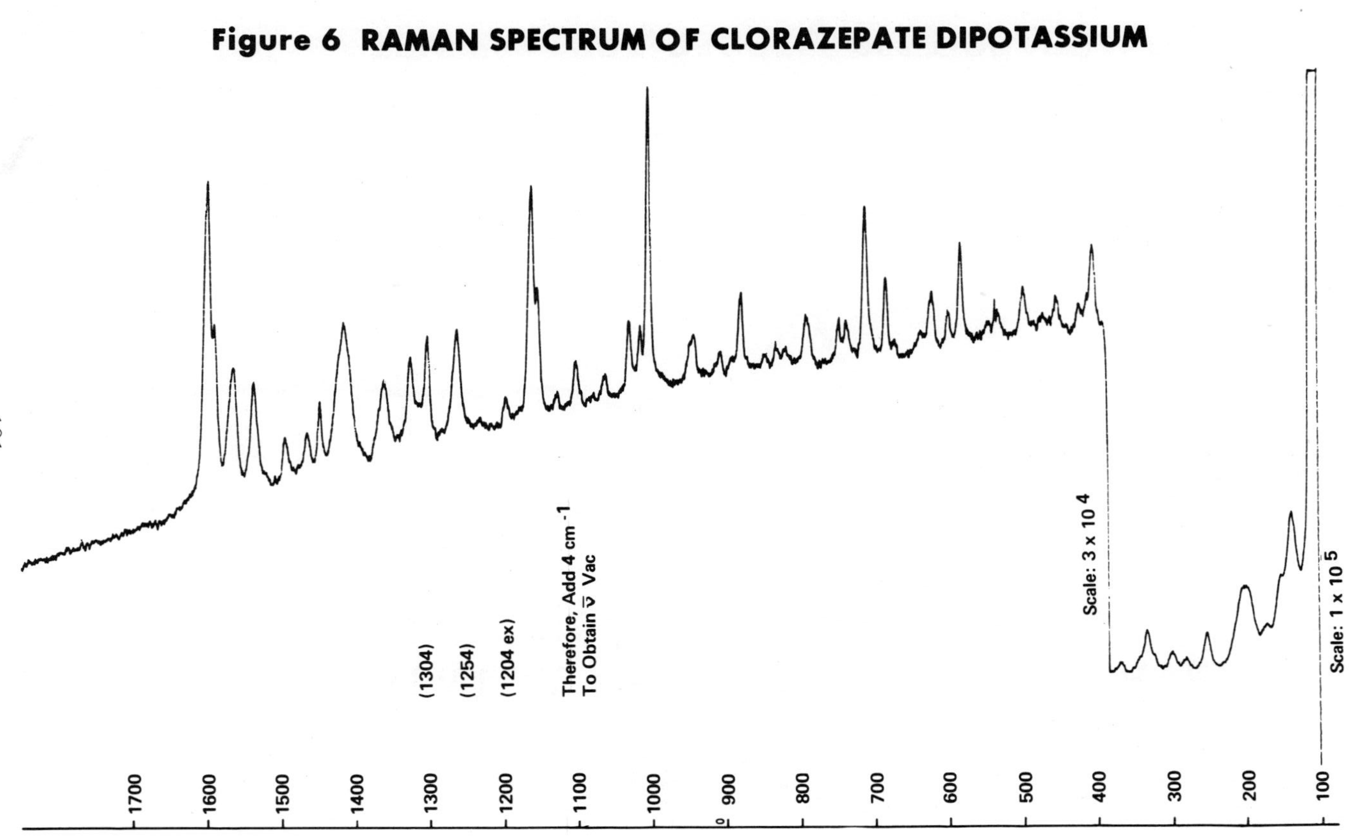

FIGURE 7.

DIFFERENTIAL THERMAL ANALYSIS CURVE OF CLORAZEPATE DIPOTASSIUM

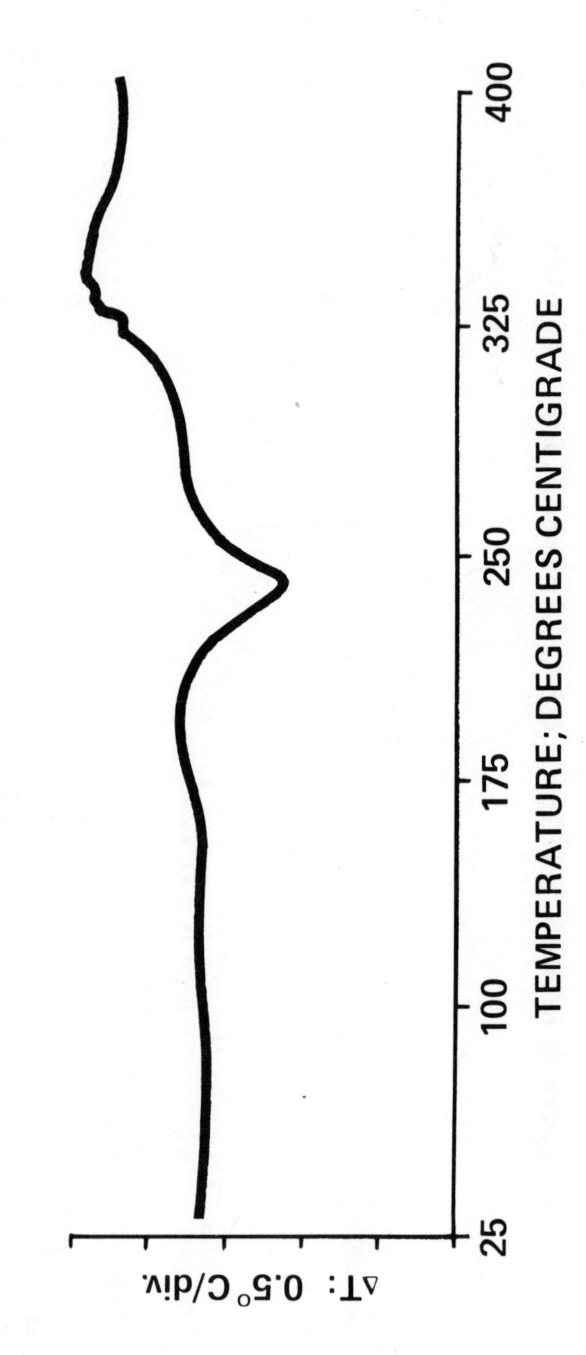

2.9 Solubility

Approximate solubility data obtained at room temperature are given in the following table:

Solvent	Solubility (mg/ml)
Water	> 100 < 200
Absolute Ethanol	0.6
Chloroform	< 0.5
Ether	< 0.5
Acetone	< 0.5
Benzene	< 0.5
Isopropanol	0.7
Methylene Dichloride	< 0.1

2.10 Crystal Properties

The X-ray powder diffraction pattern of clorazepate dipotassium was determined by visual observation of a film obtained with a 143.2 mm Debye-Scherrer Powder Camera. An Enraf-Nonius Diffractis 601 Generator; 38 KV and 18 MA with nickel filtered copper radiation; λ = 1.5418, were employed. (5)

2.11 Dissociation Constant

Attempts to measure the pKa of the carboxyl group by titration in water with hydrochloric acid were unsuccessful. Only the potassium hydroxide which is liberated on dissolving clorazepate dipotassium in water is titrated. (6)

2.12 Fluorescence

Clorazepate dipotassium does not exhibit fluorescent properties in an aqueous solution, however, it does exhibit fluorescence at 508 nm when excited at 388 nm in 9 *N* alcoholic sulfuric acid. (7)

2.13 Hygroscopic Behavior

Clorazepate dipotassium was not hygroscopic when exposed to a relative humidity of 30%-40% for 4½ months.

3. Synthesis

Clorazepate dipotassium may be prepared by the reaction scheme shown in Figure 8 with the reaction of (2-amino-5-chlorophenyl)phenyl methane imine and diethyl 2-amino malonate to form ethyl 7-chloro-1,3-dihydro-2-oxo-5-phenyl-1H-

TABLE II

X-Ray Powder Diffraction Pattern

d-Spacings and Intensities

dA	I/I_1	dA	I/I_1
17.5	100	2.99	5
8.6	10	2.93	1
7.7	25	2.88	2
7.0	10	2.82	10
6.18	20	2.78	3
5.7	20	2.67	5
5.05	30	2.62	5
4.82	75	2.57	1
4.35	10	2.46	10
4.20	5	2.42	2
4.11	1	2.27	1
4.05	1	2.17	5
3.95	5	2.13	2
3.80	5	2.07	1
3.66	25	2.03	1
3.52	40	1.99	1
3.48	1	1.97	1
3.40	10	1.93	1
3.30	60	1.87	1
3.25	65	1.81	1
3.14	1	1.78	1
3.06	15	1.71	1

FIGURE 8.
SYNTHESIS OF CLORAZEPATE DIPOTASSIUM

Cl, NH_2, C=NH + H_2N-CH, $C(=O)-OC_2H_5$, $C(=O)-OC_2H_5$ →

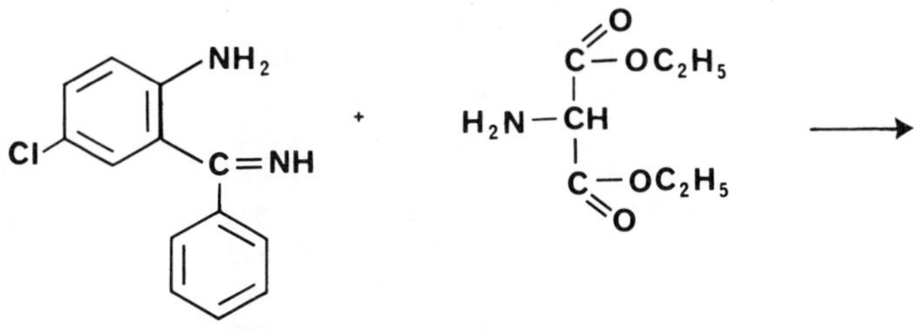

(2-Amino-5-chlorophenyl) phenyl methane imine

Diethyl-2-aminomalonate

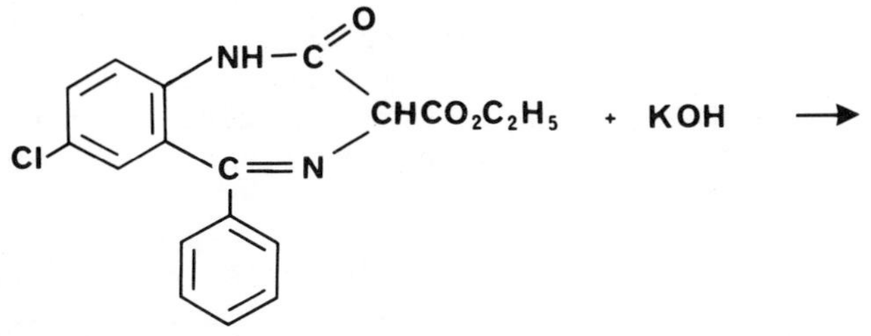

Cl, NH—C(=O), $CHCO_2C_2H_5$, C=N + KOH →

Ethyl 7-chloro-1,3-dihydro-2-oxo-5-phenyl-1H-1,4-benzodiazpine-3-carboxylate

Potassium Hydroxide

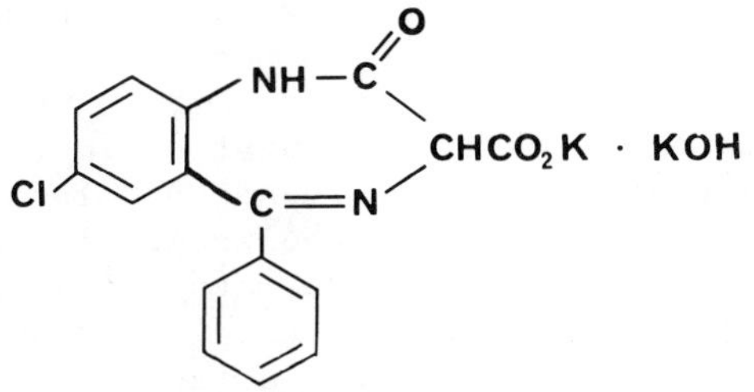

Cl, NH—C(=O), $CHCO_2K \cdot KOH$, C=N

Clorazepate Dipotassium

1,4-benzodiazepine-2-carboxylate. This intermediate is then converted to the drug substance with alcoholic potassium hydroxide.(8)

4. Stability - Degradation

The hydrolysis products for clorazepate dipotassium are shown in Figure 9. The final hydrolysis product is the same as that reported for the acid hydrolysis of clordiazepoxide and the major metabolite of diazepam.(4) The kinetics of the decomposition of clorazepate dipotassium in buffered aqueous solution at different temperatures was studied over the pH range of 2-11.(9) The extent of degradation was determined by a dichloromethane extraction and spectrophotometric measurement. Ring opening was negligible under the conditions of pH and temperature used.

The degradation to N-desmethyl diazepam is first order with respect to clorazepate dipotassium. The relationship $\log_e (C_t/C_o) = Kt$ was verified where C_t is the concentration of clorazepate dipotassium at time t, C_o is the concentration of clorazepate dipotassium at time 0, and k is the reaction rate constant. Plots of $\log_e (C_t/C_o)$ as a function of time t were linear. The slopes of these plots give the reaction rate constants k. Some rate constants, sec^{-1}, at different pH and temperature values are shown in Table III.

The data shows that in the degradation of clorazepate dipotassium the reaction rate constant, k, increases with temperature. The activation energy of the degradation reaction, Ea (obtained from the slope of the plot of $\log_e k$ as a function of 1/T where T is the absolute temperature), is about 20.3 kilocalories/mole and is independent of pH.

5. Drug Metabolic Products and Pharmacokinetics

The major metabolites of clorazepate dipotassium are shown in Figure 10. The drug substance is decarboxylated in acidic media to N-desmethyl diazepam, which can undergo hydroxylation to oxazepam and conjugation in the urine to the glucuronide.(10) Some clorazepate dipotassium (2-6%) is excreted into the urine unmetabolized.(11) Acid hydrolysis of the drug substance and its metabolites affords 2-amino-5-chlorobenzophenone. Analytical procedures have been published for the metabolites using gas chromatography (12), thin layer chromatography (7), high pressure

TABLE III

Rate Constant vs pH for Clorazepate Dipotassium

t°C → / pH ↓	5°	15°	22°	27.5°	37°
2.0	$7.60X10^{-4}$	$1.32X10^{-3}$	$3.60X10^{-3}$	$6.46X10^{-3}$	$1.93X10^{-2}$
3.0	$4.75X10^{-4}$	$1.12X10^{-3}$	$3.40X10^{-3}$	$5.96X10^{-3}$	$1.42X10^{-2}$
4.0	$8.74X10^{-5}$	$4.08X10^{-4}$	$1.16X10^{-3}$	$1.98X10^{-3}$	$6.33X10^{-3}$
5.0	$1.65X10^{-5}$	$8.50X10^{-5}$	$2.17X10^{-4}$	$4.10X10^{-4}$	$1.14X10^{-3}$
6.0	---	---	$2.23X10^{-5}$	---	$1.18X10^{-4}$
6.5	---	---	$3.90X10^{-6}$	---	$8.33X10^{-5}$
7.0	---	---	$2.33X10^{-6}$	---	$2.14X10^{-5}$
7.5	---	---	$1.05X10^{-6}$	---	$1.32X10^{-5}$
8.0	---	---	---	---	$1.27X10^{-5}$
8.5	---	---	$1.05X10^{-6}$	---	$5.61X10^{-6}$
9.5	---	---	$9.10X10^{-7}$	---	$4.83X10^{-6}$
11.0	---	---	$8.64X10^{-7}$	---	$4.55X10^{-6}$

FIGURE 9.
DEGRADATION PATHWAYS OF CLORAZEPATE DIPOTASSIUM

$CHCOK \cdot KOH$

Clorazepate Dipotassium

↓

CH_2 + K_2CO_3

N-Desmethyl Diazepam

↓

NH_2 + H_2NCH_2COH

2-Amino-5-chlorobenzophenone Glycine

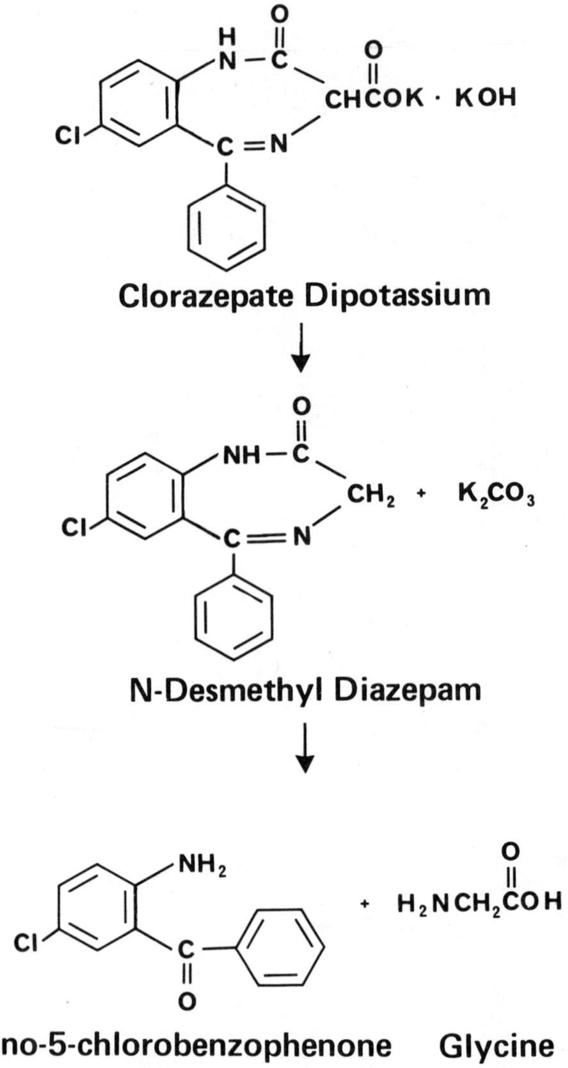

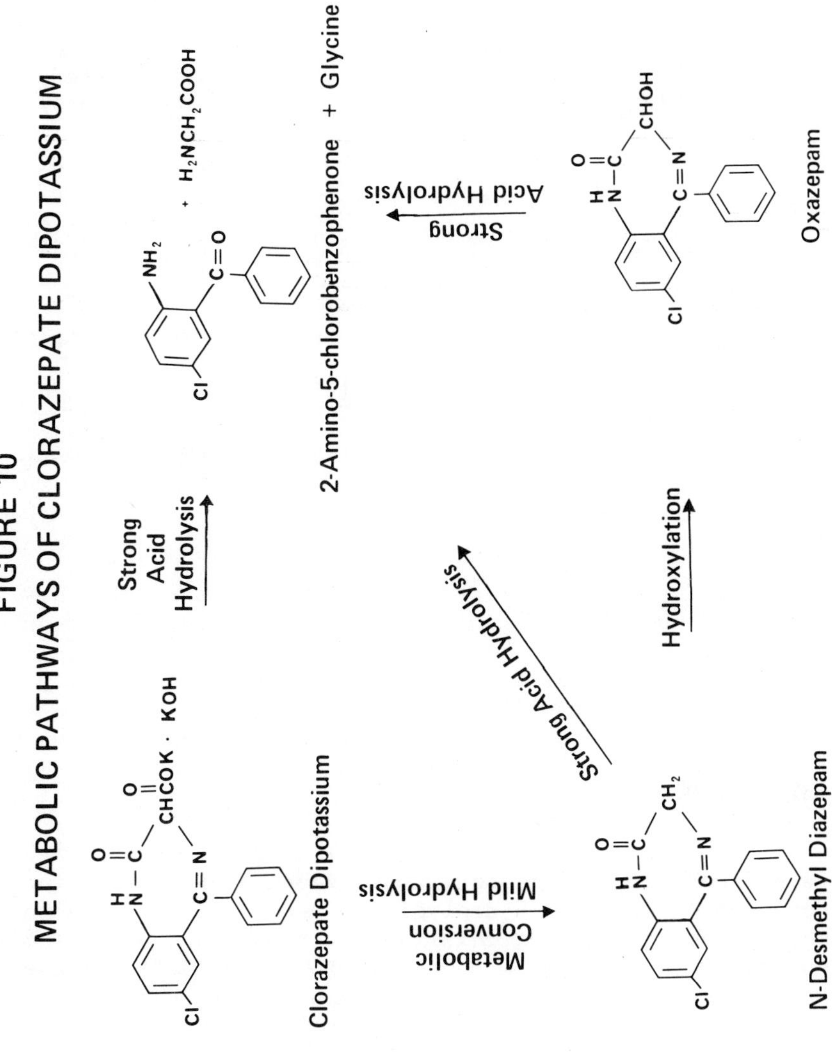

FIGURE 10
METABOLIC PATHWAYS OF CLORAZEPATE DIPOTASSIUM
Clorazepate Dipotassium
Strong Acid Hydrolysis
2-Amino-5-chlorobenzophenone + Glycine
Metabolic Conversion
Mild Hydrolysis
Strong Acid Hydrolysis
N-Desmethyl Diazepam
Hydroxylation
Oxazepam
Strong Acid Hydrolysis

liquid chromatography (13) and colorimetry. (14)

6. Methods of Analysis

6.1 Elemental Analysis

Element	% Theory	Typical Result Reported for Lot 849-648
Carbon	46.99	46.82
Hydrogen	2.71	2.65
Nitrogen	6.85	6.95
Chlorine	8.67	8.85
Potassium	19.12	19.48, 19.34*

* Values determined by atomic absorption analysis

6.2 Phase Solubility Analysis

Data is not available due to the instability of clorazepate dipotassium in the solvent systems screened.

6.3 Chromatographic Analysis

6.31 Thin Layer Chromatography

Clorazepate dipotassium is partially degraded in numerous systems to N-desmethyl diazepam, therefore, thin layer chromatography is not considered a reliable indicator of purity. The drug substance is readily identified by ultraviolet light on Silica Gel GF_{254} in the system methanol:acetone (1:1). Clorazepate dipotassium has an R_f value of 0.15 while N-desmethyl diazepam has an R_f value of 0.90.

LaFargue, et al (7) differentiated clorazepate dipotassium from other 1,4-benzodiazepines in urine and gastric fluid by hydrolysis to 2-amino-5-chlorobenzophenone and subsequent thin layer chromatography on aluminum oxide in the system benzene:chloroform (3:1). The R_f value of the hydrolysis product is 0.55. If benzophenone is used as a reference standard, the relative R_f value of 2-amino-5-chlorobenzophenone is 0.79.

Clorazepate may be directly extracted from gastric fluid as N-desmethyl diazepam and identified on aluminum oxide in the system chloroform:ethanol (29:1). The absolute R_f is reported as 0.39 or 0.48 relative to a diazepam reference standard.(7)

6.32 Gas Chromatography

Clorazepate dipotassium can not be directly chromatographed, however, gas chromatography is readily applied to the metabolites and acid hydrolysis products.(7, 12) LaFargue (15) has employed 3% OV-17 (methylphenylsilicone) on Gas Chrom Q to separate 2-amino-5-chlorobenzophenone from the acid hydrolysis products of five major benzodiazepins.

6.4 Direct Spectrophotometric Analysis

Direct spectrophotometric analysis of clorazepate dipotassium is applicable provided significant quantities of interfering contaminants are not present. The drug substance may be examined directly in an aqueous carbonate buffered media at 230 nm (ϵ = 35,000) or indirectly in alcoholic sulfuric acid by ultraviolet absorption (16) at 388 nm or fluorescence (7) with an excitation maximum of 388 nm and an emission maximum at 508 nm.

The degradation products may be quantitated in the drug substance by a solid-liquid extraction into dichloromethane. The solvent is separated from the drug and evaporated. The residue is redissolved in alcohol and compared to N-desmethyl diazepam at 230 nm.

6.5 Colorimetric Analysis

Clorazepate dipotassium and its metabolites may be determined by the Bratton-Marshall reaction after hydolysis to 2-amino-5-chlorobenzophenone.(14)

6.6 Non-Aqueous Titration

Clorazepate dipotassium may be potentiometrically titrated in glacial acetic acid using perchloric acid in glacial acetic acid and glass-calomel (0.1 N $LiClO_4$ in HOAc) electrodes. Each ml of 0.1 N $HClO_4$ is equal to 136.31 mg of clorazepate dipotassium.

7. References

1. Washburn, W., Abbott Laboratories, Personal Communication.
2. Egan, R., Abbott Laboratories, Personal Communication.
3. Mueller, S., Abbott Laboratories, Personal Communication.
4. MacDonald, A., Michaelis, A. R., and Senkowski, B. Z., "Analytical Profiles of Drug Substances," Vol. I, K. Florey, Ed., Academic Press, New York, (1972).
5. Quick, J., Abbott Laboratories, Personal Communication.
6. Wimer, D. C., Abbott Laboratories, Personal Communication.
7. LaFargue, P., Meunier, J., and Lemontey, Y., J. Chromatog., 62, 423, (1971).
8. Schmitt, J., (Establissements Clin-Byla, 4306CB) U.S. Patent 3,516,988.
9. Raveux, R., and Briot, M., Chem. Therap., 4, 303, (1969).
10. Beyer, K., Deut. Apoth. Ztg., 111, 1503, (1971).
11. LaFargue, P., Pont, P., and Meunier, J., Am. Pharm. Fran., 28, 343, (1970).
12. Viala, A., Cano, J. P., and Angeletti-Philippe, A., J. Eur. Toxicol., 3, 109, (1971).
13. Scott, C. G., and Bommer, P., J. Chrom. Sci., 8, 446, (1970).
14. Gros, P., and Raveux, R., Chim. Ther., 4, 312, (1969).
15. Ibid 11, 28, 477, (1970).
16. Laguleau, J., Crockett, R., and Mesnard, P., Bull. Soc. Pharm. Bordeaux, 110, 10, (1971).

CLOXACILLIN SODIUM

David L. Mays

TABLE OF CONTENTS

1. Description

1.1 Name, Formula, Molecular Weight

Sodium cloxacillin is found in Chemical Abstracts under 4-Thia-1-azabicyclo [3.2.0] heptane-2-carboxylic acid, 6-[[[3-(2-chlorophenyl)-5-methyl-4-isoxazolyl] carbonyl] amino]-3,3-dimethyl-7-oxo-, monosodium salt (1). It is more commonly known as 3-o-chlorophenyl-5-methyl-4-isoxazolyl penicillin sodium salt (2).

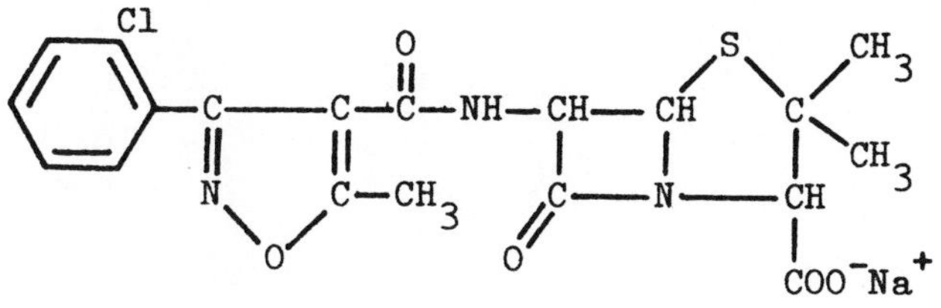

$C_{19}H_{17}ClN_3NaO_5S$ Molecular Weight 457.89

1.2 Appearance

Sodium cloxacillin is a white, odorless, crystalline powder (3).

2. Physical Properties

2.1 Infrared Spectra

Infrared absorption frequencies were reported for oil suspensions of cloxacillin and other penicillins (3a). An infrared spectrogram of sodium cloxacillin monohydrate obtained on a Perkin-Elmer Model 21 Spectrophotometer is included in the compilation of Wayland and Weiss (4). A spectrogram of Bristol Laboratories Primary Reference Standard recorded as a potassium bromide disk using a Beckman Model IR9

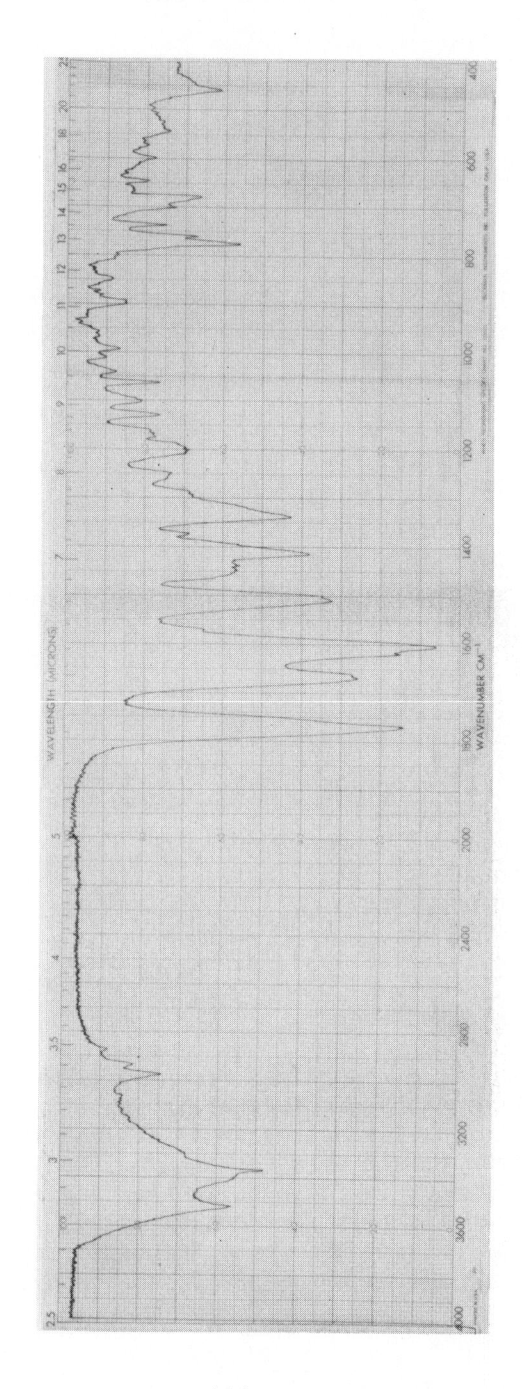

Figure 1 Infrared absorption spectrum of sodium cloxacillin monohydrate

Spectrophotometer is shown in Figure 1. Characteristic absorption frequencies (cm^{-1}) are as follows:

a.	H_2O:	3519
b.	N-H stretching band:	3370
c.	beta-lactam carbonyl:	1770
d.	secondary amide carbonyl:	1669
e.	aromatic ring:	1619
f.	carboxylate carbonyl:	1604

Within the carbonyl stretching region, the beta-lactam frequency is most characteristic of penicillins. Opening of the beta-lactam can be indicated by changes in this part of the spectrum.

2.2 Nuclear Magnetic Resonance Spectra

Proton nuclear magnetic resonance spectra for a number of penicillins were reported by Pek and coworkers (5) and the chemical shifts useful for identification were tabulated. An nmr spectrogram of Bristol Laboratories Primary Reference Standard sodium cloxacillin monohydrate as obtained on a Varian HA-100 spectrometer is shown in Figure 2. Proton resonance lines were measured in D_2O solution with deuterated sodium trimethylsilyl propionate as the internal reference. Structural assignments are as follows:

Assignment	Chemical Shift, (Coupling Constant, Hz)
aromatic	7.50 m
6-H	5.62 d (J=4.0)
5-H	5.46 d (J=4.0)
3-H	4.81 s
5-CH_3 (isoxazole)	2.63 s
2-β-CH_3	1.43 s
2-α-CH_3	1.39 s

2.3 Ultraviolet Spectra

The ultraviolet spectrum of an aqueous solution of sodium cloxacillin monohydrate obtained on a Cary-14 spectrophotometer is shown in Figure 3. An absorption maximum at 194 nm and a weak shoulder with

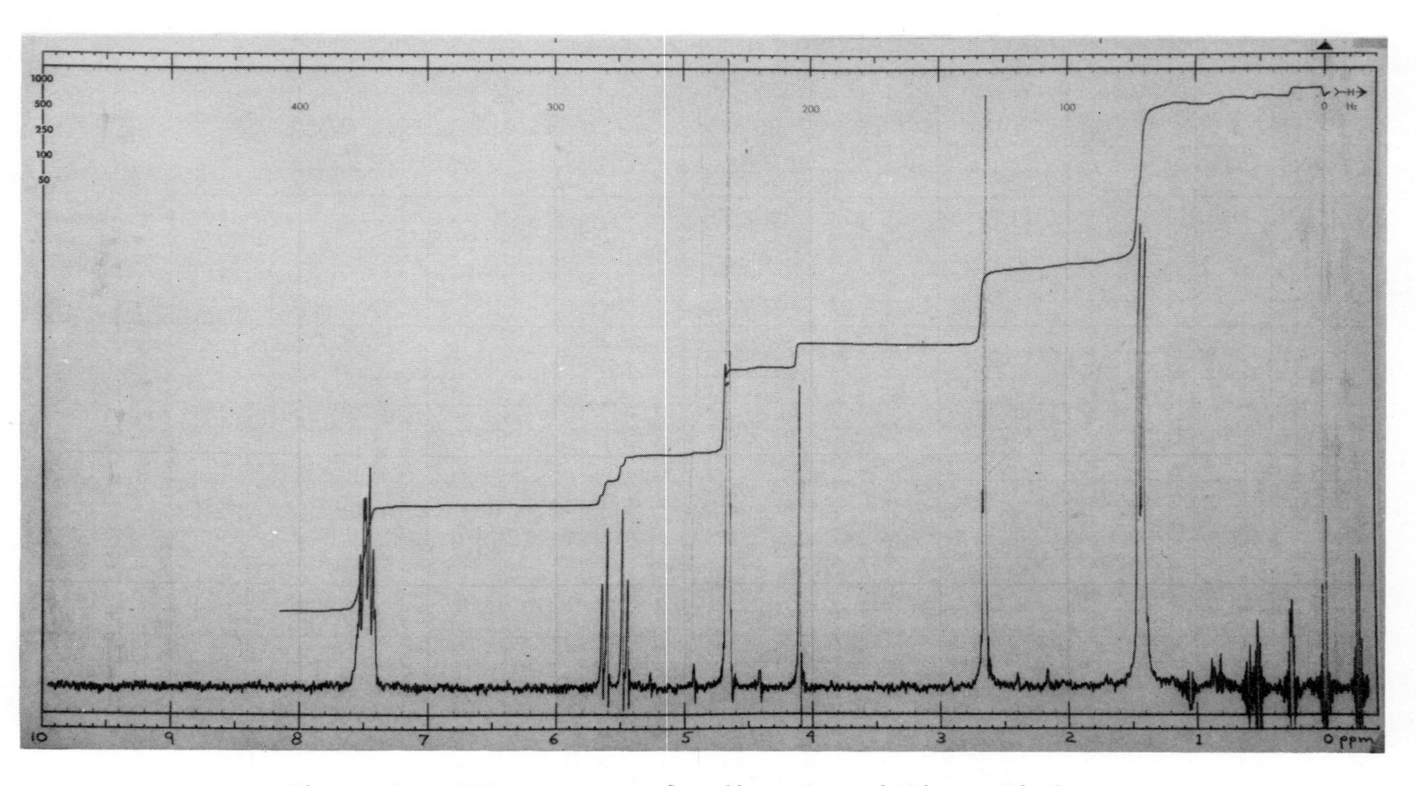

Figure 2 NMR spectrum of sodium cloxacillin monohydrate

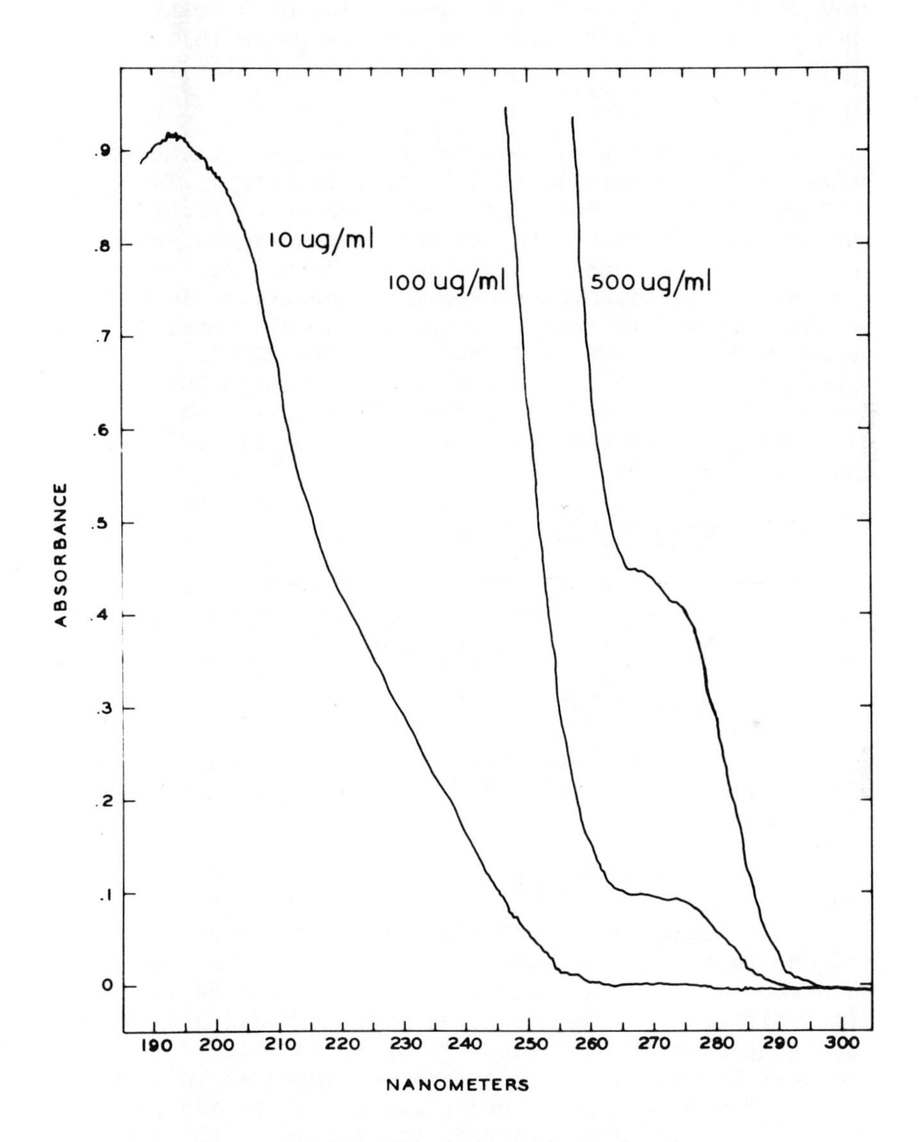

Figure 3 Ultraviolet absorption spectrum of sodium cloxacillin monohydrate

maxima at 267 and 273 nm are observed. The effect of halogen substitution on the phenyl ring of 5-methyl-3-phenyl isoxazole has been discussed by Doyle (6).

2.4 Mass Spectra

With few differences the mass spectrometric behavior of cloxacillin methyl ester (Figure 4, Table I) follows the fragmentation pattern deduced by Richter and Biemann (7) from high resolution mass measurements. These fragmentations have previously been discussed in this series (8,9). The presence of the isoxazole ring in the acyl moiety results in diagnostically useful peaks at m/e 43, 178, 193, and 220. M/e 220 is presumed to result from cleavage of the amide bond. The origin of the other three peaks follows from the discussion of the mass spectra of isoxazoles by Ohashi and coworkers (10).

2.5 Crystal Properties

Sodium cloxacillin is a microcrystalline powder exhibiting birefringence and extinction positions under a light polarizing microscope (11-13).

2.6 Melting Range

Sodium cloxacillin melts with decomposition at 170° (6,11). Cloxacillin free acid melts with decomposition at 126-127° (14).

2.7 Thermal Analysis

Differential thermal analysis curves of sodium cloxacillin monohydrate and cloxacillin free acid were recorded over a range of 0 to 250° on a Perkin Elmer Differential Scanning Calorimeter Model DSC-1B (15). No significant transitions were observed with cloxacillin free acid; decomposition appeared to occur over several broad temperature ranges. With sodium cloxacillin, broad endothermic transitions with peak temperatures at 176° and 193° were recorded. Dehydration probably occurred during the first transition, since no loss of water was observed at lower temperature.

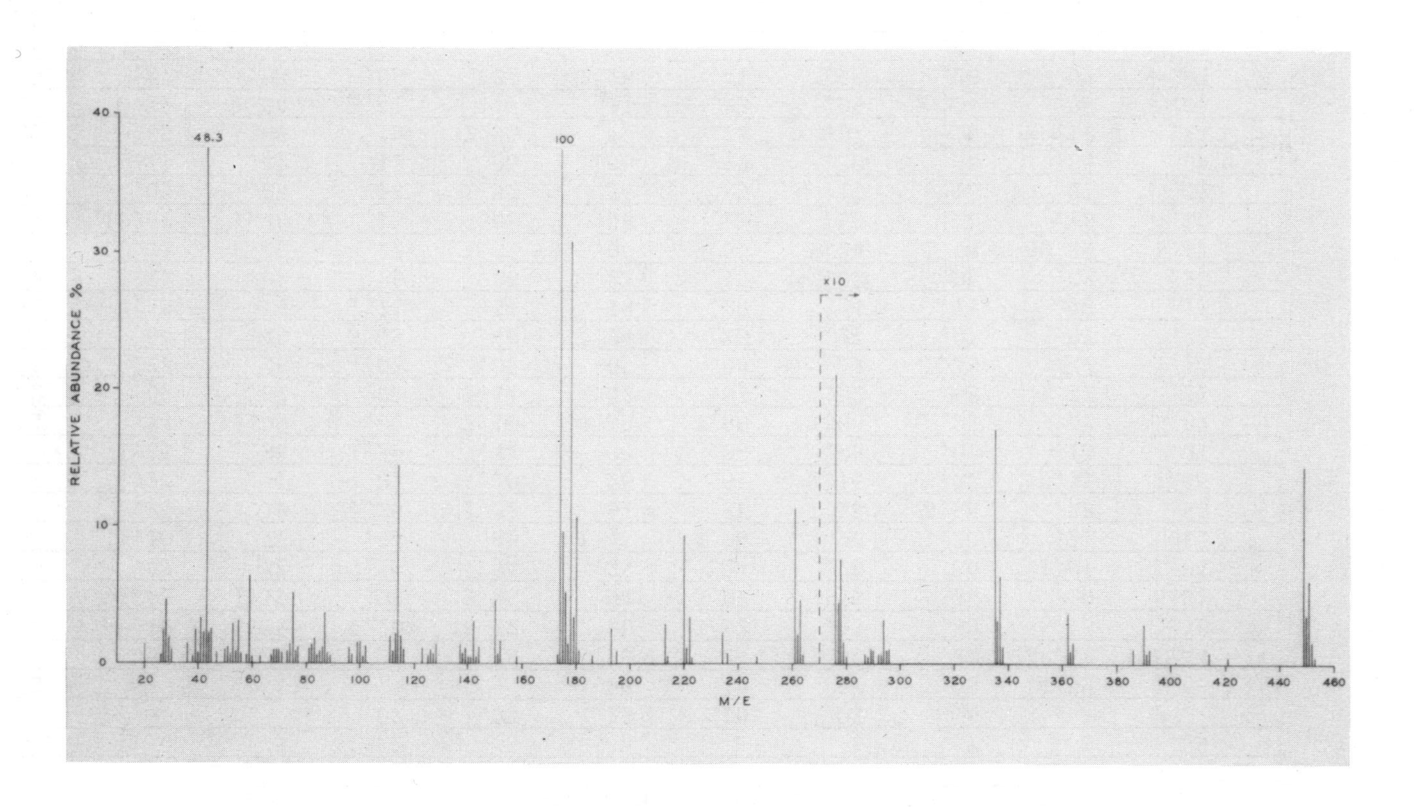

Figure 4 Mass spectrum of cloxacillin methyl ester

TABLE I

Low Resolution Mass Spectrum of Cloxacillin Methyl Ester

m/e	0	1	2	3	4	5	6	7	8	9
1			.07	.05	.23	.10	.03	.16		
2					.06	.10	.62	2.28	4.59	1.86
3	.98	.36	.26	.18	.36	.24	1.36	.24	.52	2.44
4	.72	3.25	2.20	48.29	2.20	2.52	.43	.80	.13	.18
5	1.02	1.19	.50	2.85	.94	3.09	.73	.35	.60	6.31
6	.47	.38	.36	.54	.25	.24	.25	.60	1.01	1.01
7	1.10	.77	.34	.88	1.41	4.96	.90	1.20	.21	.11
8	.25	1.12	1.25	1.82	.61	.88	1.15	3.58	.78	.56
9	.20	.11	.07	.10	.26	.31	1.16	.72	.37	1.56
10	1.63	.54	1.29	.44	.20	.08	.08	.08	.15	.21
11	.34	1.99	.90	2.15	14.35	2.00	1.06	.23	.07	.07
12	.07	.07	.07	1.14	.42	.63	1.02	.68	1.37	.31
13	.28	.08	.03	.05	.05	.08	.12	1.38	.72	1.14
14	.54	.45	3.05	.49	1.24	.15	.21	.07	.07	.23
15	4.55	.72	1.71	.31	.28	.20	.24	.11	.48	.13
16	.21	.08	.16	.23	.21	.16	.08	.05	.08	.20
17	.21	.40	.22	.65	100.00	9.59	5.24	1.54	30.89	3.41
18	10.68	1.40	.18	.08	.08	.23	.55	.20	.18	.10
19	.05	.05	.18	2.57	.36	.85	.13	.10	.07	.32
20	.12	.42	.12	.06	.06	.10	.10	.06	.06	.04
21	.05	.05	.10	2.93	.45	.13	.07	.10	.11	.31
22	9.27	1.24	3.41	.47	.21	.23	.05	.24	.08	.05
23	.07	.03	.05	.23	2.28	.41	.76	.20	.07	.03
24	.07	.15	.21	.13		.06	.05	.48	.14	.21
25	.07									

TABLE I (cont.)

m/e	0	1	2	3	4	5	6	7	8	9
26		13.30	2.11	4.63	.70	.07				
27							2.11	.44	.76	.16
28	.05							.07	.50	.11
29	.18	.03	.07	.07	.32	.10	.11			
30										
31										
32										
33						1.71	.31	.63	.12	.03
34										
35										
36			.37	.09	.16					
37										
38							.11			
39	.29	.08	.11	.02						
40										
41					.09	.04				
42		.05								
43										
44										1.44
45	.35	.60	.16	.05						

Base Peak: m/e 174 = 100.00% Relative Abundance

2.8 Solubility

Sodium cloxacillin is very soluble in cold water (11,12,16). The ratio of sodium cloxacillin concentration in chloroform versus pH 6 buffer solution was determined to be 0.118 (17). The solubility of penicillin salts in nonpolar solvents is significantly increased by the presence of a small amount of water, even water of hydration (18). The tabulation of sodium cloxacillin solubility shown below is taken from Marsh and Weiss (19).

Solvent	Solubility (mg/ml)	Solvent	Solubility (mg/ml)
Water	> 20	Methyl ethyl ketone	1.771
Methanol	> 20	Diethyl ether	0.086
Ethanol	> 20	Ethylene chloride	0.260
Isopropanol	9.158	1,4-Dioxane	4.224
Isoamyl alcohol	5.865	Chloroform	1.820
Cyclohexane	0.028	Carbon disulfide	0.062
Benzene	0.044	Pyridine	> 20
Petroleum ether	0.0	Formamide	> 20
Isooctane	0.0	Ethylene glycol	> 20
Carbon tetrachloride	0.010	Propylene glycol	> 20
Ethyl acetate	0.598	Dimethyl sulfoxide	> 20
Isoamyl acetate	0.421	0.1 N NaOH	> 20
Acetone	2.723	0.1 N HCl	4.526

2.9 Ionization Constant, pKa

Budgaard and Ilver (20) reported an apparent pKa of 2.68 ± 0.05 at 35°C., determined by measuring the pH of a partially neutralized 0.0025 *M* solution of sodium cloxacillin. Rapson and Bird (21) obtained replicate apparent pK values of 2.73 ± 0.04 and 2.70 ± 0.03 at 25°C. by titrating 0.0025 *M* sodium cloxacillin solutions.

2.10 Optical Rotation

Cloxacillin has 3 asymmetric carbon atoms and is strongly dextrorotatory. Specific rotation values in the literature are shown below:

Cloxacillin	$[\alpha]_D$	Temp.	Conc. and Solvent	Ref.
sodium salt	+163°	20°	1% in water	6,11,12
sodium salt	+159°	20°	1% in water	14
free acid	+122°	20°	1% in acetone	14

3. Synthesis and Purification

Cloxacillin is a semisynthetic beta-lactam antibiotic prepared by acylation of 6-amino penicillanic acid with 3-o-chlorophenyl-5-methyl isoxazolyl chloride (6,22, 23). The route of preparation of isoxazole acid chlorides has been described by Doyle (6). More recently, the conversion of Penicillin G to other penicillins without isolation of 6-amino penicillanic acid has been reported (24-26).

A method of purification by deionized water elution through a G-25 Sephadex column has been described (27). Recrystallization of the sodium salt may be accomplished by methyl isobutyl ketone extraction of the free acid from an acidic aqueous solution and precipitation with sodium salts of carboxylic acids.

4. Stability

The general pattern of penicillin degradation has been described in several reviews (18, 28). Sodium cloxacillin is stable in water for one week at 5°C. The half-life in a solution of 50% aqueous alcohol at pH 1.3 and 35°C. was reported to be 160 minutes, approximately the same as Penicillin V (12, 18). At a concentration of 4 mg/ml in water or saline at pH 5.5 to 6.0, cloxacillin loses 15-25% of its biological activity in 7 hours (29).

The rates of hydrolysis of cloxacillin and other penicillins were studied by Kinget and Schwartz (30). By following the amount of acid formed, they showed that alkaline hydrolysis was more rapid for cloxacillin than other penicillins tested. The presence of aminoalkyl catechols doubled the rate of hydrolysis at pH 8 but reduced the rate at pH 6.5. The rate of cloxacillin hydrolysis was also increased above pH 8 by the presence of carbohydrates (31) and at neutral pH by aminoglycosides (32).

The kinetics of cloxacillin degradation in aqueous solution has been studied as a function of pH, buffer species, ionic strength, and temperature (20). Decomposition appears first order with respect to sodium cloxacillin content at any given pH. A rate-of-degradation versus pH profile over a range of pH from 1 to 11 showed that decomposition is accelerated as the pH is moved above or below 6. The rate of hydrolysis is significantly affected by the type of buffer ions. Citrate buffers show the least catalytic effect and phosphate show the greatest.

Degradation by enzymes has been studied extensively. Cloxacillin is resistant to enzymes which catalyze hydrolysis of the beta-lactam to produce the corresponding penicilloic acid (penicillinase or beta-lactamase enzymes). Citri and Zyk (33) studied the effect of different penicillin side chains on penicillinase activity. They showed the rate of cloxacillin inactivation by penicillinase (penicillin amidohydrolase EC 3.5.2.6) to be about 3% the rate of penicillin G inactivation. Smith and coworkers (34) investigated the stability of some penicillins to penicillinase. The initial rate of hydrolysis of cloxacillin was 0.3% and 1.1% of penicillin G hydrolysis by the penicillinases produced by B. cereus and Staph. aureus, respectively. Chapman (35) used the infrared absorption of the beta-lactam at 5.6 μ to observe penicillin inactivation by beta-lactamase. Cloxacillin was more stable to staphylococci beta-lactamase than to coliform beta-lactamase. Combined treatment of penicillin amidase (enzymes which deacylate penicillins to 6-amino penicillanic acid) and beta-lactamase caused considerable inactivation of cloxacillin in 2 hours.

5. Methods of Analysis

5.1 Analysis of Impurities

The penicillenic acid of cloxacillin and other thiol-containing products such as penamaldic acid and penicillamine can be determined by reaction of the free thiol group with Ellman's reagent (36, 36a). The bright yellow anion formed is measured at the 412 nm absorption maximum. Other thiols and reducing agents interfere. If present in sufficient quantity, penicillenic acid may be measured directly by the natural absorbance at 337 nm (36a).

The penicilloic acid of cloxacillin has been separated from cloxacillin by thin-layer chromatography (37). The penicilloic acids (and other iodine consuming substances) of some penicillins have been estimated by direct consumption of iodine (36a, 38).

5.2 Identification Tests

Sodium cloxacillin is identified by the infrared absorption spectrum and by the penicillin characteristic purple color formed upon treatment with chromotropic acid in sulfuric acid at 150°C. (38).

In dosage forms, cloxacillin has been identified by the infrared absorption spectrum after extraction from aqueous phosphoric acid solution into chloroform and evaporation to a concentration of 2 mg/ml (39).

Weiss and coworkers (17) identified penicillins in dosage forms by determining the amount of penicillin partitioned between pH 6 buffer and organic solvent.

5.3 Quantitative Methods

5.31 Volumetric Methods

Iodine is not consumed by penicillins but is consumed by the hydrolysis products. The difference in iodine consumption before and after alkaline hydrolysis is used as a standard method of determining cloxacillin content (13).

Hydrolysis with a measured excess of alkali generates an additional carboxyl group in the penicilloic acid. Back titration of the excess alkali with hydrochloric acid is used for determination of cloxacillin content. (38)

5.32 Colorimetric Methods

Cloxacillin has been determined by measurement of the side chain absorbance at 275 nm (39a). Penicillins have been determined by measurement of the absorption maximum near 340 nm produced by acid degradation to the corresponding penicillenic acids (38, 40-44). Formation of the penicillenic acid is catalyzed by copper and other metals and by imidazole (44).

Penicillins have long been determined by reaction with hydroxylamine. The hydroxamic acids generated produce a red-colored chelate with iron (III). Details of the procedure are available in the Federal Register (13). Automated versions of the method have also been used (45).

5.33 Polarography

Penicillins in serum were determined by polarographic techniques (46). The analysis required about two hours. Cloxacillin was determined at levels of 5 μg/ml in sulfuric acid solution.

5.34 Gas Chromatography

Organic acid side chains produced by vigorous alkaline hydrolysis of penicillins were converted to methyl esters for gas chromatographic separation on a 3.5% SE-30 column (47). Intact penicillins were gas chromatographed as the corresponding methyl esters on a 2% fluoro-silicone phase by Martin and coworkers (48). Separation of the trimethylsilyl esters of several penicillins on a 2% OV-17 column was reported by Hishta, et. al. (49). Cloxacillin was separated from several other penicillins and quantitation was indicated by reproducibility of response factors on reference samples.

5.35 Infrared Spectroscopy

The infrared absorption due to the beta-lactam has been used to quantitate penicillins after extraction into a suitable solvent (39). The cloxacillin beta-lactam band near 1760 cm^{-1} was used to measure cloxacillin inactivation by beta-lactamase as differentiated from amidase (35). For this work solutions were lyophilized in the presence of potassium bromide and infrared absorption measurements were made from the solid disks.

5.36 Optical Rotation

The change in optical rotation upon treatment with penicillinase has been used to quantitate penicillins (50). Penicillins which are more susceptible to penicillinase can be determined in the presence of resistant penicillins. Ampicillin and cloxacillin were determined in combination.

5.37 Biological Methods

The cylinder-plate agar diffusion method is the official microbiological method of determination (13). Staph aureus (ATCC 6538P) is the organism of choice (13, 51). Cloxacillin has been microbiologically determined in the presence of ampicillin using agar impregnated with penicillinase to destroy the ampicillin activity (52) or after ion exchange separation on a column of IRA-402 (53). Cloxacillin is routinely measured by the turbidimetric method using Staph aureus FDA-209P (ATCC 6538P) or Staph aureus BL-A9596 (54).

A microbiological paper disk procedure has been described for measuring cloxacillin and other antibiotics in as little as 10 μl of plasma (55). Hooke and Ball (56) used an agar plate method and Sarcina lutea NCIB 8553 to measure cloxacillin at levels below 10 μg/ml in milk.

In a test to detect trace levels in milk, penicillin inhibits growth of Streptococcus thermophilus B.C., which otherwise causes the dye 2,3,5-triphenyltetrazolium chloride, to turn from colorless to red (57). Other antibiotics interfere.

5.38 Automated Methods

Several chemical methods for penicillin determination have been automated, including the hydroxylamine method (45, 58-60), the iodometric method (61-63), the penicillenic acid method (64), and a colorimetric method based on enzyme deacylation to 6-amino penicillenic acid and detection with p-dimethylaminobenzaldehyde (65). Cloxacillin, specifically, is mentioned in the hydroxylamine (45) and penicillenic acid (64) automated procedures.

5.4 Thin-layer Chromatography

Reviews on chromatography (66) and analysis (67) of antibiotics are available. Cloxacillin has been chromatographed on paper (68), kieselguhr G (68), cellulose (69-71), polyamide (72), silica gel G (37, 70, 73-75), commercial silica gel plates (76, 77) and silica gel G predeveloped with silicone (5% DC 200 in ether) (78-80).

Cloxacillin has been separated from other penicillins (37, 66, 75, 77-80), and from impurities (37), from cephalosporins (79), from other antibiotics (70, 74), and from constituents of body fluids (76, 77). Chromatography has been used to study solvent partitioning (78) and structure-activity relationships (79, 80).

Visualization spray reagents used include iodine-azide solution followed by aqueous starch to give white spots on blue purple (37, 68, 74); 10% acetic acid in acetone followed by starch-iodine to give white spots on blue (69); ammoniated copper sulfate (73); 0.5% bromine solution (72); 0.25% fluorescein (72); ferric chloride and potassium ferricyanide with sulfuric acid to give blue spots on green (70, 71); alkaline potassium permanganate and heat to give yellow spots on pink (78-80); chloroplatinic acid with potassium iodide in acetone to give white spots on red-purple (75); and alkaline silver nitrate (68). Bacillus subtilis ATCC 6633 impregnated in agar has been used to detect penicillins by observation of the zone of inhibition after contact with the plates (76, 77).

6. Protein Binding

The reversible binding of protein to penicillins (81), and drugs in general (82), has been reviewed. Schwartz's penicillin review (28) includes a section on reversible and irreversible protein binding. Schwartz postulated that the high local concentration of penicillin reversibly bound to protein accelerates aminolysis, particularly in the case of cloxacillin which is highly bound to serum protein. Batchelor (83) also presented evidence that penicillins, including cloxacillin, can become irreversibly bound to protein.

Several workers have measured the proportion of cloxacillin reversibly bound to serum protein. Values of 94% (29), 95% and 96% (84), 85% (85) and 62% (86) have been reported. Rolinson and Sutherland (87) made a thorough study of the degree of binding of several penicillins as a function of concentration of penicillin and protein, of sera from different animal species, and of temperature. Binding was shown to be essentially reversible and competitive with other drugs. Cloxacillin at 50 μg/ml (considered approximate blood level from a normal dose) was 94% bound to human serum.

Affinity for Sephadex gel of a number of penicillins was studied as a function of protein binding (88).

7. Metabolism

After oral administration of 100 mg/kg to rats, 18.7% of the cloxacillin was recovered in the urine after 24 hours, and 10.2% was recovered in the bile. Paper chromatographic examination of the urine indicated small amounts of two unidentified metabolites (12). Cloxacillin was reported to be approximately 10% metabolized in man (16, 89). One unidentified metabolite was found which had biological activity similar to the parent compound. The rate of elimination of cloxacillin from young men given oral doses was studied. (90)

8. References

1. Chem. Abs. Index Guide 76 Jan-June (1972).
2. J. Amer. Med. Ass. 185 (8), 656 (1963).
3. United States Pharmacopeia XVIII, Mack Publishing Co., Eaton, Pa., 1970, p. 620.

3a. Rudzit, E. A., et. al., Antibiotiki 17 (11), 978-981 (1972).

4. Wayland, L., and Weiss, P. J., J. Ass. Offic. Anal. Chem., 48 (5), 965-972 (1965).
5. Pek, G. Yu., et. al., Izv. Akad. Nauk SSSR, Ser. Khim. 1968, (10), 2213-22 through Chem. Abs. 70:28229a (1969).
6. Doyle, F. P., et. al., J. Chem. Soc. 5838-5845 (1963).
7. Richter, W., and Biemann, K., Monatsh. Chem. 95, 766-778 (1964).
8. Dunham, J. M., Analytical Profiles of Drug Substances, Vol. 1, K. Florey, ed., Academic Press, New York, New York (1973) p. 258.
9. Ivashkiv, E., Analytical Profiles of Drug Substances, Vol. 2, K. Florey, ed., Academic Press, New York, New York (1973), pp. 12-13.
10. Ohashi, M., et. al., Org. Mass Spectrom. 2, 195-207 (1969).
11. The Merck Index, 8th ed., Merck and Co., Inc., Rahway, N. J., p. 271.
12. Nayler, J. H. C., et. al., Nature (London) 195, 1264-1267 (1962).
13. Code of Federal Regulations, Title 21, April 1973 Revision, Chapter I, Section 149j.1.
14. Koenig, R., et. al., Hung. Patent 151,377 (May 23, 1964) through Chem. Abs. 61:5658f (1964).
15. Marr, T. R., Bristol Laboratories, personal communication.
16. Knudsen, E. T., et. al., Lancet 2, 632-634 (1962).
17. Weiss, P. J., et. al., J. Ass. Offic. Anal. Chem. 50 (6), 1294-1297 (1967).
18. Hou, J. P., and Poole, J. W., J. Pharm. Sci. 60 (4), 503-532 (1971).

19. Marsh, J. R., and Weiss, P. J., J. Ass. Offic. Anal. Chem. 50 (2), 457-462 (1967).
20. Bundgaard, H., and Ilver, K., Dan. Tidsskr., Farm. 44, 365-380 (1970).
21. Rapson, H. D. C., and Bird, A. E., J. Pharm. Pharmacol. 15 (Suppl.) 222T-231T (1963).
22. Axerio, P., Farm. Nueva 28 (318), 315-320 (1963). through Chem. Abs. 59:15269c (1963).
23. Doyle, F. P., and Naylor, J. H. C., U.S. Patent 2,996,501 (August 15, 1961).
24. Fosker, G. R., et. al., J. Chem. Soc. C., 1917-1919 (1971).
25. Heuser, L. J., Fr. Patent 1,596,495 (July 31, 1970) through Chem. Abs. 74:141781 (1971).
26. Jinnosuke, A., et. al., Ger. Patent 1,943,667 (May 14, 1970) through Chem. Abs. 73:45505s (1970).
27. Feinberg, J. G., and Weston, R. D., Brit. Patent 1,131,741 (October 23, 1968) through Chem. Abs. 70:14410r (1969).
28. Schwartz, M. A., J. Pharm. Sci 58 (6), 643-661 (1969).
29. Sidell, S., Clin. Pharmacol. Ther. 5, 26-34 (1964).
30. Kinget, R. D., and Schwartz, M. A., J. Pharm. Sci. 58 (9), 1102-1105 (1969).
31. Simberkoff, M. S., et. al., N. Engl. J. Med. 283 (3), 116-119 (1970).
32. Lynn, B., and Jones, A., Advances in Antimicrobial and Antineoplastic Chemotherapy, Vol. I/2, University Park Press, Baltimore, Md., (1972) pp. 701-705.
33. Citri, N., and Zyk, N., Biochim. Biophys. Acta 99, 427-441 (1965).
34. Smith, J. T., et. al., Nature 195, 1300-1301 (1962).
35. Chapman, J. J., et. al., J. Gen. Microbiol. 36, 215-223 (1964).
36. Vermeij, P., Pharm. Weekbl. 107, 249-259 (1972).
36a. Seitzinger, Ing. R. W. Th., Pharm. Weekbl. 108, 961-968 (1973).
37. Vandamme, E. J., and Voets, J. P., J. Chromatog. 71, 141-148 (1972).
38. British Pharmacopeia, Her Majesty's Stationery Office, London, 1973, pp. 81, 117.
39. Coclers, L., et. al., J. Pharm. Belg. 24, 475-491 (1969).

39a. Davidson, A. G., and Stenlake, J. B., J. Pharm. Pharmacol. 25, Suppl., 156P-157P (1973).
40. Herriott, R. M., J. Biol. Chem. 164, 725-736 (1946).
41. Weaver, W. J., and Reschke, R. F., J. Pharm. Sci. 52 (4), 362-364 (1963).
42. Saccani, F., and Pitrolo, G., Boll. Chim. Farm. 108, 29-33 (1969).
43. Yasuda, T., and Shimada, S., J. Antibiot. 24 (5), 290-293 (1971).
44. Bundgaard, H., and Ilver, K., J. Pharm. Pharmacol. 24, 790-794 (1972).
45. Lane, J. R., and Weiss, P. J., Presented at the Technicon Symposium, "Automation in Anal. Chem.", New York, New York, October 17, 1966.
46. Benner, E. J., Presented at the 10th Interscience Conference on Antimicrobial Agents and Chemotheraphy Chicago, October 18-21, 1970.
47. Kawai, S., and Hashiba, S., Bunseki Kagaku 13 (12), 1223-1226 (1964).
48. Martin, J. B., et. al., presented at the 17th Annual Pittsburgh Conference on Applied Spectroscopy, February 21, 1966.
49. Hishta, C., et. al., Anal. Chem. 43 (11), 1530-1532 (1971).
50. Rasmussen, C. E., and Higuchi, T., J. Pharm. Sci. 60 (11), 1608-1616 (1971).
51. Arret, B., et. al., J. Pharm. Sci. 60 (11), 1689-1694 (1971).
52. Sabath, L. D., et. al., Appl. Microbiol. 15 (3), 468-470 (1967).
53. Saccani, F., et. al., Boll. Chim. Farm. 108 (12), 777-780 (1969).
54. Tylec, F., and Kianka, J., Bristol Laboratories, private communication.
55. Jalling, B., et. al., Pharmacol. Clin. 4, 150-157 (1972).
56. Hooke, E. J., and Ball, G. M., J. Appl. Bacteriol. 26 (2), 216-218 (1963).
57. Scarlett, C. A., Agriculture (London) 73 (9), 423-426 (1966).
58. Avanzini, F., et. al., Automation in Analytical Chemistry, 1966, Vol. II, Mediad, Inc., White Plains, New York, pp. 31-34.
59. Niedermeyer, A. O., et. al., Anal. Chem. 32, 664-666

(1960).
60. Stevenson, C. E., Automation in Analytical Chemistry 1969, Vol. II, Mediad, Inc., White Plains, New York, pp. 251-256.
61. Bomstein, J., et. al., Ann. N. Y. Acad. Sci. 130, 589-595 (1965).
62. Ferrari, A., et. al., Anal. Chem. 31 (10), 1710-1717 (1959).
63. Russo-Alessi, F. M., et. al., Ann. N. Y. Acad. Sci. 87, 822-829 (1960).
64. Celletti, P., et. al., Farmaco, Ed. Prat. 27 (12), 688-698 (1972) through Analyt. Abs. 25:427 (1973).
65. Evans, W. G., et. al., Automation in Analytical Chemistry, 1969, Vol. II, Mediad, Inc., White Plains, New York, pp. 257-259.
66. Chromatography of Antibiotics, Wagman, G. H., and Weinstein, J. J., Elsevier Sci. Pub., New York (1973), pp. 140-146.
67. Schmitt, J. P., and Mathis, C., Pharma. Int., Engl. Ed., pp. 17-28 Mar. (1970).
68. Hellberg, H., J. Ass. Offic. Anal. Chem. 51 (3), 522-556 (1968).
69. Bird, A. E., and Marshall, A. C., J. Chromatog. 63, 313-319 (1971).
70. McGilvery, I., and Strickland, R. D., J. Pharm. Sci. 56, 77-79 (1967).
71. Wayland, L. G., and Weiss, P. J., J. Pharm. Sci. 57 (5), 806-810 (1968).
72. Wang, J. T., et. al., Tai-Wan K'o Hsueh, 24 (1-2) 19-20 (1970) through Chem. Abs. 74:115942d (1971).
73. Guven, D. C., and Ari, A., Eczacilik Bul. 13 (2), 20-23 (1971).
74. Saccani, F., Boll. Chem. Farm. 106, 625-628 (1967).
75. Pokorny, M., et. al., J. Chromatog. 77, 457-460 (1973).
76. Nishida, M., et. al., Nippon Kagaku Ryohogakukai Zasshi 17 (10), 1973-1976 (1969) through Chem. Abs. 73:12731j (1970).
77. Murakawa, T., J. Antibiot. 23 (5), 250-251 (1970).
78. Biagi, G. L., et. al., J. Chromatog. 41, 371-379 (1969).
79. Biagi, G. L., et. al., J. Chromatog. 51, 548-552 (1970).
80. Biagi, G. L., et. al. J. Med. Chem. 13 (3), 511-516

(1970).
81. Scholtan, W., Antibiot. Chemother. (Basel) 14 53-93 (1968).
82. Meyer, M. C., and Guttman, D. E., J. Pharm. Sci. 57 (6), 895-918 (1968).
83. Batchelor, F. R., et. al., Nature 206, 362-364 (1965).
84. Kunin, C. M., Antimicrob. Ag. Chemother., 1025-1034 (1965).
85. Sidell, S., et. al., Arch. Intern. Med. 112, 21-28 (1963).
86. Kato, Y., Nippon Kagaku Ryohogakukai Zasshi 13 (2), 89-93 (1965) through Chem. Abs. 64:7214e (1966).
87. Rolinson, G. N., and Sutherland, R., Brit. J. Pharmacol. 25, 638-650 (1965).
88. Murakawa, T., J. Antibiot. 23 (10), 481-487 (1970).
89. Lynn, B., Antibiot. Chemother. 13, 125-226. (1965).
90. Kislak, J. W., et. al., Amer. J. Med. Sci. 249 (6), 636-646 (1965).

I appreciate the help of Mr. N. Muhammad, Dr. R. D. Brown, and Mr. T. R. Marr of the Chemical Control Department of Bristol Laboratories for assistance in obtaining and interpreting the spectrograms and thermal analysis data used in this profile.

DIATRIZOIC ACID

Hyam Henry Lerner

TABLE OF CONTENTS

1. Description

1.1 Name, Formula, Molecular Weight

Diatrizoic Acid is 3,5- diacetamido - 2,4,6 triiodbenzoic acid. Chemical Abstracts listings are under the heading benzoic acid, 3,5 - diacetamido - 2,4,6 - triiodo-.

Among the generic and trivial names for this compound are urografin acid. Common trade names include Renografin and Hypaque.

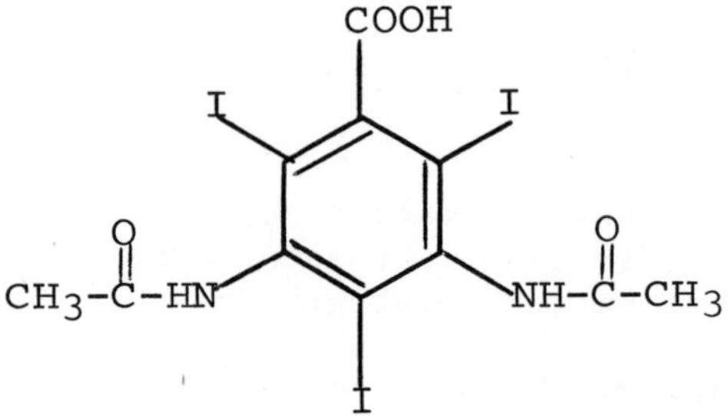

$C_{11}H_9I_3N_2O_4$ Mol. Wt. 613.928

1.2 Appearance, Color, Odor

Diatrizoic Acid is a white, odorless and tasteless crystalline powder. The sodium salt is colorless and odorless[1,4]; its taste reported as both weakly salty[1] and slightly bitter[4]. The methylglucamine (meglumine) salt is colorless, odorless and has a slight sweetish taste[1].

2. Physical Properties

2.1 Spectra

2.11 Infrared

The infrared spectrum in Fig. 1 was obtained on a Perkin-Elmer Model 21 infrared spectrophotometer, from a mineral oil dispersion. The following spectral assignments were made by Toeplitz[5].

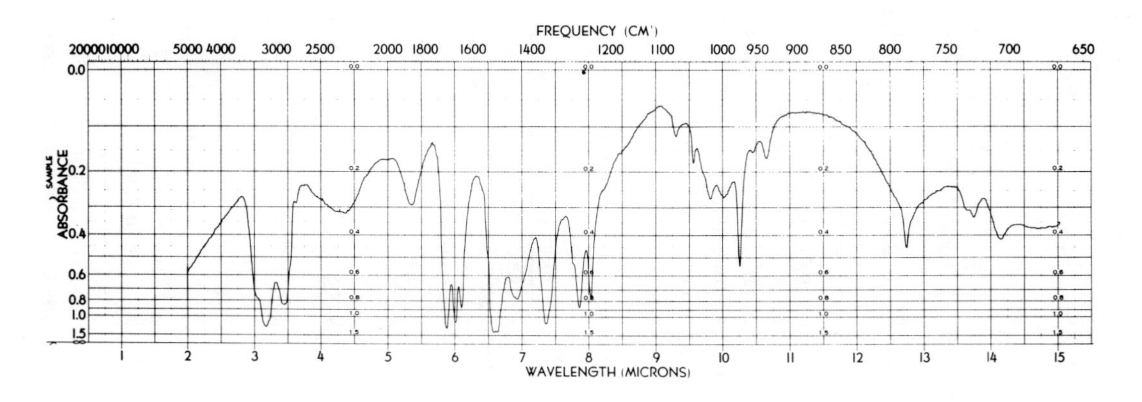

Fig. 1 Infrared Spectrum of Diatrizoic Acid

Wavelength		Assignment
cm^{-1}	μ	
2985	3.35	N-H stretch
1700	5.88	Acid C=0
1661	6.02	Amide C=0
1515	6.60	Secondary amide and aromatic C=C stretch

This spectrum is in basic agreement with a published spectrum for sodium diatrizoate[4].

2.12 Nuclear (Proton) Magnetic Resonance

The NMR spectrum of diatrizoic acid shown in Fig. 2, was determined on a Varian XL-100 NMR spectrometer at ambient probe temperature (ca.31°). The sample was dissolved in deuterated dimethylsulfoxide, containing tetramethylsilane as an internal reference. Spectral assignments are recorded in Table I[2].

TABLE I

NMR Spectral Assignments

Chemical Shift (ppm,δ)	No. of Protons	Assignment
2.02	6 (m)	$-CH_3$
9.88	1 (s)	$-\overset{O}{\overset{\|}{C}}-OH$
9.96	2 (s)	$-NH-\overset{O}{\overset{\|}{C}}-$

s = singlet
m = multiplet

2.13 Ultraviolet

The following ultraviolet spectral data have been reported for diatrizoic acid:

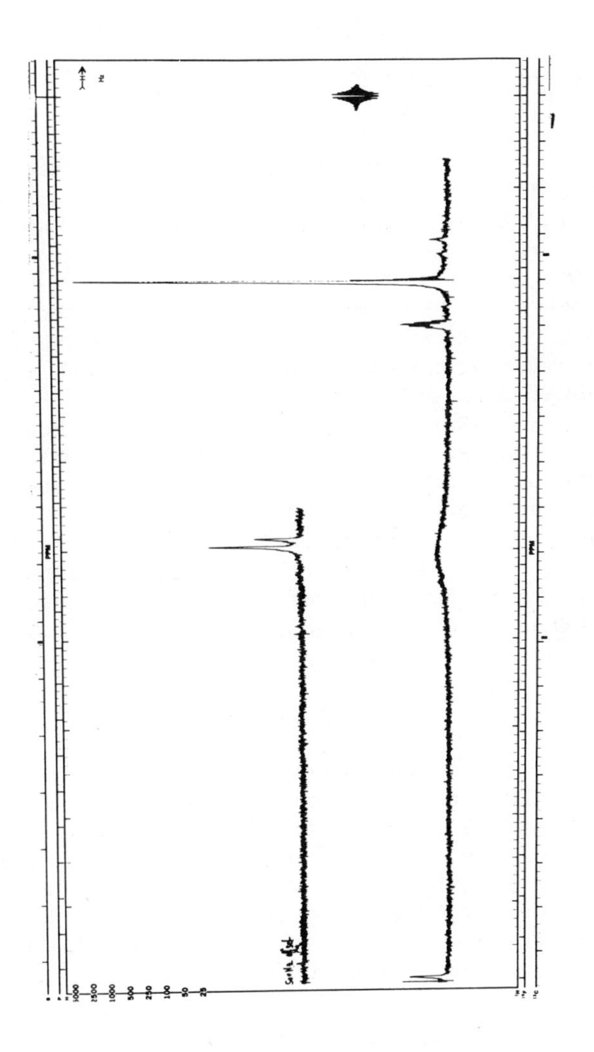

Fig. 2 NMR Spectrum Diatrizoic Acid

Form	Solvent	λmax, nm	$E^{1\%}_{1cm}$	ε	Reference
Free Acid	0.1N NaOH	238	521	32,000	1
Sodium Salt	Ethanol	239	525	33,400	4
Free Acid	Ethanol	238	589	36,200	12
Free Acid	Methanol	238	539	33,100	32
Free Acid	.01N Methanolic NaOH	238	538	33,000	32
Free Acid	.01N Methanolic HCl	238	531	32,600	32

Although none of the references explicity state it, it appears the above results are reported on an "as is" basis, except for reference 12, which appears to be on the dried basis.

Purkiss *et.al.*[26], reported that the absorbance of sodium diatrizoate is due to the presence of the acetamido group and is not dependent on iodine content. Removal of iodine from the compound did not affect absorbance at 238 nm. Neudert and Ropke[33], however, determined the molecular extinction coefficients (ε) of related iodinated compounds at their peaks. They reported the ε for 1,3,5,-triiodobenzene at 232 nm,,is 36,600 and that for 2, 4,6-triiodobenzoic acid at 238 nm, to be 33,600. They also reported the extinction of 3-acetamidobenzoic acid at 243 nm to be 13,500. Therefore it seems reasonable to suggest that the strong ultraviolet absorption maximum of diatrizoic acid at 238 nm, is due, to a large extent, to iodine.

2.14 Mass Spectrometry

No molecular ion is observed for diatrizoic acid. Weak peaks at m/e 569, corresponding to the loss of -COOH at m/e 572, for the loss of $-COCH_2$, from the unobserved molecular ion, do occur. Major fragment ions at m/e 487 and 486 occur, due to loss of I and HI, respectively. Loss of CH_3CO from the m/e 486 fragment results in a peak at m/e 443 and loss of a second CH_3CO results in the observed peak at m/e 400[3].

2.2 Crystal Properties

2.21 Differential Thermal Analysis

Valenti[34] determined the DTA of diatrizoic acid on a DuPont 900 Thermoanalyzer at a temperature rise of 15° per minute. The thermogram showed endotherms at 166° and 321°, exotherms at 220° (small) and 330°, and a shoulder at 306°. Kabadi[35] reprecipitated diatrizoic acid from alkaline solution and dried it at room temperature, under vacuum. He reported DTA data under the same conditions referred to above as follows: endotherms at 119° (large), 157° (small), 316° broad; exotherm at 234°; shoulder at 294°. The moisture content of this specimen was $<0.4\%$.

The work of Kabadi[35] indicates that other polymorphic forms of diatrizoic acid exist.

2.22 Thermal Gravimetric Analysis

Valenti[34] determined the TGA of diatrizoic acid on a DuPont Thermogravimetric Analyzer. The compound was heated at a rate of 15° per minute under nitrogen sweep. Weight loss of 5% was observed, with all volatile material driven off before 140°.

2.23 X-Ray Powder Diffraction

Ochs[13] obtained the X-ray powder diffraction spectrum of diatrizoic acid on a Phillips X-Ray Powder Diffractometer, at a voltage of 45 kv and a current of 15 ma. The sample was irradiated by a copper source at 1.54Å. Data are recorded in Table 2.

2.24 Water of Crystallization

Langecker *et.al.* reported that the diatrizoic acid crystal can contain up to 2 moles of water of crystallization.

TABLE II

X-Ray Powder Diffraction Pattern of Diatrizoic Acid Squibb Lot 1987

d(Å)*	Relative Intensity**	d(Å)*	Relative Intensity**
13.2	0.23	3.13	0.45
9.10	0.40	3.12	0.49
7.20	0.19	3.07	0.71
6.50	0.34	3.00	0.23
5.66	0.51	2.98	0.27
5.50	0.19	2.93	0.21
5.00	0.15	2.88	0.13
4.50	0.20	2.68	0.11
4.42	0.31	2.60	0.19
4.28	1.00	2.56	0.16
4.18	0.43	2.54	0.24
4.02	0.32	2.50	0.27
3.90	0.49	2.46	0.21
3.75	0.69	2.37	0.12
3.68	0.19	2.28	0.19
3.60	0.19	2.24	0.19
3.48	0.16	2.18	0.17
3.44	0.53	2.16	0.16
3.36	0.43	2.13	0.15
3.30	0.53	2.09	0.17
3.23	0.29	1.93	0.13
3.17	0.17	1.89	0.15
		1.87	0.16

* d = (interplanar distance) = $\frac{n\lambda}{2 \sin \theta}$

where λ = 1.539Å

** Based on highest intensity of 1.00

2.25 Melting Range

Diverse values have been reported in the literature for the melting range of diatrizoic acid. They are shown in the table below. A possible explanation for the wide diversity in reported values may be that different (unidentified) polymorph forms were used in the work.

Melting Range	Remarks	Ref.
>300	--	6
from 260	Decompositon, I_2 Vapors reported	1
>340	Crystal form: needles	11
260° - 290	Using Köfler Microblock (Reichert)	12

2.3 Solution Data

2.31 Solubility

The data in Table III were reported for the solubility of diatrizoic acid, at room temperature.

TABLE III - SOLUBILITY DATA

	Solubility (mg/100 ml)		
Solvent	Ref. 1	Ref. 12	Ref. 34
Acetone	-	-	<100
Benzene	<10	-	<10
Chloroform	<10	0.1	<10
Ethanol	700	1,018	-
Ether	<10	0.3	<10
Hexane	-	-	<10
Methanol	2,400	7,205	-
Propylene Glycol	-	-	<10
Water at 25°	100	-	-
Water at 50°	150	-	-
Water at 90°	270	-	-
0.1N Sodium hydroxide	-	-	6,200

Langecker *et.al.*[1] reported the solubility at 20° of the sodium salt of diatrizoic acid to be 60 g/100 ml in water and that of the methylglucamine salt to be 89 g/100 ml. Drug Standards[4] reported that the sodium salt is freely soluble in water and dimethylformamide and very slightly soluble in chloroform and ether. The solubility in ethanol was reported to be 2 g/100.

2.32 pKa

The pKa of diatrizoic acid is 3.4[1].

2.33 pH

The pH of a 1% suspension of the free acid in water is 2.1[12]. The pH of a 50% solution of the sodium salt is between 7 and 9[4]. Langecker *et.al.*[1] reported the pH of a solution of the sodium salt to be 7.3 and that of the meglumine salt to be 6.0.

2.34 Osmotic Properties and Ionic Strength

Berdalen *et.al.*[36] determined the molal osmotic coefficients and ionic strength of some commercial preparations of diatrizoate.

2.35 Index of Refraction

The refractive index (n_D^{25}) of sodium diatrizoate in water, at 25°, is given in Table IV.

2.36 Specific Gravity

Langecker *et.al.*[1] reported the specific gravities (ρ) of water solutions of sodium diatrizoate at 25°. Data are recorded in Table IV.

2.37 Freezing Point Depression

The freezing point depression ($-\Delta T$) of various concentrations of sodium diatrizoate were reported by Langecker *et.al.*[1], and are reported in Table IV.

2.38 Viscosity

Schmid[59] studied the viscosity of 0.75M aqueous solutions of diatrizoate at pH 7, and 20°. He reported viscosities by Rheomat 16 and Ostwald viscometers for sodium diatrizoate (3.29 cps), meglumine diatrizoate (6.19 cps), and tris-(hydroxymethyl)-aminomethane diatrizoate (4.76 cps.).

TABLE IV

Physicochemical Data of Sodium Diatrizoate

g/100 ml	Molarity	n_D^{25}	ρ (g/ml)	$-\Delta T°$
0.0	0.0	0.997	-	-
4.74	0.074	1.338	1.025	0.27
14.20	0.223	1.353	1.085	0.78
28.40	0.446	1.376	1.175	1.70
37.90	0.595	1.392	1.235	2.38
47.40	0.744	1.407	1.295	3.20

3. Manufacturing Procedures

3.1 Synthesis

Diatrizoic acid may be prepared from 3,5-dinitrobenzoic acid by catalytic reduction with *e.g.* Raney nickel in methanol[10] or palladium (on carbon support)[11], to yield the 3,5 diaminobenzoic acid. Reaction of dilute hydrochloric acid solutions of the latter compound with potassium iodochloride or iodine monochloride yields the 3,5-diamino-2,4,6-triidobenzoic acid[7,10,11]. Acylation of this compound is accomplished by dissolving it in acetic anhydride, heating and then acidifying with concentrated sulfuric acid, added dropwise. After a brief exothermic reaction the solution is heated on a steam bath and cooled in ice[6]. Alternatively 3,5-diacetamidobenzoic acid can be iodinated in pH 5-6 methanolic solution with iodine monochloride, at 80°C, to yield the triiodo compound[8,10,11]. (See Fig. 3 for stepwise synthetic route)

Davidson[28] acetylated the diamino compound with acetic ^3H anhydride to obtain tritiated diatrizoic acid.

Figure 3

Synthetic Route for Diatrizoic Acid

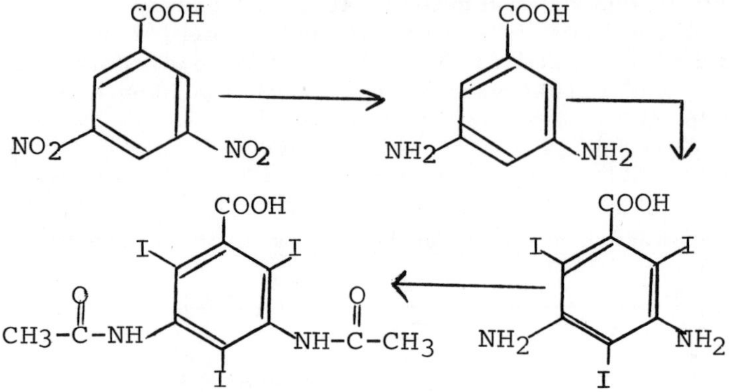

COOH
NO_2 NO_2
COOH
NH_2 NH_2
COOH
I I
CH_3-C(=O)-NH NH-C(=O)-CH_3
I
COOH
I I
NH_2 NH_2
I

3.2 Purification

Crude diatrizoic acid may be purified in any of the following ways:

The precipitate is dissolved in dilute ammonia or sodium hydroxide, charcoaled and reprecipitated by acidification[6,8,10,11].

Recrystallization from a 50% aqueous dimethylformamide solution with charcoaling[6].

Filtration of a solution in dilute sodium hydroxide containing sodium hydrosulfite and charcoal. The filtrate is passed through a medium basic anion exchange resin (in the OH form). The eluate is acidified to reprecipitate the pure product[9].

4. Stability

Diatrizoic acid is chemically stable at room temperature. Extrapolation of the activation energy and frequency constant data, between 70° and 100° to room temperature, indicates decomposition of the N-acyl bond to be 0.1% in 50 years[37]. Evolution of iodine at 90° from diatrizoic acid in solution at pH 9, is less than 3% in 75 hours. The presence of impurities e.g. under iodinated and free amino compounds, enhance degradation[37].

Commercial preparations of diatrizoic acid neutralized with meglumine, and containing sodium citrate (3.2 mg/ml) and disodium edetate (0.4 mg/ml), show no significant loss of potency after storage for 5 years, at room temperature[52].

5. Separation Technique and Analysis for Impurities

Many of the chromatographic procedures described in Section 6.6 are used to separate, detect, and estimate impurities in diatrizoic acid. In addition, the following methods have been described in the literature to detect and quantify impurities and decomposition products, and to isolate pure diatrizoic acid.

5.1 Free Amino Compounds

Hartmann and Röpke[37] determined free amino compounds of iodinated contrast agents by a kinetic method based on the greater reactivity of these impurities with elemental bromine in acetic acid. Under these conditions, the free amino compound quantitatively splits off iodine, which is oxidized to iodate by the bromine. After destruction of the excess bromine, the iodate is reduced to iodine and titrated with thiosulfate, to a starch endpoint. Free iodides in the sample also react with bromine and give positive results, by this procedure.

Hoevel-Kestermann and Muhlemann[12] and Hartmann and Röpke[37] described a Bratton-Marshall colorimetric reaction to quantitate free amino compound impurities.

5.2 Free Iodine and Free Halide

Gonda[38] detected free iodine in diatrizoic acid by dispersal in water, filtering and adding starch to the filtrate. A blue coloration indicates the presence of free iodine. An alternate procedure[4,12] is to acidify the filtrate, add chloroform and sodium nitrite and observe for reddish coloration in the lower chloroform layer. Free halide was detected[38] by acidification of a water suspension filtrate of the drug substance with nitric acid, addition of silver nitrate and observing for the presence of opalescence or precipitation.

Poet[39] quantitated free halide coulometrically, with an Aminco-Cotlove Chloride Titrator. Hartmann and Röpke[37] quantitated free halide in contrast media by potentiometric titration with 0.001N silver nitrate, under a protective cover of nitrogen.

5.3 Complexometric Method of Separation

Hentrich and Pfeifer[40] described methods for the precipitation of ten contrast agents as their metallic salts or metallic complex salts. Diatrizoic acid can be precipitated quantitatively by cadmium sulfate with thiourea, copper sulfate with thiourea, silver nitrate, cadmium sulfate with pyridine and copper sulfate with pyridine. Chelatometric methods are also described for the titration of excess precipitant, after separation of the precipitated salt by filtration. The molecular formulae, weights, melting range and equivalent weight of diatrizoic acid precipitated by 1 ml of 0.1M solution of the precipitant, for the salts and complexes, are given in Table V below.

TABLE V

Metallic Salt and Metallic Complexes of Diatrizoic Acid[40]

Precipitant	Mol. Formula of Salt or Complex	Mol. Wt.	Melting Range	Equiv. Weight of Diatrizoic acid to 1 ml of 0.1M Precipitant
$AgNO_3$	$C_{11}H_8I_3N_2O_4Ag$	720.8	281-284	0.0614 g
$CdSO_4$-Thiourea	$\{Cd[(NH_2)_2CS]_4\}(C_{11}H_8I_3N_2O_4)_2$	1,642.7	226-227	0.0307 g
$CdSO_4$-	$[Cd(C_6H_5N)_4]\cdot(C_{11}H_8I_3N_2O_4)_2$	1,656.6	288-290	0.0307 g
$CuSO_4$-Thiourea	$\{Cu[(NH_2)_2CS]\}C_{11}H_8I_3N_2O_4$	828.7	245	0.0614
$CuSO_4$-Pyridine	$[Cu(C_6H_5N)_2]\cdot(C_{11}H_8I_3N_2O_4)_2$	1,447.5	266-267	0.0307

5.4 Countercurrent Distribution

Strickler *et.al.*[42] separated diatrizoic acid and other contrast agents from sera by countercurrent distribution. They used a solvent system composed of *sec*-butanol: dil. aqueous ammonia (1:1). Both a 30-tube manual procedure and a 200-tube automatic procedure are described. Evaluation of other methods *e.g.* separation by ion exchange on Dowex 1 column (not successful) and 2-dimensional paper chromatography, are also discussed.

5.5 Phase Solubility Analysis

Kallos[47] reported on the use of phase solubility analysis to determine the purity of diatrizoic acid. He used a solvent system composed of acetone: ethanol (17:3), with equilibration over a 23 hour period at 25°. The extrapolated solubility in this system is 3.5 mg per g.

6. Methods of Analysis

6.1 Elemental Analysis

The elemental analysis[60] results were calculated to the anhydrous basis.

Element	% Theoretical	% Reported[60]
C	21.52	21.98
H	1.478	1.78
I	62.01	61.69
N	4.562	4.26
O	10.42	10.29 (CALC)

6.2 Identification Tests

Infrared (section 2.11), paper chromatography (section 6.61), and thin-layer chromatography (section 6.62) have been used to identify diatrizoic acid.

The evolution of intense violet fumes of liberated iodine can be observed by heating a sample of diatrizoic acid over an open flame[4,12].

Identification of the amide funcionality can be made by first ascertaining the absence of free amino groups (section 5.1) and then cleaving the acyl group, by refluxing in base, and repeating the Bratton-Marshall reaction[12].

The pH (section 2.33) of a water suspension and the ability to titrate a suspension with base, can be used to identify the presence of the carboxylic acid[12].

6.3 Direct Spectrophotometric Analysis

Purkiss *et.al.*[26], determined diatrizoic acid directly in urine and heparinized protein-free plasma at the 238 nm maximum. Samples taken from the subject prior to administration of diatrizoic acid were used as blanks. Dilution with water decreased blank interference.

Hoevel-Kestermann and Mühlmann[12] used spectrophotometric analysis at the 238 nm peak to determine solubility in various solvents.

6.4 Polarography

Kabasakalian and Mc Glotten[25] studied 10 iodinated radiocontrast agents polarographically in ethanol-water (1:1) solutions between pH 1.3 and 10.5. The effects of pH, temperature and sample concentration on wave form and height are examined and the analytical possibilities of polarography as a quantitative procedure for diatrizoic acid are discussed. Diatrizoic acid in phosphate buffer, pH 8.9, exhibits two waves: $-E_{1/2}$ at 0.90 and $-E^{I}_{1/2}$ at 1.22.

6.5 Organically Bound Iodine

Hoevel Kestermann and Mühlemann[12] reviewed three methods for liberating organically bound iodine in contrast agents: A. Parr bomb (fusion with sodium peroxide); B. catalytic dehalogenation with sodium borohydride; and C. reductive dehalogenation with zinc-sodium hydroxide. They recommend method B because of its simplicity, short assay time and high precision. This method was originally proposed by Egli[45]. Quantitation is accomplished by potentiometric titration with silver nitrate.

Yakatan and Tuckerman[43] reviewed four methods for decomposing iodinated contrast agents to liberate organically bound iodine: A. Parr bomb (fusion with sodium peroxide); B. alkaline permanganate reduction; C. Zinc-sodium hydroxide reduction; and D. oxygen flask (Schoniger) combustion. Method D is recommended as a general technique for all iodinated organic compounds because of its high reproducibility, simplicity, and rapidity of analysis (20 min. per test). For compounds that have all iodine atoms ortho or para to the electronegative carboxylic acid on the aromatic ring e.g. diatrizoic acid, method C, zinc-sodium hydroxide reduction, is recommended. In method C, the iodine is reduced under reflux conditions and replaced by hydrogen, generated by the reaction of powdered zinc with sodium hydroxide. The iodide thus formed, is determined by titration with silver nitrate in acidic solution, in the presence of the absorption indicator, tetrabromophenolphtalein ethyl ester[4].

Ates and Amal[14] decomposed diatrizoic acid with alkaline permanganate. After decoloration of the permanganate with sodium nitrite and acid, they titrated the liberated iodine with thiosulfate.

Zak and Boyle[44] used chloric acid as an oxidizing digestion reagent to wet ash iodine-containing organic compounds. The iodine present in this solution as iodate may then be determined by titrimetry, spectrophotometry or by polarography.

Strauss and Erhardt[46] determined diatrizoic acid directly, in urine, by splitting off the organic iodine, followed by thiosulfate determination. They boiled urine samples with sulfuric acid at 150° for 5 minutes, cooled the sample, added potassium iodide, and titrated.

6.6 Chromatographic Analysis

6.61 Paper Chromatography

Many systems have been described in the literature for separation and detection of diatrizoic acid. These are summarized in Table VI. The drug substance may be quantitated from the developed chromatogram by elution in aqueous solution and subsequent spectrophotometric analysis by densitometry, or by other suitable means.

Pileggi *et.al.*[17], described a paper chromatographic precedure for separation of 19 organic iodide compounds from blood serum. Ates and Amal[14] describe systems for separation of 8 contrast agents. Bl fox *et.al.*[24], and Turula[19] described 2 dimensional paper chromatographic procedures for separation of various iodinated contrast agents and iodinated proteins, from bile and serum.

Key To Table VI.

Solvent Systems

- I Ethanol: 25% ammonia (ratio of solvents not given).
- II 0.25*M* sodium citrate, pH 8.2.
- III *sec*-butanol: 4% ammonia (3:1).
- IV toluene: ethanol: formic acid (10:3:3).
- V *tert*.-amyl alcohol: 2*M* ammonia (1:1)
- VI water: formic acid (5:1)
- VII *n*-butanol: 95% ethanol: water (4:1:5); upper phase used for development.
- VIII *n*-butanol: 1*N* ammonium hydroxide: ethanol USP (5:2:1)
- IX water: *n*-butanol: ethanol USP (5:4:1); upper phase used for development.
- X *n*-butanol: dioxane: ammonia (ratios not reported).

Methods of Detection

A. Spray with 10% ceric sulfate and 5% sodium arsenite, both prepared in 1*N* sulfuric acid[15].

B. Compound assayed was labeled with I^{131}. Detection accomplished by autoradiography with Kodak Radiography film.

C. Viewed under ultraviolet light.

D. Spray with mixture of ceric ammonium sulfate and arsenious acid, followed by spraying with 0.5% solution of methylene blue[18].

E. Spray on both sides with a reagent containing 5 parts of a 2.7% solution of $FeCl_3 \cdot 6H_2O$ in 2N HCl, 5 parts of a 3.5% solution of $K_3Fe(CN)_6$ and 1 part

TABLE VI

Paper Chromatographic Systems for Separating Diatrizoic Acid

Solvent System	Paper	Method of Development	Method of Detection	Rf	Reference
I	Whatman No. 3	descending	A,C	Not reported	14
II	Whatman AE 30	ascending	B	0.76	16
II	DE 20	ascending	B	0.66	16
II	Whatman No. 1	ascending	B	0.73	16
III	Whatman No. 3	descending	D	0.57	17
IV	Whatman No. 3	ascending	E	Not reported	19
V	Whatman No. 3	descending	E	0.00	19,24
VI	Whatman No. 3	descending	E	1.00	19
VII	Whatman No. 1	descending	C	0.27	21
VIII	Whatman No. 1	descending	C	Not reported	22
IX	Whatman No. 4	descending	C	Not reported	23
X	Not reported	descending	B	Not reported	24

of an acified 5% $NaAsO_2$. The color is allowed to develop for 15 minutes, between glass plates, the paper washed with water and dried with a current of warm air. Spraying and development are carried out in a dim-lit or dark room[20].

6.62 Thin-Layer Chromatography

Thin-layer chromatographic methods found suitable for the separation and detection of diatrizoic acid, are summarized in Table VII. Some of the references cited, present sample preparation techniques and methods for eluting the drug from the developed plate, for quantitation by other means.

Hollingsworth *et.al.*, separated iodoamino acids and related compounds on cellulose plates with a solvent system composed of *tert*.-butanol: 2*N* ammonia: chloroform (376:70:60). They did not report this technique for diatrizoic acid, however, it seems reasonable to expect that this system will separate iodinated contrast agents. Stahl and Pfeifle[29] reported on 6 systems to separate 17 iodinated contrast agents and Hoevel-Kestermann and Muhlemann[12] separated 8 contrast agents, with a single system.

6.63 Electrochromatography

Dennenberg[27] separated diatrizoic acid on Whatman No. 3 paper by high voltage paper electrophoresis. He used a pyridine: acetic acid water (100:10:890), pH 6 buffer, at a potential gradient of 45 v/cm, for 1 hour. Diatrizoic acid in this system has a mobility of 3 cm/1800 v/hr.

Turula[19] used a pH 8.6 diethylbarbiturate buffer system, Whatman Nos. 1 and 3 paper, a potential gradient of 6 v/cm, for 17 hours, to separate iodinated contrast agents.

Gopal[16] separated a number of iodinated compounds including diatrizoic acid, by paper electrophoresis on Whatman AE 30, Whatman No. 1 and DE 20 (W and R Balston Ltd) paper. Electrophoresis was accomplished in .025*M* sodium citrate buffer, at a potential gradient of 7 v/cm,

Key To Table VII

Solvent System

I	ethanol: ammonia, 25% (2:1)
II	benzene: acetone: formic acid (5:5:1)
III	ethylacetate: isopropanol: ammonia, 25% (11:7:4)
IV	acetone: isopropanol: ammonia, 25% (
V	isopropanol: ammonia 25% (4:1)
VI	acetic acid: chloroform (1:19)
VII	ethylacetate: methanol: diethylamine (5:4:2)
VIII	chloroform: methanol: pyridine (17:1:2)
IX	methanol: chloroform: ammonia (10:20:2)
X	methanol: chloroform: ammonia (4:5:2)

Plate

A	Silica gel G (Merck)
B	Silica gel HF 254-366 (Merck)
C	Silica gel GF (Brinkman)

Detection System

a. Spray with 1:1 solution of 10% ceric sulfate and 5% sodium aresenite, both in 1N sulfuric acid[15].

b. Autoradiography of H^3 labeled diatrizoic acid.

c. Ultraviolet light.

d. Spray with 50% acetic acid, followed by irradiation at 254 nm for 10 min., to give blue violet spots.

TABLE VII

Thin Layer Systems for Separation and Detection of Diatrizoic Acid

Solvent System	Plate	Detection System	Rf	Reference
I	A,B	a	Not reported	14
II	A	b	Not reported	28
III	A,B	c,d	0.12	12,29
IV	A	c,d	0.30	29
V	A	c,d	0.37	29
VI	A	c,d	0.40	29
VII	A	c,d	.37	29
VIII	A	c,d	.00	29
IX	C	c	0.45	30

in 1 hour.

6.7 Titrimetric Methods of Analysis

Sodium diatrizoate dissolved in dimethylformamide can be titrated potentiometrically with perchloric acid in dioxane[4].

Although no references appear in the literature, it seems reasonable that a simple acid-base titration to a phenolphtalein endpoint, of diatrizoic acid in aqueous methanolic solution, would be successful.

6.8 Radiometric Methods of Analysis

6.81 X-Ray

Holynaka and Jankiewicz[48] used X-ray fluorescence and absorption techniques to determine iodine in various commercial contrast agent preparations. In the X-ray fluorescence work, excitation of the sample was accomplished by irradiation from a ^{241}Am source, of 5-mCi activity. The energy of excitation was 60 keV. Fluorescence of the K series of iodine (28.5 keV) was measured in a scintillation counter having a 6-mm thick NaI/Tl crystal. The characteristic radiation of iodine was separated by means of a single-channel, pulse-amplitude analyzer covering the total width of the K-peak iodine. Measurement time was 1 min. A calibration curve was made using HIO_3 as a standard.

A ^{241}Am source at 60 keV was used in the X-ray absorption procedure. The detector was a scintillation counter equipped with a 6-mm thick NaI/Tl crystal. Absorption measurements were made by means of a single-channel, pulse-amplitude analyzer in the energy channel of 5keV, covering 60 keV. Calibration curves were prepared, using HIO_3 as a standard. The authors[48] preferred the X-ray fluorescence technique because of its higher precision and accuracy.

Schiller and Synek[49] used radiometric methods to determine iodine in various preparations of contrast agents. They used a measuring probe with a GM counter and scaler to determine reflected radiation.

Buersch et.al.[50] determined the continuous spectrum of X-rays on aqueous solutions of diatrizoic acid. They reported that Beer's law with respect to linear absorption, holds, provided the excitation beam is sufficiently monochromatic. The application to biological systems is also discussed.

6.82 β- Particle Dispersion

Mikolajek et.al.[51] used the method of retrograde dispersion of beta particles to assay contrast media. ^{208}Tl with about 3-μCi activity, deposited on a ring was used as the source of radiation. A calibration showing the number of scattered electrons vs concentration was established. Results by this method are in good agreement with determinations made by more conventional methods.

7. Pharmacology

7.1 Drug Metabolism

Diatrizoic acid is excreted rapidly and unchanged largely through the kidneys by glomular filtration[28,46,53,54,55]. Tauxe et.al[53] suggest that there is a minor degree of plasma protein binding, in man. Woodruff and Malvin[56] inferred about 5-10% protein binding of diatrizoate, but did not present data. Stokes et.al.[57] reported that 50% diatrizoate was protein bound, but binding was easily reversed. No reports were uncovered in the literature suggesting metabolites of diatrizoic acid or concentration of diatrizoic acid in organs of man, other than the kidney.

7.2 Toxicity

Langecker et.al.[1] reported the LD_{50} of diatrizoic acid in the rat, to be 14.7 g/kg. Miller[58] reported the LD_{50} of diatrizoic acid to be about 10.1 g/kg, in the mouse. Schmid[59] reported the LD_{50} of sodium diatrizoate to be 14.0 g/kg in the mouse and 11.4 g/kg in the rat. The LD_{50} of meglumine diatrizoate was reported as 14.7 g/kg in the rat[59].

Acknowledgement

The author acknowledges with gratitude the cooperation of Ms. Cheryl Pernell for her patience in the preparation and correction of this manuscript.

8. References

1. Langecker, H., A. Harwart and K. Junkmann, Arch. Exp. Pathol. Pharmakol., 222, 584 (1954).

2. Puar, M., Squibb Institute, personal communication.

3. Funke, P., Squibb Institute, personal communication.

4. Drug Standards, 25, 84 (1957).

5. Toeplitz, B., Squibb Institute, personal communication.

6. Larsen, A., C. Moore, J. Sprague, B. Cloke, J. Moss and J. Hoppe, J. Am. Chem. Soc., 78, 3210 (1956).

7. Lugens, J., Ber., 29, 2836 (1896).

8. Lenkowski, P., T. Dzieciak, H. Cieslik and M. Drzejszczak, Polish Patent 62,290 (1967); Chem. Abstr., 75, 151549 k (1971).

9. Kulhanek, J., J. Chromik, K. Skaba and K. Mejstrick, Czech. Patent 134,140 (1969); Chem. Abstr., 74, 53288 j (1971).

10. Cassebaum, H. and K. Dierbach, German (East) Patent 43,994 (1965); Chem. Abstr., 64, 17494 f (1966).

11. British Patent, assigned to Mallinckrodt Chemical Works, No. 782,313 (1957); Chem. Abstr., 52, 2916 b (1958).

12. Hoevel-Kestermann, H., and H. Mühlemann, Pharm. Acta. Helv., 47, 394 (1972).

13. Ochs, Q., Squibb Institute, personal communication.

14. Ates, O., and H. Amal, Istambul Univ. Eczacilik Fak. Mecm., 2, 82 (1966).

15. Lissitzky, S., Bull. Soc. Chim. biol., 37, 89 (1955).

16. Gopal, N., J. Appl. Radiation Isotopes, 17, 75 (1966).

17. Pileggi, V., R. Henry, M. Segalove and G. Hamill, Clin. Chem., 8, 647 (1962).

18. Mandl, R. and R. Block, Arch. Biochem. Biophys, 81, 25 (1959).

19. Turula, K., Acta Endocrinol. 48, 31 (1965).

20. Gmelin, R. and A. Virtanen, Acta Chem. Scand., 13, 1469 (1959).

21. Brewer, G., Squibb Institute, personal communication.

22. Soh, T., Squibb Quality Control, personal communication.

23. Chow, L., Squibb Quality Control, personal communication.

24. Blaufox, M., D. Sanderson, W. Tauxe, K. Wakim, A. Orvis and C. Owens, Amer. J. Physiol., 204, 536 (1963).

25. Kabasakalian, P., and J. Mc Glotten, Anal. Chem., 31, 513 (1959)

26. Purkiss, P., R. Lane, W. Cattell, I. Fry, and A. Spencer, Invest. Radiol., 3, 271 (1968).

27. Dennenberg, T., Acta Med. Scand. Suppl, 442, 17 (1966).

28. Davidson, A., Invest. Radiol., 1, 271 (1966).

29. Stahl, E. and J. Pfeifle, Z. Anal. Chem., 200, 377 (1964).

30. Jacob, Z., Squibb Quality Control, personal communication.

31. Hollingsworth, D., M. Dillard and P. Bondy, J. Lab. Clin. Med., 62, 346 (1963).

32. Willis, S., Squibb Institute, personal communication.

33. Neudert, W., and H. Röpke, Chem. Ber., 87, 659 (1954).

34. Valenti, V., Squibb Institute, personal communication.

35. Kabadi, B., Squibb Quality Control, personal communication.

36. Boerdalen, B., H. Wang and H. Hotterman, Invest. Radiol., 5, 595 (1970); Chem. Abstr., 74, 123391 p (1971).

37. Hartmann, E. and H. Röpke, Z. Anal. Chem., 232, 268 (1967).

38. Gonda, H., Squibb Quality Control, personal communication.

39. Poet, R., Squibb Institute, personal communication.

40. Hentrich, K. and S. Pfeifer, Pharmazie, 21, 296 (1966).

41. Targos, F., Squibb Institute, personal communication.

42. Strickler, H., E. Saier, E. Kelvington, J. Kempic, E. Campbell and R. Grauer, J. Clin. endocrin. Metabol., 24, 15 (1964).

43. Yakatan, G. and M. Tuckerman, J. Pharm. Sci., 55, 532 (1966).

44. Zak, B. and A. Boyle, J. Amer. Pharm. Assn., sci. ed., 41, 260 (1952).

45. Egli, R., Z. Anal. Chem., 247, 39 (1969).

46. Strauss, I., F. Erhardt and K. May, Fortschr. Med., 87, 1296 (1969).

47. Kallos, P., Squibb Quality Control, personal communication.

48. Holynska, B. and J. Jankiewicz, Chem. Anal. (Warsaw), 14, 219 (1969); Chem. Abstr., 71, 24774 X (1969).

49. Schiller, P., and J. Synek, Cesk. Farm., 15, 508 (1966); Chem. Abstr., 66, 79647 k (1967).

50. Buersch, J., R. Johs and P. Heintzen, Fortschr. Geb. Roentgenstr. Nuklearmed., 112, 259 (1970); Chem. Abstr., 72, 126781 k (1970).

51. Mikolajek, E., J. Kormicki and Z. Kowalczyk, Diss. Pharm., Pharmacol, 18, 523 (1966); Chem. Abstr. 64, 14905 u (1967).

52. Zatz, L., Squibb Quality Control, personal communication.

53. Tauxe W., M. Burbank, F. Maher and J. Hunt, Proc. Mayo Clinic, 39, 761 (1964).

54. Mc Chesney, E. and J. Hoppe, Amer. J. Roentgen., 78, 137 (1957).

55. Knoefel, P., Ann. Rev. Pharmacol., 5, 321 (1965).

56. Woodruff, M. and R. Malvin, J. Urol., 84, 677 (1960).

57. Stokes, J., J. Conklin and H. Huntley, J. Urol., 87, 630 (1962).

58. Miller, M., Squibb Institute, personal communication.

59. Schmid, H., Pharm.ActaHelv., 46, 134 (1971); Chem. Abstr., 75, 3915 m (1971).

60. Alicino, J., Squibb Institute, personal communication.

DISULFIRAM

Norris G. Nash and Raymond D. Daley

CONTENTS

1. DESCRIPTION

1.1 Name, Formula, Molecular Weight

The Chemical Abstracts name for disulfiram is tetraethylthioperoxydicarbonic diamide, starting with Volume 76. Previously the Chemical Abstracts name was bis(diethylthiocarbamoyl)disulfide. The CAS Registry No. is [97-77-8]. Other common names are tetraethylthiuram disulfide and TTD.

$$(CH_3CH_2)_2N-\overset{S}{\overset{\|}{C}}-S-S-\overset{S}{\overset{\|}{C}}-N(CH_2CH_3)_2$$

$C_{10}H_{20}N_2S_4$ Mol. Wt.: 296.54

1.2 Appearance, Color, Odor

Disulfiram is a light yellow odorless crystalline powder.

2. PHYSICAL PROPERTIES

2.1 Infrared Spectra

The infrared spectrum of disulfiram (Figure 1) was obtained with a Beckman IR-12 spectrophotometer. The spectrum was run in three sections: (a) from 200 to 650 cm^{-1} as a mineral oil mull on polyethylene; (b) from 500 to 1365 cm^{-1} as a mineral oil mull between potassium bromide plates; (c) from 1360 to 4000 cm^{-1} as a perfluorinated oil mull between potassium bromide plates. The spectrum matches those published elsewhere (1,2).

Some of the absorption bands can be assigned to the alkyl groups: 2980 to 2870 cm^{-1}, C-H stretching; 1460 cm^{-1}, C-H bending; 1378 cm^{-1}, methyl C-H bending (symmetrical). The assignment of bands to the various vibrations of the remainder of the molecule is complicated by the extensive coupling of vibrations of the N-C-S bonds and the apparent conjugation of the nitrogen unshared electron pair with the C=S bond (2,3). The 1500 cm^{-1} band has been assigned to a C-N stretching vibration (3) and the 1000 cm^{-1} band to a C=S stretching vibration (4). Other suggested assignments (5) are: 926 cm^{-1}, $N-\overset{\|}{C}-S$; 976 cm^{-1}, C=S; 1149 and 1206 cm^{-1}, $N-\underset{S}{\overset{\|}{C}}-S$; 1274 cm^{-1}, $N-\overset{\|}{C}$.

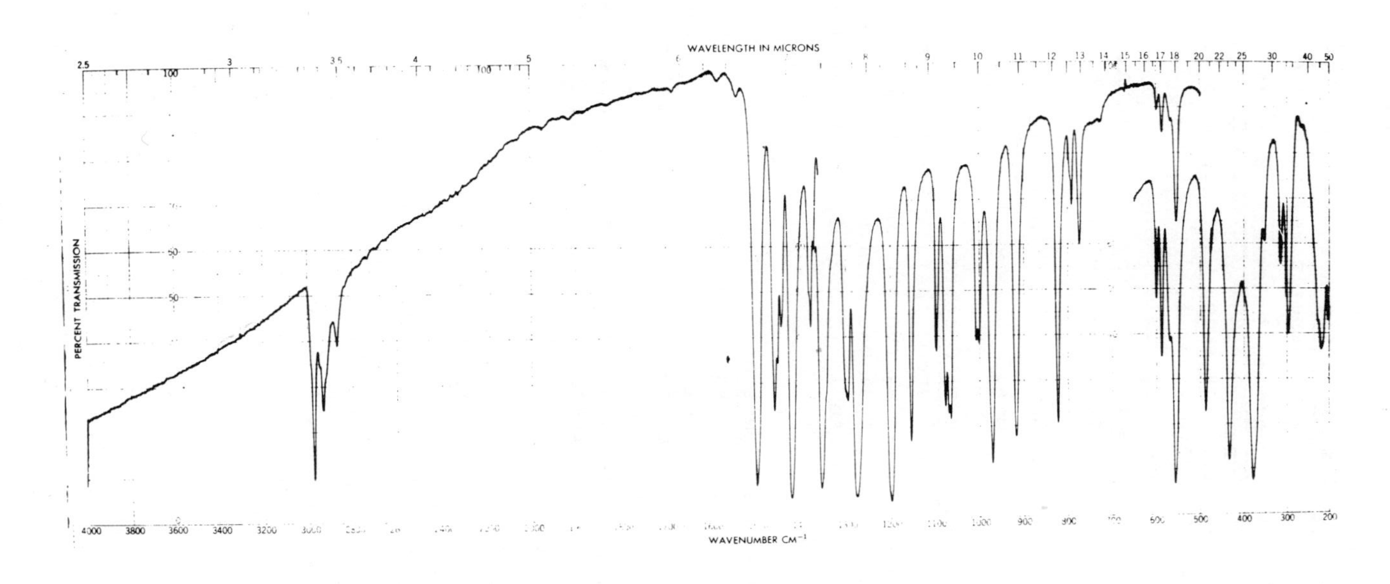

Figure 1. Infrared spectrum of disulfiram, mineral oil and perfluorinated oil mulls.

2.2 Nuclear Magnetic Resonance Spectra

The proton magnetic resonance spectrum of disulfiram shown in Figure 2 was run in $CDCl_3$, on a Varian A-60A 60 MHz NMR spectrometer, with a tetramethylsilane reference (6). The sweep width is 0 to 500 Hz.

The triplet at 1.40 ppm (J=7.3 Hz) is assigned to the methyl protons, while the quartet at 4.06 ppm (J=7.3 Hz) is assigned to the methylene protons. The small bands at $\pm$ 27 Hz of each line are spinning sidebands (6).

The temperature dependence of the NMR spectrum of disulfiram and similar compounds has been investigated (7,8).

2.3 Ultraviolet Spectra

Figure 3 shows the ultraviolet absorption spectrum of disulfiram, run on a Cary Model 14 spectrophotometer. The spectrum was obtained at two concentrations, 100 μg per ml and 10 μg per ml, in methanol solution, in 1 cm cells.

2.4 Mass Spectra

The low resolution mass spectrum of disulfiram is shown in Figure 4. This spectrum was obtained on an LKB 9000S mass spectrometer, with an ionization voltage of 70 eV, source temperature 250°C (6). This spectrum is similar to one reported and interpreted by Madsen et al (9); their suggested fragmentation scheme is shown in Figure 5.

2.5 Differential Thermal Analysis

The only thermal event in the differential thermal analysis curve of this compound is the melting endotherm at 70°C (10).

2.6 Solubility

The solubility of disulfiram at room temperature is as follows:-

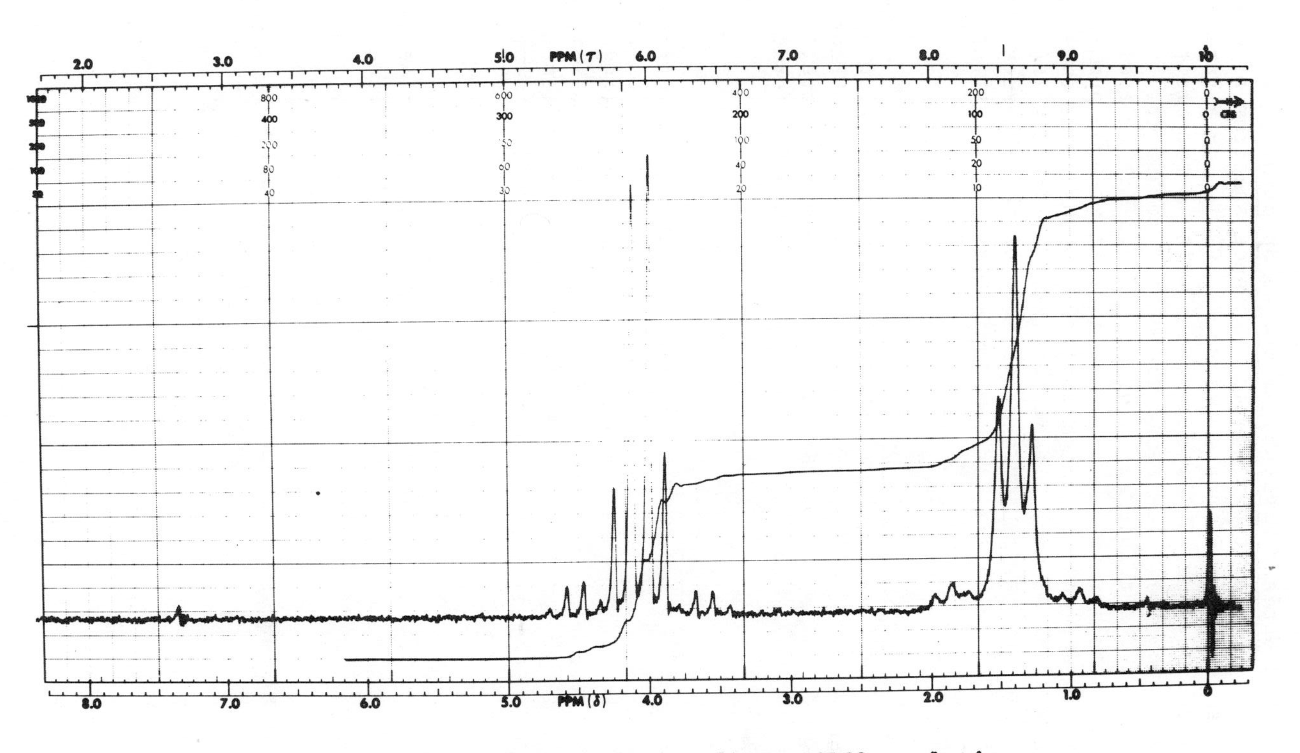

Figure 2. NMR spectrum of disulfiram, $CDCl_3$ solution.

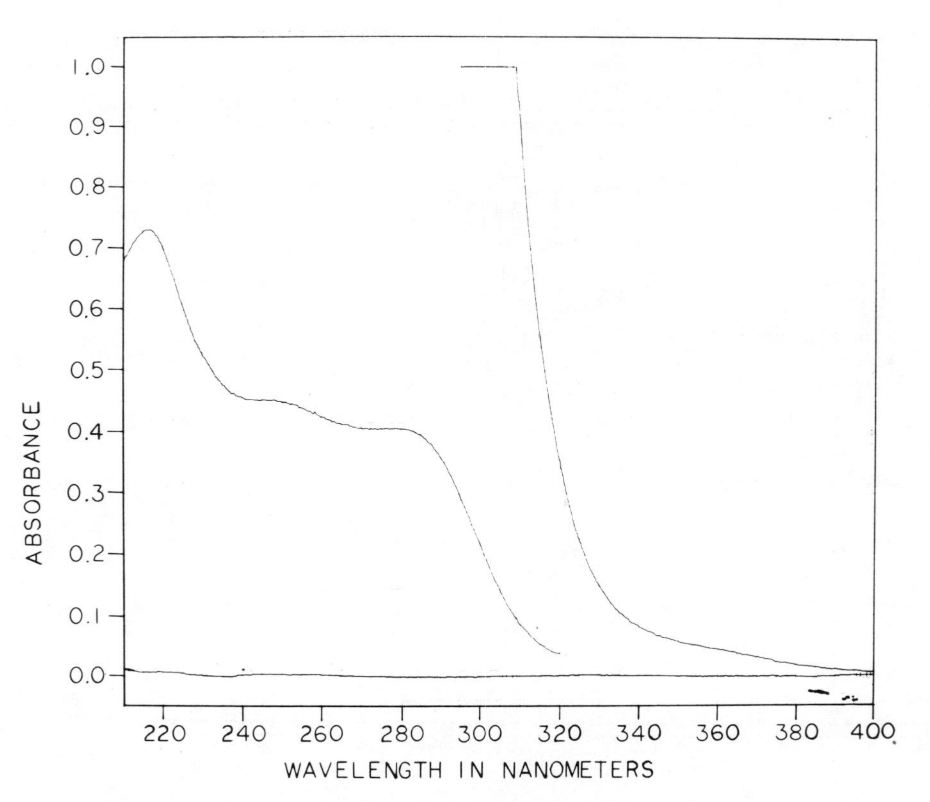

Figure 3. Ultraviolet spectrum of disulfiram, 100 μg per ml and 10 μg per ml, in methanol, 1 cm cells.

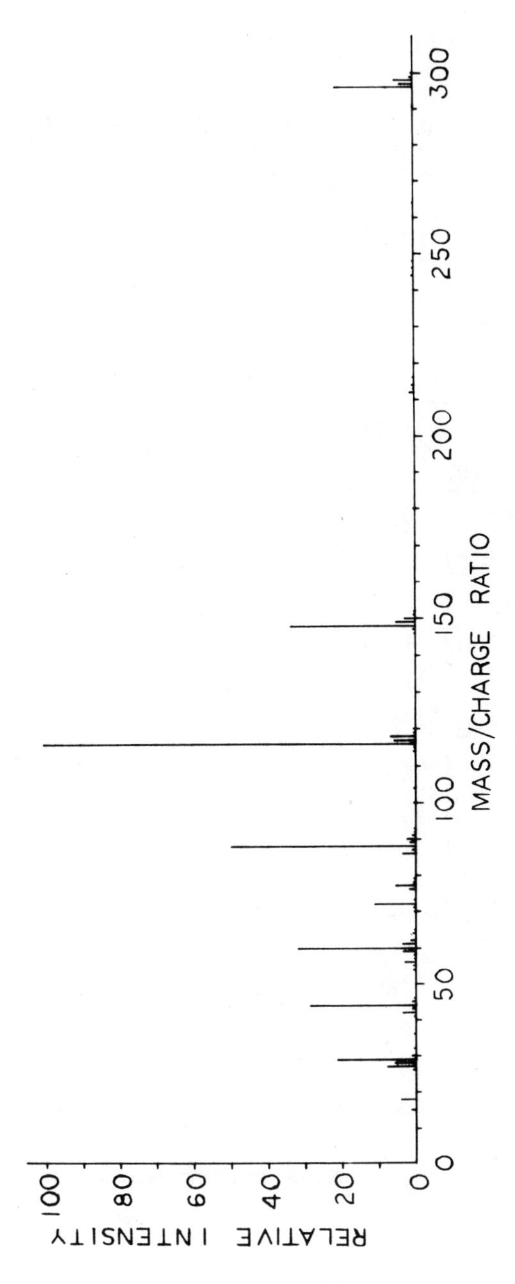

Figure 4. Mass spectrum of disulfiram.

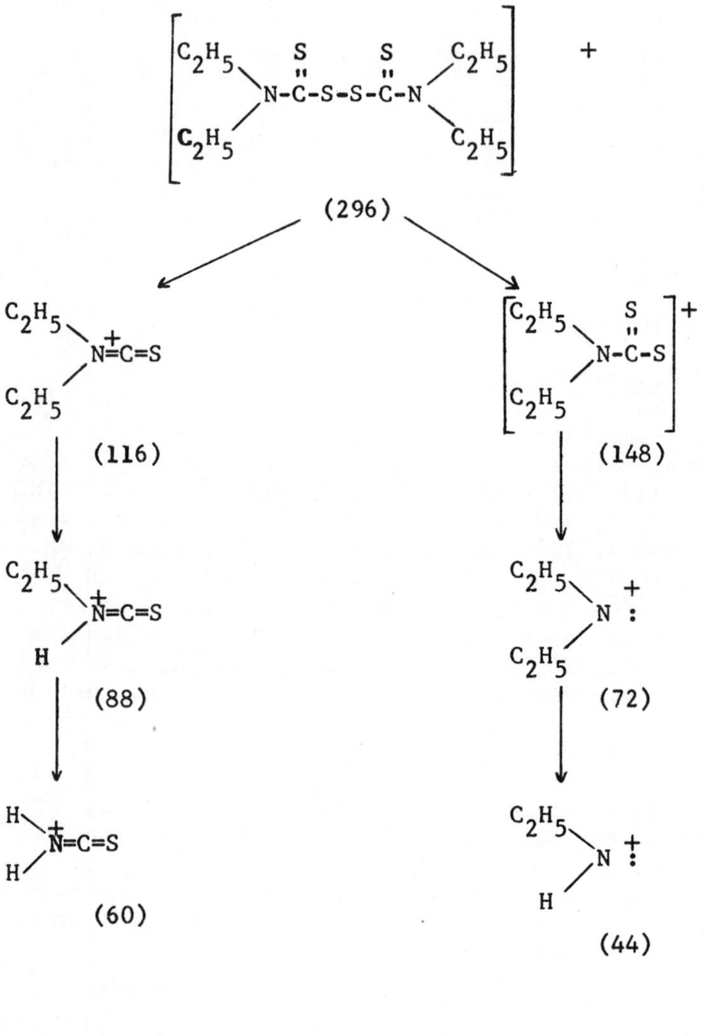

$$\left[(C_2H_5)_2N-\overset{S}{\overset{\|}{C}}-S-S-\overset{S}{\overset{\|}{C}}-N(C_2H_5)_2\right]^{+}$$

(296)

$$(C_2H_5)_2\overset{+}{N}=C=S$$

(116)

$$(C_2H_5)(H)\overset{+}{N}=C=S$$

(88)

$$H_2\overset{+}{N}=C=S$$

(60)

$$\left[(C_2H_5)_2N-\overset{S}{\overset{\|}{C}}-S\right]^{+}$$

(148)

$$(C_2H_5)_2N^{+}:$$

(72)

$$(C_2H_5)(H)N^{+}:$$

(44)

Figure 5. Mass spectrometric fragmentation (9).

Solvent	Approximate Solubility, mg/ml
Methanol	25
Ethanol (95%)	22
Acetone	440
Water	0.3
Chloroform	740
Ether	80
Benzene	560
Petroleum ether	2.9

2.7 Crystal Properties

Grabar and McCrone reported the crystal properties of disulfiram (11). The crystal system is monoclinic, with axial ratios a:b:c of 0.870 : 1 : 0.545, interfacial angles $(011 \wedge 0\bar{1}1) = 47.5°$, profile angle $(011 \wedge 0\bar{1}1$ in 100 plane$) = 123°$, and beta angle = 126°. The refractive indices (5893 A, 25°C) are $\alpha=1.590 + 0.005$, $\beta=1.67 \pm 0.01$, and $\gamma=1.740 \pm 0.005$. The optic axial angle is $2V = 84 \pm 5°$, dispersion $v > r$, optic axial plane 010, sign of double refraction negative. The molecular refraction is 84.2 (calculated), 85.0 (observed). Some additional information regarding crystal habit, fusion behavior, and x-ray diffraction is also given in this paper. No polymorphism was observed.

Grabar and McCrone (11) also reported x-ray diffraction data for disulfiram. They found the cell dimensions to be a=13.84A, b=15.90A, and c=8.66A, with four formula weights per cell. The d-spacings and relative intensities of the principal diffraction lines are tabulated.

Karle et al (12) made a complete crystal structure analysis. The molecule contains two planar

$$-S-\overset{\overset{S}{\|}}{C}-N\langle{}^{C-}_{C-}$$

groups, nearly perpendicular to each other. The configuration about the nitrogen atoms is planar, not pyramidal. The space group is $P2_1/c$, with four molecules in the unit cell. Their unit cell parameters were $a=11.11 \pm .02$, $b=15.90 \pm .03$, $c=8.66 \pm .02$, $\beta=92° 42' \pm 15'$. The unit

TABLE I

X-RAY POWDER DIFFRACTION PATTERN OF DISULFIRAM

d(A)	I/I_1	d(A)	I/I_1
9.11	10	3.18	23
7.94	8	3.17	23
7.59	27	3.08	5
6.36	100	3.05	6
6.15	37	3.00	16
5.55	9	2.91	10
5.24	14	2.83	3
5.11	49	2.80	4
4.77	7	2.77	12
4.55	21	2.70	4
4.50	11	2.65	3
4.40	1	2.61	2
4.32	12	2.56	2
4.21	9	2.52	10
4.16	60	2.48	8
4.08	8	2.44	5
3.96	13	2.38	3
3.83	3	2.33	3
3.79	4	2.29	4
3.73	1	2.26	4
3.62	27	2.22	10
3.54	1	2.20	5
3.46	19	2.15	7
3.41	11	2.06	5
3.34	7	1.88	4
3.28	8	1.71	2
3.23	5		

cell reported by Grabar and McCrone can be mathematically transformed into this one.

The x-ray powder diffraction pattern for disulfiram is shown in Table I. This pattern was obtained on a Norelco x-ray diffractometer with nickel-filtered copper Kα radiation. This pattern closely resembles that reported by Grabar and McCrone (11).

2.8 Melting Point

The following melting points have been reported:

67-71°C	(13)
70.5	(14)
68.4	(15)
74	(16)
69-70	(17)
72	(73)
69.5-70	(72)

3. SYNTHESIS

Disulfiram has been prepared in a variety of ways. The usual method involves oxidation, by various means, of diethyldithiocarbamic acid, its sodium salt, or some other metal salt, to give the disulfiram product directly. Oxidants used include hydrogen peroxide (18, 19, 20, 72), sodium hypochlorite (21, 22), chlorine or bromine (13, 15, 23), sodium nitrite and hydrochloric acid (24, 73), potassium ferricyanide (14), air (25), and ammonium persulfate (16). An alternative method is the reaction of diethylamine monosulfide with carbon disulfide (17).

These reactions are shown in Figure 6.

4. STABILITY-DEGRADATION

No reports of stability studies or of the formation of degradation products were found in the literature, except for metabolic products (Section 5). Several unidentified degradation products are formed if an alkaline solution of disulfiram in methanol is heated to 50°C for 2 hours, however, and at least one unidentified product is formed when an acidic methanol solution is heated similarly (26).

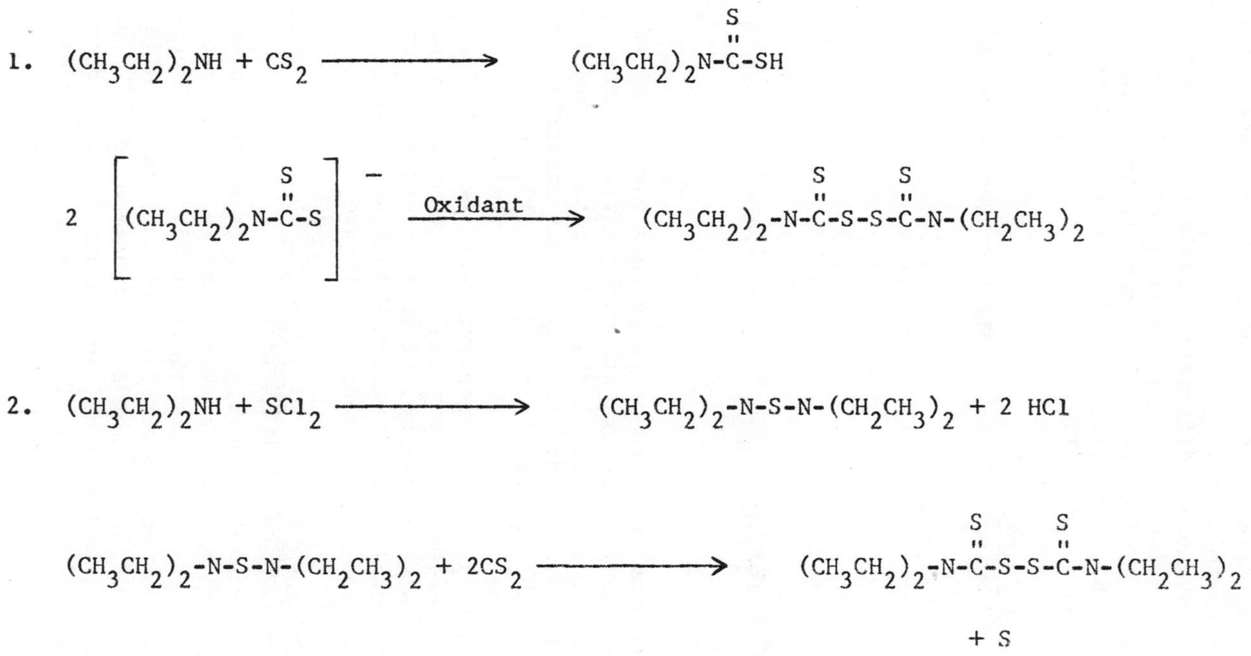

1. $$(CH_3CH_2)_2NH + CS_2 \longrightarrow (CH_3CH_2)_2N\text{-}\overset{S}{\overset{\|}{C}}\text{-}SH$$

$$2\left[(CH_3CH_2)_2N\text{-}\overset{S}{\overset{\|}{C}}\text{-}S\right]^- \xrightarrow{\text{Oxidant}} (CH_3CH_2)_2\text{-}N\text{-}\overset{S}{\overset{\|}{C}}\text{-}S\text{-}S\text{-}\overset{S}{\overset{\|}{C}}\text{-}N\text{-}(CH_2CH_3)_2$$

2. $$(CH_3CH_2)_2NH + SCl_2 \longrightarrow (CH_3CH_2)_2\text{-}N\text{-}S\text{-}N\text{-}(CH_2CH_3)_2 + 2\ HCl$$

$$(CH_3CH_2)_2\text{-}N\text{-}S\text{-}N\text{-}(CH_2CH_3)_2 + 2CS_2 \longrightarrow (CH_3CH_2)_2\text{-}N\text{-}\overset{S}{\overset{\|}{C}}\text{-}S\text{-}S\text{-}\overset{S}{\overset{\|}{C}}\text{-}N\text{-}(CH_2CH_3)_2 + S$$

Figure 6. Synthesis of Disulfiram.

5. DRUG METABOLIC PRODUCTS

The reported metabolic products of disulfiram in various species are as follows:

Metabolite	Species	References
Diethyldithiocarbamic Acid	Man	(27,28,29,30)
	Rat	(31,32)
	Rabbit	(27,29,30)
	Cattle	(30)
	Mice	(33,34)
Carbon Disulfide	Man	(30,35)
	Rat	(31,32)
	Rabbit	(30)
	Cattle	(30)
	Guinea Pig	(35)
N,N-Diethyldithiocarbamoyl-1-thio-β-D-glucopyranosiduronic Acid	Man	(36)
	Rat	(31,32)
Diethyldithiocarbamic Acid, Methyl Ester	Mice	(34)
Hydrogen Sulfide	Man	(30)
	Rabbit	(30)
	Cattle	(30)
Sulfate	Rat	(31)

Gessner and Jakubowski have proposed the metabolic scheme in Figure 7 (34).

6. METHODS OF ANALYSIS

6.1 Identification Tests

Disulfiram is easily identified by the physical properties described in section 2 above. Where identification of disulfiram in formulations is necessary, it can be extracted by solvents such as carbon disulfide. If identification of very small amounts is necessary, the colorimetric tests described in section 6.3 or polarog-

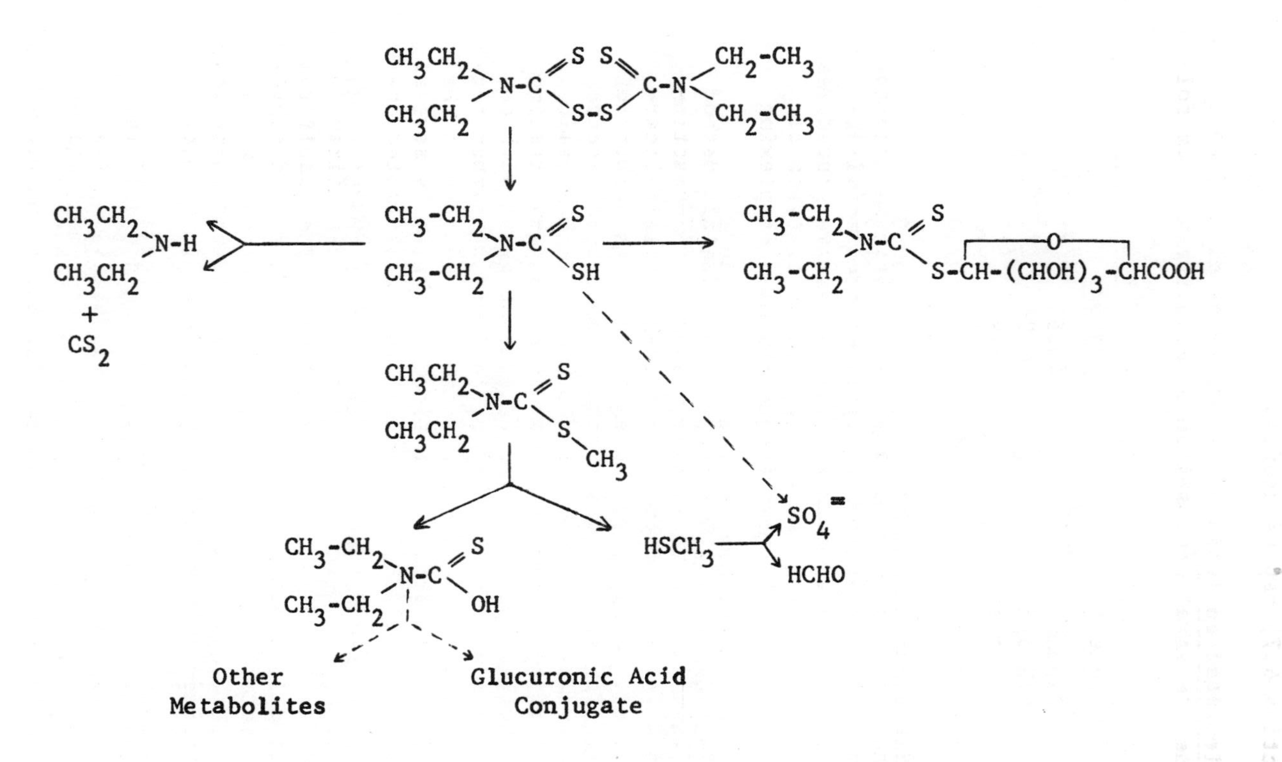

Figure 7. Metabolic products of disulfiram (34).

raphy (section 6.7) may be useful.

6.2 Elemental Analysis

The elemental composition of disulfiram is as follows:

Element	% Theory
Carbon	40.50
Hydrogen	6.80
Nitrogen	9.45
Sulfur	43.25

6.3 Colorimetric Methods

The basis for many of the colorimetric methods for disulfiram is the formation of highly colored metal-diethyldithiocarbamate complexes. Very sensitive procedures have been developed for use with a variety of types of samples. Copper is the most frequently used complexing metal.

Domar et al (27) developed a colorimetric method for disulfiram in excreta samples, based on the reaction with cuprous iodide to form the copper-diethyldithiocarbamate complex. Disulfiram was extracted with carbon tetrachloride. The extract was treated with cuprous iodide to yield a brown-yellow complex. Diethyldithiocarbamic acid, a disulfiram metabolite, was determined by treating the carbon tetrachloride extracted aqueous phase with cupric sulfate in a buffer solution, followed by carbon tetrachloride extraction. Domar's method was adapted to the determination of metabolites in blood and urine by several investigators (29,37,38,39). Patzch (40) substituted cupric chloride for the cuprous iodide used by Domar.

Parquot and Mercier (41) determined disulfiram in fat samples at the 5 to 50 μg per g level. The disulfiram was extracted into petroleum ether and reduced to diethyldithiocarbamate. The copper-dithiocarbamate complex was formed by treatment with a cupric sulfate-hydroquinone reagent, extracted into chloroform, and determined colorimetrically.

Egorova and Trukhina (42) determined disulfiram in suppositories by carbon tetrachloride extraction, reaction with cuprous iodide, and colorimetric measurement against a blank containing no cuprous iodide.

Vignoli and Cristau (43) determined disulfiram by reduction of phosphomolybdotungstate in alkaline solution, to form a blue color.

Fried et al (44) adapted the disulfiram identification color reaction in Clarke's manual of drug identification (45) to develop a quantitative procedure. Reaction of disulfiram with ethanol and sodium cyanide produces a blue color. The response is linear in the range 2 to 20 μg disulfiram per ml. Diethyldithiocarbamate and carbon disulfide do not interfere.

Kofman and Arzamastsev (46) and Piotrowska (47) reported determining disulfiram by combustion by the oxygen flask method. The sulfur dioxide produced is bound as a nonvolatile complex by sodium tetrachloromercurate. The complex is determined colorimetrically by reaction with formaldehyde and pararosaniline in hydrochloric acid solution.

6.4 Titration Methods

Disulfiram is not sufficiently basic to be titrated with acid even in non-aqueous media. However, several titration methods based on complex formation or oxidation have been reported.

Ferreira (48) titrated an acidified carbon tetrachloride solution, containing from 15 to 25 mg of disulfiram, with a 0.1$\underline{N}$ bromate-bromide solution. The following reaction is given:

$$(C_2H_5)_2\text{-N-}\overset{\displaystyle S}{\overset{\|}{C}}\text{-S-S-}\overset{\displaystyle S}{\overset{\|}{C}}\text{-N-}(C_2H_5)_2 + 13\ Br_2 + 20\ H_2O \longrightarrow$$

$$26\ HBr + 4\ H_2SO_4 + 2\ HCOOH + 2\ (C_2H_5)_2NH$$

Analyses of the reaction mixture for sulfate and diethylamine, as well as the amount of bromine consumed, support the stoichiometry given above.

Varga (49) determined disulfiram in the drug substance and tablets by oxidation with excess 0.1$\underline{N}$ ceric sulfate and determination of the excess reagent by back titration with 0.1$\underline{N}$ ferrous sulfate. Sandri (50) oxidized disulfiram with an excess of periodic acid; after 30 minutes the excess periodic acid was determined volumetrically by addition of potassium iodide and titration of the iodine formed.

Roth (51) determined disulfiram in crude products and fungicide formulations after conversion to carbon disulfide. The sample was refluxed in formic acid and the carbon disulfide evolved was trapped in an alkaline cadmium acetate solution. The cadmium acetate solution was neutralized and titrated with 0.1N methanolic iodine to a starch-iodine end point.

Bayer and Posgay (52,53) determined that the addition of mercuric acetate to a glacial acetic acid solution of disulfiram formed a complex that consumed 2 moles of perchloric acid per mole of disulfiram. The sample was titrated to a gentian violet end point with perchloric acid in glacial acetic acid.

Bukreev et al (54) treated disulfiram with 0.1N nickel sulfate in an alcohol-ammonia solution. After 1 hour the excess nickel ion was titrated with 0.1N tetrasodium edetate. The accuracy of the method was stated to be within $\pm$ 0.6%. Li and Li (55) treated a dioxane solution of tetramethylthiuram disulfide with an excess of silver nitrate solution, separated the precipitate, and determined the excess silver ion by titrating with thiocyanate solution. Two moles of silver reacted with 1 mole of tetramethylthiuram disulfide.

Fournier et al (56) determined disulfiram metabolites and other compounds with C-S bonds in urine samples by the catalytic effect of the metabolites on the reaction of iodine with iodazide. Scalicka (57) used sodium azide instead of iodazide to determine disulfiram metabolites by the catalytic effect; quantitation was by titration of excess sodium azide with sodium arsenate after the reaction was completed.

Stefek et al (58) determined the nitrogen content by a Kjeldahl method. The ammonia was collected in a 2% boric acid solution, cooled in an ice-salt mixture, and titrated with 0.05N hydrochloric acid. The standard deviation for the determination of nitrogen was less than 0.2%.

6.5 Infrared Spectroscopy

Salvesen et al (59) determined disulfiram in tablets by infrared spectroscopy. The strength was determined in a carbon tetrachloride solution using the absorption bands at 7.41, 8.32, 9.92, and 10.31 microns with a relative error of 2%.

The band at 818 cm^{-1} (carbon disulfide solution)

can also be used. Diethyldithiocarbamic acid has no absorption bands near 818 cm^{-1}.

6.6 Proton Magnetic Resonance Spectroscopy

Sheinin et al (60) determined disulfiram in the drug substance and tablets by proton magnetic resonance spectroscopy. The average results on a tablet composite were 100.8 ± 1.4% of claim by magnetic resonance spectroscopy and 100.7 ± 0.4% by a colorimetric procedure (61).

6.7 Polarography

Belitskaya (62) analyzed disulfiram in rubber samples by micropolarographic procedure at a 0.003M level using the wave at -2.5V in an ammonia-ammonium chloride buffer. Taylor (63) used cathode-ray polarography to make the determination in the 1.5 to 15 ppm range on rubber samples; at this level a linear response was obtained. Porter et al (64) used cathode-ray polarography on urine samples and increased the sensitivity and specificity by the addition of copper to the sample. The wave for thiuram at -0.5V was replaced by a wave at -0.65V and that of the excess copper at -0.23V. The method gave 85% recovery on urine samples at the 0.5 μg per ml level and 95% recovery on samples extracted with chloroform to give a 2 μg per ml solution.

Prue et al (65) used pulse polarography for determining disulfiram in tablet samples. The pulse polarographic method is more sensitive than d.c. polarography and the peak is easier to identify and quantitate. The use of an aqueous-ethanol acetate buffer rather than an aqueous-ethanol ammonia buffer eliminates the interference of diethyldithiocarbamate, increasing the specificity of the method.

6.8 Gravimetric Methods

Three papers describe oxidation of disulfiram in various ways to produce sulfate, determined gravimetrically as barium sulfate. Ferreira (48) used bromine, Wojahn and Wempe (66) used strong alkali, and Parrak (67) used nitric acid in the presence of ammonium vanadate.

6.9 Chromatography

Parker and Berriman (68) developed a separation system utilizing silica gel-Celite adsorbents and 4 binary solvent mixtures to separate 32 rubber additives, one of which was disulfiram.

Stromme (32) developed a method for the separation of proteins, diethyldithiocarbamate, and disulfiram in diluted plasma, liver homogenate, and diluted urine using a Sephadex G-25 gel filtration column.

Goeckeritz (69) used paper chromatography to separate disulfiram from 4 other antimycotic drugs using paper impregnated with one of the following: (a) heptane; (b) heptane-xylene (1:1); (c) cyclohexane; (d) methanol-ethanol (1:1); (e) acetone with dimethylformamide; (f) paraffin. As stationary solvents acetone or hexane was used. For detection, iodoazide or mercury fluorescein was used.

Gessner and Jakubowski (34) used descending paper chromatography on Whatman No. 1 or No. 4 paper to separate disulfiram metabolites. The following systems were employed: (a) n-hexane on paper impregnated with 50% dimethylformamide solution in ethanol; (b) cyclohexane-n-butanol (1:1) on paper impregnated with 50% ethanol solution of dimethylformamide. Detection was by observation of quenching spots under ultraviolet light.

Farago (70) chromatographed biological extracts on silica gel G plates. The R_f values for disulfiram were (a) 0.82, using methanol-acetone-triethanolamine (1:1:0.03); and (b) 0.77, using ethylene dichloride-benzene-88% formic acid-H_2O. Palladium chloride was used to detect disulfiram. The unsprayed spot can be eluted from the silica gel with chloroform-ethanol (2:1) and the absorbance determined at 272 nanometers.

Drost and Reith (71) chromatographed biological extracts on Kieselgel G plates. Two solvent systems were used: (a) toluene-ethyl acetate-diethylamine (7:2:1), and (b) cyclohexane-diethylamine (9:1). The spots were detected by ultraviolet light irradiation or by spraying with a mixture of 3 ml 10% H_2PtCl_6, 100 ml 6% KI, and 47 ml of water. Unsprayed spots were scraped off and extracted with 1,2-dinitrochloroethane; the solution was evaporated and the residue taken up in methanol for ultraviolet determination.

7. REFERENCES

1. C. J. Pouchert, "The Aldrich Library of Infrared Spectra", Aldrich Chemical Company, Milwaukee, Wisconsin, 1970, spectrum 369D.
2. A. T. Pilipenko and N. V. Mel'nikova, Zh. Neorg. Khim. 14 (7), 1843-6 (1969); C.A. 71, 75815u.
3. L. J. Bellamy, "Advances in Infrared Group Frequencies", Methuen & Co. Ltd., London, 1968, pp. 212-214.
4. H. C. Brinkhoff and A. M. Grotens, Rec. Trav. Chim. Pays-Bas 90 (3), 252-7 (1971).
5. M. L. Shankaranarayana and C. C. Patel, Spectrochim. Acta 21 (1), 95-103 (1965).
6. G. Schilling, Ayerst Research Laboratories, private communication.
7. H. C. Brinkhoff, A. M. Grotens, and J. J. Steggerda, Rec. Trav. Chim. Pays-Bas 89 (1), 11-17 (1970).
8. A. M. Grotens and F. W. Pijpers, Rec. Trav. Chim. Pays-Bas 92 (5), 619-27 (1973).
9. J. O. Madsen, S.-O. Lawesson, A. M. Duffield, and C. Djerrasi, J. Org. Chem. 32, 2054-8 (1967).
10. C. E. Orzech, Ayerst Laboratories Inc., private communication.
11. D. J. Grabar and W. C. McCrone, Anal. Chem. 22 (4), 620-1 (1950).
12. I. L. Karle, J. A. Estlin, and K. Britts, Acta Crystallogr. **22** (2), 273-80 (1967).
13. R. H. Cooper, U.S. Patent No. 2,375,083, May 1, 1945; C.A. 40, 1875.
14. R. Rothstein and K. Binovic, Rec. Trav. Chim. Pays-Bas 73, 561-2 (1954).
15. M. M. Miville, Ger. Patent No. 1,023,030, Jan. 23, 1958; C.A. 54, 17274c.
16. B. L. Richards, Belg. Patent No. 648,878, Sept. 30, 1964; C.A. 63, 14714a.
17. E. S. Blake, J. Am. Chem. Soc. 65, 1267-9 (1943).
18. H. S. Adams and L. Meuser, U.S. Patent No. 1,782, 111; C.A. 25, 303.
19. Span. Patent No. 281,099, Nov. 6, 1962; C.A. 59, 7378a.
20. J. P. Zumbrunn, Fr. Patent No. 2,038,575, Jan. 8, 1971; C.A. 75, 87658r.

21. G. C. Bailey, U.S. Patent No. 1,796,977; C.A. 25, 2598.
22. S. L. Kemichrom, Span. Patent No. 354,918, Nov. 16, 1969; C.A. 72, 110847a.
23. J. C. Counts, R. T. Nelson, and W. R. Trutna, U.S. Patent No. 2,777,878, Jan. 15, 1957; C.A. 51, 9681i.
24. J. L. Eaton, U.S. Patent No. 2,464,799, Mar. 22, 1949; C.A. 43, 4690h.
25. W. L. Cox and A. Gaydash, Fr. Patent No. 1,322,579, Mar. 29, 1963; C.A. 59, 9812.
26. G. L. Boyden, Ayerst Laboratories Inc., private communication.
27. G. Domar, A. Eredga, and H. Linderholm, Acta Chem. Scand. 3, 1441-2 (1949); C.A. 44, 10025f.
28. L. Eldjarn, Scand. J. Clin. Lab. Invest. 2, 202-8 (1950); C.A. 45, 6298c.
29. H. Linderholm and K. Berg, Scand. J. Clin. Lab. Invest. 3, 96-102 (1951); C.A. 45, 10286e.
30. R. Fischer and H. Brantner, Naturwissenschaften 50, 551 (1963).
31. J. H. Stromme, Biochem. J. 92, 25P (1964).
32. J. H. Stromme, Biochem. Pharmacol. 14, 393-410 (1965).
33. J. H. Stromme, Biochem. Pharmacol. 15, 287-97 (1966).
34. T. Gessner and M. Jakubowski, Biochem. Pharmacol. 21, 219-30 (1972).
35. E. Merlevede and J. Peters, Arch. Belg. Med. Soc., Hyg., Med. Trav. Med. Leg. 23, 513-51 (1965); C.A. 64, 18224c.
36. J. Kaslander, Biochim. Biophys. Acta 71, 730-2 (1963).
37. K. J. Divatia, C. H. Hine, and T. N. Burbridge, J. Lab. Clin. Med. 39, 974-82 (1952).
38. S. L. Tompsett, Acta Pharmacol. Toxicol. 21, 20-2 (1964).
39. G. Bors and N. Ioanid, Farmacia (Bucharest) 13, 593-8 (1965); C.A. 64, 8784h.
40. H. Patzsch, Deut. Apoth. Ztg. 94, 1284-6 (1954); C.A. 53, 20229i.
41. C. Parquot and J. Mercier, Rev. Franc. Corps Gras 6, 695-9 (1959); C.A. 54, 6158g.
42. N. M. Egorova and V. I. Trukhina, Nauch. Tr. Perm. Farm. Inst. 4, 96-7 (1971); C.A. 79, 149338e.

43. L. Vignoli and B. Cristau, Congr. Soc. Pharm. Fr. C. R., 9th, 221-8 (1957); C.A. 53, 20690f.
44. R. Fried, A. N. Masoud, and M. Francis, J. Pharm. Sci. 62, 1368-9 (1973).
45. E. G. C. Clarke, Editor, "Isolation and Identification of Drugs", The Pharmaceutical Press, London, England, 1969, p. 319.
46. M. D. Kofman and A. P. Arzamastsev, Farm. Zh. (Kiev) 26, 52-6 (1971); C.A. 75, 91338d.
47. A. Piotrowska, Diss. Pharm. Pharmacol. 24, 93-7 (1972); C.A. 76, 158430u.
48. P. C. Ferreira, Arquív. biol. (Sao Paulo) 34, 103-5 (1950).
49. E. Varga, Acta Pharm. Hung. 28, 38-43 (1958); C.A. 53, 2541bc.
50. G. Sandri, Atti Accad. Sci. Ferrara 35, 17-24 (1957-58); C.A. 54, 18160d.
51. H. Roth, Angew. Chem. 73, 167-9 (1961); C.A. 55, 13161e.
52. I. Bayer and Mrs. G. Posgay, Acta Pharm. Hung. 31, 43-50 (1961); C.A. 56, 7431f.
53. I. Bayer and E. Posgay, Pharm. Zentralh. 100, 65-71 (1961); C.A. 55, 14819a.
54. A. I. Bukreev, V. V. Minakova, and V. F. Soinikova, Med. Prom. SSSR 18, 32-3 (1964).
55. M. Li and K. Li, Chung-Shan Ta Hsueh Hsueh Pao-Tzu Jan K'o Hsueh, 1959 (3), 26-8; C.A. 56, 6667.
56. E. Fournier, L. Petit and A. Lecorsier, J. Eur. Toxicol. 3, 337-40 (1971); C.A. 76, 149631h.
57. B. Scalicka, Prac. Lek 19, 408-12 (1967); C.A. 68, 58318e.
58. S. Stefak, M. Blesova, and M. Zahradnicek, Cesk. Farm. 21, 64-6 (1972); C.A. 77, 39330x.
59. B. Salvesen, L. Domange, J. Guy, Ann. Pharm. Fr. 13, 208-15 (1955); C.A. 49, 11959a.
60. E. B. Sheinin, W. R. Benson and M. M. Smith, Jr., J. Ass. Offic. Anal. Chem. 56, 124-7 (1973).
61. "Official Methods of Analysis", 1970, 11th edition, Association of Official Analytical Chemists, Washington, D. C., section 29.148.
62. R. M. Belitskaya, Kauch. Rezina 19, 53-5 (1960); C.A. 55, 4025e.
63. A. F. Taylor, Talanta 11, 894-6 (1964).

64. G. S. Porter and A. Williams, J. Pharm. Pharmacol. 24, 144-145P (1972).
65. D. G. Prue, C. R. Warner, and B. T. Kho, J. Pharm. Sci. 61, 249-51 (1972).
66. H. Wojahn and E. Wempe, Arch. Pharm. (Weinheim) 285, 375-82 (1952); C.A. 47, 12126g.
67. V. Parrak, Farmacia 22, 19-22 (1953); C.A. 49, 5776g.
68. C. A. Parker and J. M. Berriman, Trans. Inst. Rubber Ind. 28, 279-96 (1952); C.A. 47, 3602i.
69. D. Goeckeritz, Pharmazie 21, 668-74 (1966); C.A. 66, 98510b.
70. A. Farago, Arch. Toxikol. 22, 396-9 (1967); C.A. 67, 79553x.
71. R. H. Drost and J. F. Reith, Pharm. Weekbl. 105, 1129-38 (1970); C.A. 74, 2316j.
72. A. D. Cummings and H. E. Simmons, Ind. Eng. Chem. 20, 1173-6 (1928).
73. J. Giral and C. Soler, Rev. Soc. Quim. Mex. 2, 167-70 (1958); C.A. 53, 11220d.

ACKNOWLEDGMENTS

The writers wish to thank Dr. B. T. Kho for his review of the manuscript, Dr. G. Schilling of Ayerst Research Laboratories for his NMR and mass spectral data, the library staff for their literature search, and Mrs. B. Juneau for typing the profile.

ESTRADIOL VALERATE

Klaus Florey

CONTENTS

1. Description

1.1 Name, Formula, Molecular Weight

Estradiol Valerate is estra-1,3,5 (10) triene-3,17β-diol-17-valerate (pentanoate).

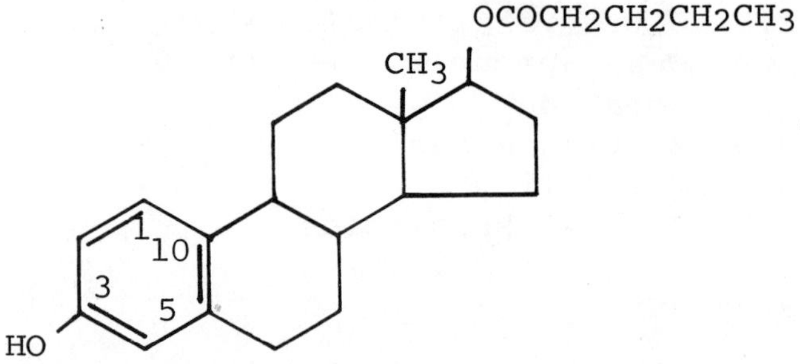

OCOCH2CH2CH2CH3
CH3
HO
1
10
3
5

$C_{23}H_{32}O_3$ Mol. Wt. 356.51

1.2 Appearance, Color, Odor

White, odorless, crystalline powder.

2. Physical Properties

2.1 Infrared Spectrum

The infrared spectrum of estradiol valerate[1] is presented in Figure 1.

2.2 Nuclear Magnetic Resonance Spectrum

The NMR spectrum is presented in Figure 2. It was obtained on a 60 MHz spectrometer in deuterochloroform containing tetramethylsilane as an internal reference. The following proton assignments were made[2]:

Protons at	Chemical Shift τ	Coupling Constants T (in Hz)
C-1H	2.86 (doublet)	1H,2H = 9
C-2H	3.38 (quartet)	1H,2H = 9;2H,4H =2.5
C-4H	3.43 (multiplet)	2H,4H = 2.5;4H,6H =1
C-18H	9.19 (singlet)	
C-17αH	5.25 (triplet)	16H,17H = 7.5
ester-CH_3	9.09 (triplet)	6.5
phenolic-OH	4.94 (singlet)	Concentration dependent

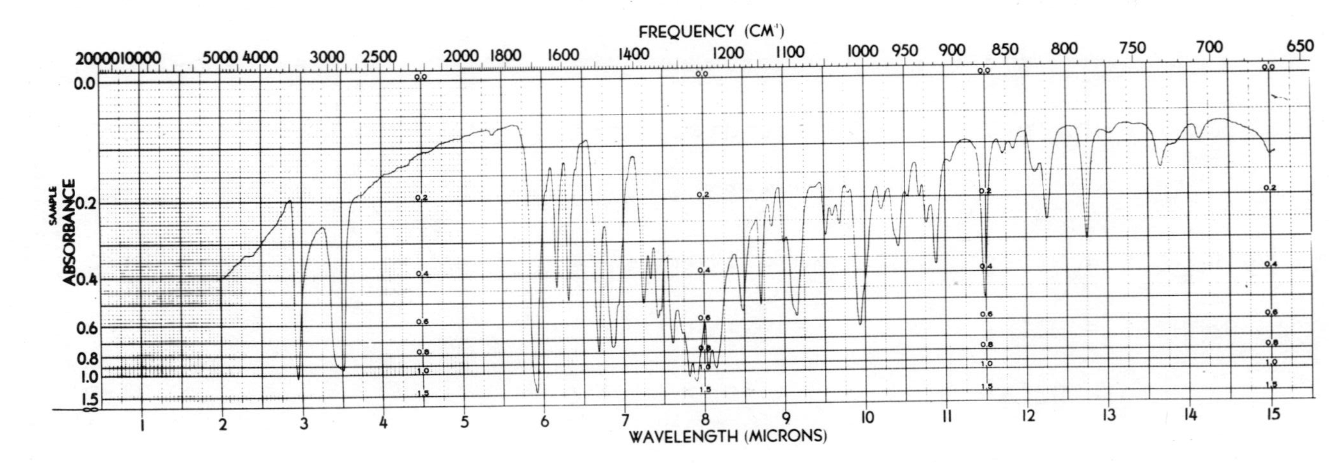

Figure 1. Infrared Spectrum of Estradiol Valerate in Mineral Mull. Instrument: Perkin-Elmer 621.

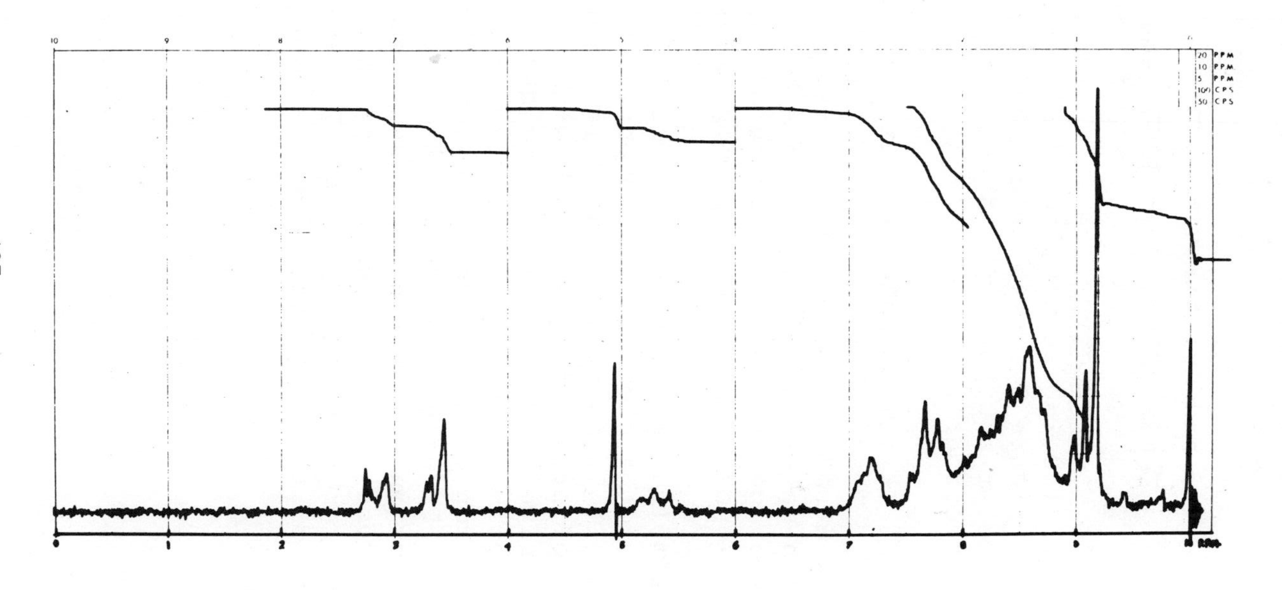

Figure 2. NMR Spectrum of Estradiol Valerate in Deuterated Chloroform. Perkin-Elmer R-12B.

2.3 Ultraviolet Spectrum

λ_{max}^{EtOH} 281 nm $E_{1cm}^{1\%}$ 61[3]

λ max 281 nm; = 2090[6]

λ max 287 nm; = 1800 (small peak)[6]

2.4 Mass Spectrum

The low resolution mass spectrum, shown in Figure 3, was obtained on an AEI MS-902 spectrometer equipped with a frequency modulated analog recorder[2].

It demonstrates the expected M^+ of m/e 356. There is also the M^+ of an acid homolog at m/e 370 present as a minor component. The high mass fragment ion at m/e 271 occurs by the cleavage of the acyl portion of the ester at the C-17 position. There is complement fragment ion at m/e 85 that represents the other portion of the molecule. The ions at m/e 253-255 represent the cleavage of the entire C-17 moiety. Because it is an aromatic steroid, there are a series of strong fragment ions with the progressive loss of first D-ring carbons, then C-ring carbons and finally B-ring carbons. The m/e 107 ion contains the A-ring and the C-6 carbon. Phenols can also eliminate the oxygen by explusion of CO or HCO. Thus a second series of progressive loss of carbons can occur but does not produce as many ions. The assignment of some of the diagnostic ions is depicted below.

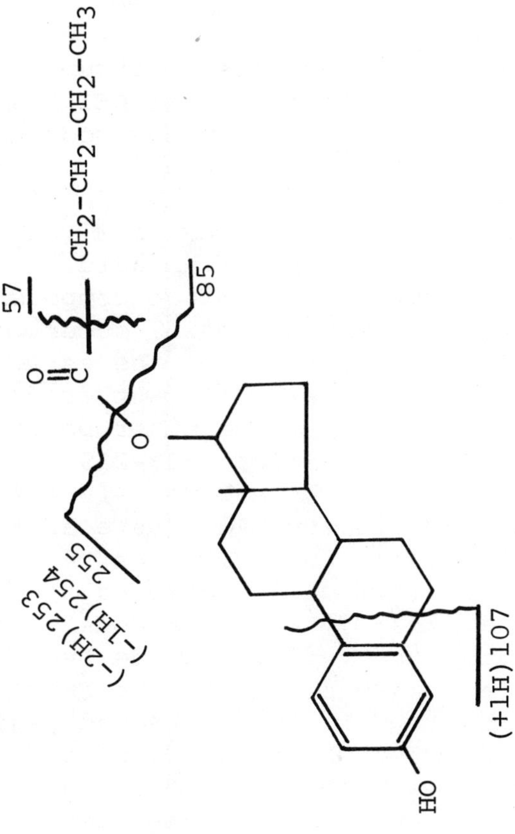
57
85
O
C
CH2-CH2-CH2-CH3
O
255
(-1H) 254
(-2H) 253
(+1H) 107
HO

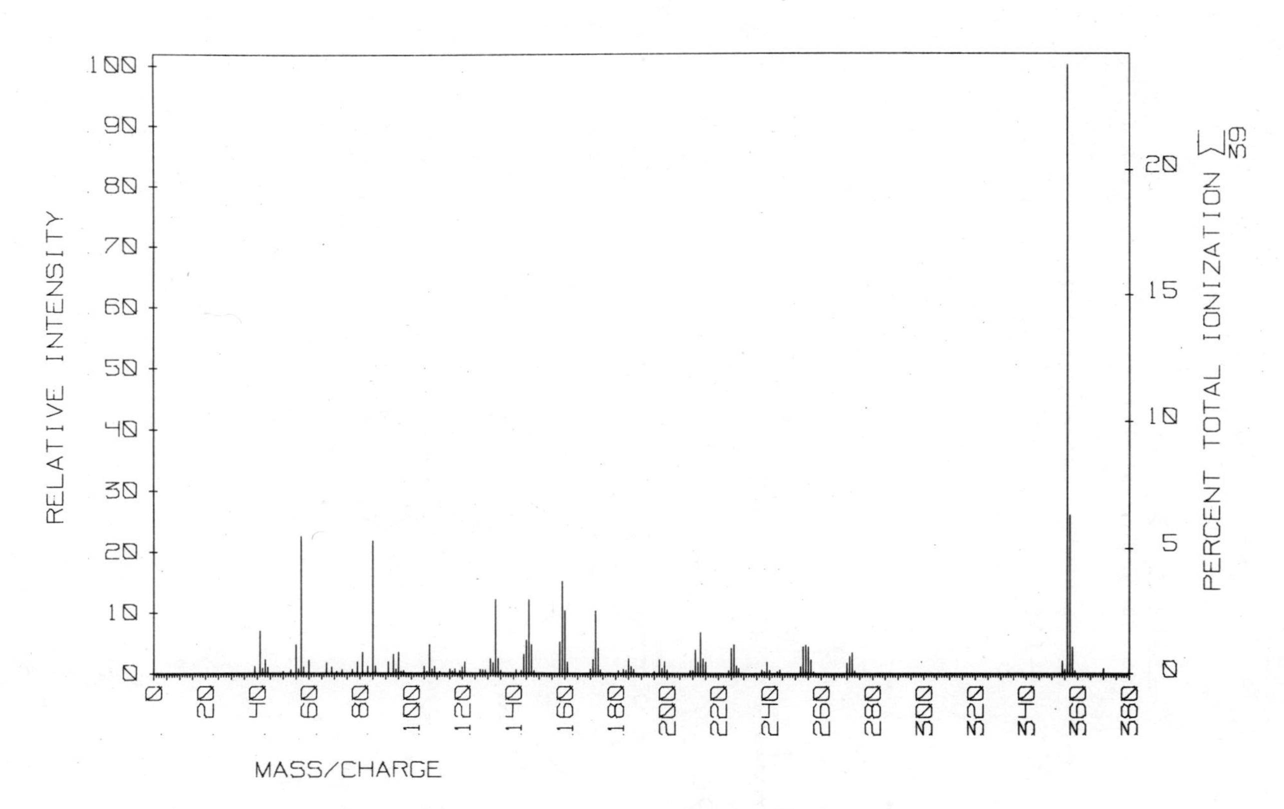

Figure 3. Low Resolution Mass Spectra of Estradiol Valerate. Instrument: AE1-902.

2.5 <u>Rotation</u>

$[\alpha]_D$ + 44° (dioxane)[3]

2.6 <u>Melting Range</u>

Like many steroids, Estradiol Valerate does not melt sharply. The following melting range temperatures (°C) were reported:

144-145[4] (from methanol-water)
150[5]
146-148[6]
147[7]
142-144[27]

The thermomicroscopic[6] and fusion properties[5] of estradiol valerate have also been described.

2.7 <u>Solubility</u>

Practically insoluble in water; soluble in caster oil, in methanol, in benzyl-benzoate and in dioxane; sparingly soluble in sesame oil and in peanut oil[8].

2.8 <u>Crystal Properties</u>

No polymorphism was found[7]. Isomorphic mixed crystals were formed with estradiol propionate[5].

The powder X-ray diffraction pattern is presented in Table I[9].

Table I
Powder X-ray Diffraction Pattern
of Estradiol Valerate

d(Å)*	I/I_1**	d(Å)*	I/I_1**
11.5	0.06	3.50	0.19
10.0	0.60	3.31	0.34
6.95	0.21	3.22	0.08
6.21	0.24	3.14	0.07
5.97	0.51	3.02	0.09
4.98	1.00	2.96	0.08
4.68	0.46	2.32	0.05
4.12	0.34		
3.96	0.32		
3.83	0.13		

*d = interplanar distance $\frac{n\lambda}{2 \sin\Theta}$

λ = 1.539A; Radiation: $K\alpha_1$ and $K\alpha_1$ Copper

** Relative intensity based on highest intensity of 1.00

3. Synthesis

The synthesis of estradiol valerate was first described by Miescher and Scholz[4] by methanlysis of estradiol 3,17-divalerate.

$CH_3-(CH_2)_3-\overset{O}{\overset{\|}{C}}O$ — $O\overset{O}{\overset{\|}{C}}-(CH_2)_3-CH_3$ $\xrightarrow[K_2CO_3]{MeOH}$ HO — $O\overset{O}{\overset{\|}{C}}-(CH_2)_3-CH_3$

Alternatively the 3-benzoate has been removed selectively with sodium borohydride[27].

4. Stability - Degradation

Estradiol Valerate is very stable as a solid. In solution under certain conditions, particularilly alkaline, saponification to valeric acid and estradiol can occur.

5. Drug Metabolic Products

An increase in blood and urine estrogen levels was noted after administration of estradiol valerate to ovarectomized [10] and postmenopausal[11] women. No metabolites have been identified so far, although estradiol valerate most likely follows the pathway of estradiol metabolism.

6. Methods of Analysis

6.1 Elemental Analysis

	% Calc.	% Found[4]
C	77.49	77.54;77.42
H	9.05	9.00;9.20
O	13.46	--

6.2 Spectrophotometric Analysis

The U.V. absorption at 282 nm can be used to determine estradiol valerate in sesame oil after extraction with 80% methanol, however, with an accuracy of not better than ± 10% because of interference from the oil[15]. In the compendial assay the absorbance of an alkaline assay solution at 300 nm is determined against an acidic assay preparation[8].

6.3 Spectroflurometric Analysis

A quantitative fluorometric method for determination of estradiol valerate in oil has been described[16]. After column clean up the steroid concentration in ethanol is determined in a spectrofluorometer (Excitation wave length: 285 nm; Fluorescence maximum ca. 328 nm).

6.4 Colorimetric Analysis

Estradiol valerate responds to colorimetric tests used for estrogens such as the phenol reagent (Folin-Ciocalteau reagent,phosphomolybdic phosphotungstic acid)[8],iron-phenol solution[17]and the Ittrich modification of the Kober reaction (see section 7).

6.5 Chromatographic Analysis

6.51 Paper

The quantitative determination of estradiol valerate by paper chromatography has been described[18]. After spotting the paper strip strips were impregnated with 20% diethylene glycol monoethylether(Carbitol) in chloroform and developed for 3 hours with methylcyclohexane saturated with diethylene glycol monoethyl ether. The steroid is located on a guide-strip by spraying with Folin-Ciocalteau reagent, eluted with ethanol and quantitated fluorometrically. (Activation wave length at 280 nm, fluorescence at 310 nm). In this system estradiol stays at the origin.

6.52 Thin-Layer

Thin-layer chromatographic systems have been tabulated in Table II.

Table II

Absorbant	Solvent System	Ref.	Ref.
Magnesium silicate	Benzene-ethanol(9:1)	0.52	19
Magnesium silicate	Chloroform	0.62	19
Silica gel G	Stationery phase:mineral oil Mobile phase:50% acetic acid	0.64	20
Silica gel G	Stationery phase:propylene glycol Mobile phase:cyclohexane-pet.ether(1:1)	0.60	22
Silica gel G	Stationery phase:tetraethylene glycol Mobile phase:xylene	--	24

Detection systems. Phosphoric acid (1:5)spray(green color);conc. sulfuric acid, p-toluenesulfonic acid.

Quantitative determination in pharmaceutical preparation has been described[23]. The tetraethylene glycol-xylene system[24] separates estradiol valerate from estradiol and is the basis for a compendial limit test for free estradiol.

6.53 Gas-Liquid

Estradiol valerate can be determined quantitatively by gas chromatography using a 3% OV-17 on 80-100 mesh Varaport 30 column at a column temperature of 260°[25]. Alternately a 3% JXR (methylsilicone polymer)on silanized 100-200 mesh Gas Chrom P column at a column temperature of 230° can be used for the tetramethylsilyl derivate[26].

7. Identification and Determination in Body Fluids and Tissues.

A method[12] has been developed to determine extremely low concentration of total estrogens in bovine tissues, making use of a shortened version of the extraction procedure of Goldzieher[13], followed by fluorometric read out using the Ittrich[14] modification of the Kober reaction.

8. References

1. B. Toeplitz, The Squibb Institute, Personal Communication.
2. A. I. Cohen, The Squibb Institute, Personal Communication.
3. N. H. Coy and C. M. Fairchild, The Squibb Institute, Personal Communication.
4. K. Miescher and C. Scholz, Helv. Chim.Acta 20,1237(1937); U.S.Patent, 2,233,035(1941).
5. J. P. Crisler, N. F. Witt and M. H. Crisler, Microchimica Acta 1962,317.
6. M. Kuhnert-Brandstatter, E. Junger and A. Kofler, Microchem. J. 9,105(1965).
7. M. Brandstatter-Kuhnert and E. Junger, Microchimica Acta 1964,238.
8. U.S.P. XVIII
9. Q. Ochs, The Squibb Institute, Personal Communication.
10. G. Ittrich and P. Potts, Abhandl. Deut. Akad. Wiss, Berlin, Kl.Med. 1965,56; C.A. 64,11501 g. (1966).
11. R. Kaiser, Symp. Deut. Ger. Endokrinol. 8,227(1962); C.A. 65,7565h(1966).
12. H. Kadin, The Squibb Institute, Personal Communication.
13. J. W. Goldzieher, R. A. Baker and E.C. Riha, J. Clin. Endocr. 21,62(1961).
14. G. Ittrich, Acta Endocr. 35,34(1960).
15. N. H. Coy and C. M. Fairchild, The Squibb Institute, Personal Communication.
16. Th. James, Journal of the AOAC 54,1192(1971); ibid. 56,86(1973).
17. British Pharmacopoeia 1973 p.330.
18. H. R. Roberts and M. R. Siino, J. Pharm.Sci. 52,370(1963).
19. V. Schwarz, Pharmazie 18,122(1963).
20. D. Sonanini and J. Anker, Pharm. Acta.Helv. 42,54(1967).

21. T. Diamanstein and K. Lorcher, J. Anal.Chem. 191,429(1962).
22. A. Vanden Bulcke, Pharm. Tijdschr. Belg. 46,221(1969) C.A. 72,103790y(1970).
23. N. Ari, Turk Hij. Tecr. Biyol. Derg. 29,200 (1969); C.A. 73,48569b(1970).
24. H. Klein and S. Hays, Drug Standards Laboratory, Personal Communication.
25. Th. James and B. Rader, F.D.A. By-lines 4, 161(1972).
26. G. Cavina, G. Moretti and P. Siniscalchi, J. Chromatog. 47,186(1970).
27. K. Tsuneda, J. Yamada, K. Yasuda and H.Mori, Chem. Pharm. Bull. (Tokyo) 11,510(1963), Japan Patent 20,163(1964);C.A.62,10484h (1965).

Literature surveyed through December 1972.

The help of H. Gonda and A. Mohr in the preparation of this profile is gratefully acknowledged.

HYDROXYPROGESTERONE CAPROATE

Klaus Florey

CONTENTS

1. Description

1.1 Name, Formula, Molecular Weight

Hydroxyprogesterone caproate is 17-[(1-oxohexyl)oxy]-4 pregnene-3,20-dione; also 17-hydroxypregn-4-one-3,20-dione hexanoate.

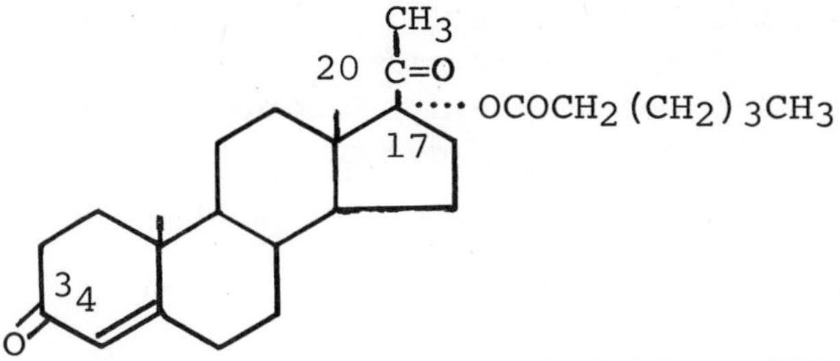

CH_3
20 C=O
17 ···· $OCOCH_2(CH_2)_3CH_3$
3 4
O

$C_{27}H_{40}O_4$ Mol.Wt.: 428.62

1.2 Appearance, Color, Odor

White or creamy white odorless crystalline powder.

2. Physical Properties

2.1 Infrared Spectrum

The infrared spectrum of hydroxyprogesterone caproate is presented in figure 1.[1]

2.2 Nuclear Magnetic Resonance Spectrum

The NMR spectrum is presented in figure 2. It was obtained on a 60 MHz NMR spectrometer in deuterochloroform containing tetramethylsilane as an internal reference. The following proton assignments were made:[6]

Protons at	Chemical Shift
C-4	4.20 singlet
C-18	9.30 singlet
C-19	8.80 singlet
C-21	7.96 singlet
Ester methyl	9.08 multiplet, coupling constant 6.5 Hz

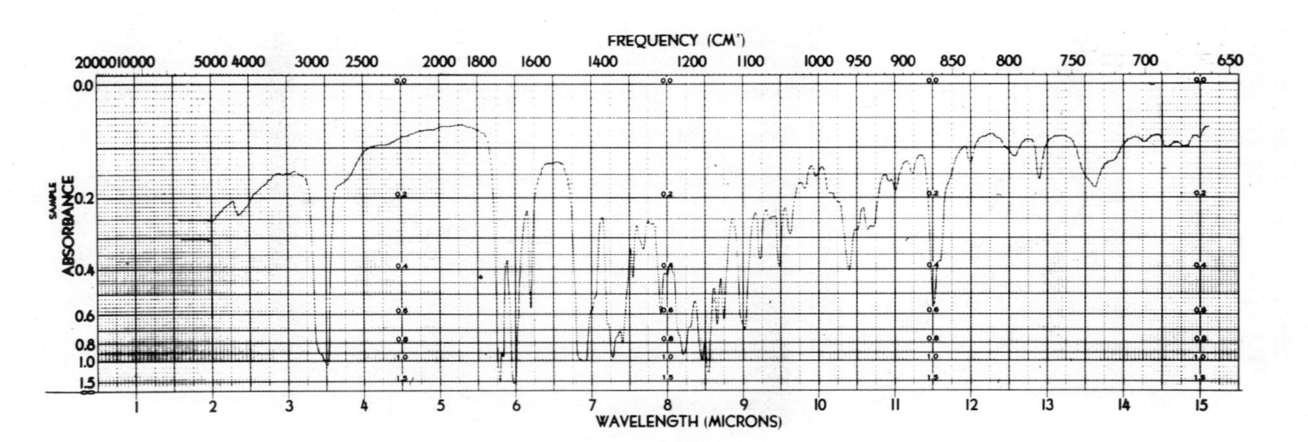

Figure 1. Infrared Spectrum of Hydroxyprogesterone Caproate in mineral oil mull. Instrument:Perkin Elmer 621.

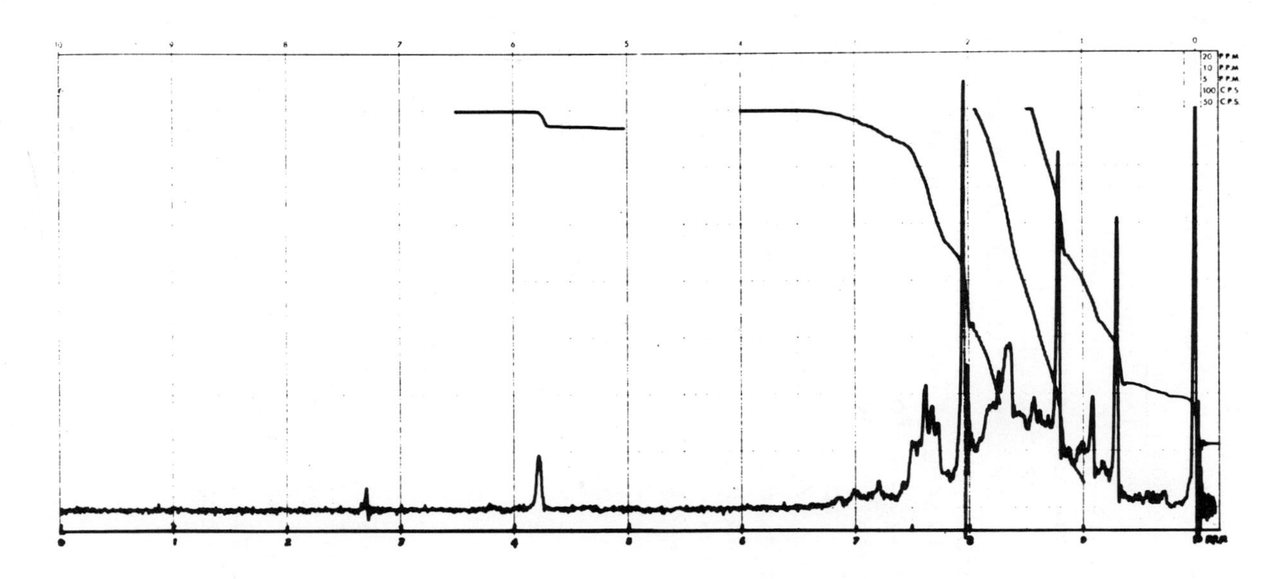

Figure 2. NMR Spectrum of Hydroxyprogesterone Caproate in Deuterated Chloroform. Instrument: Perkin-Elmer R-12B.

2.3 Ultraviolet Spectrum

λ max 241 nm; ε = 17000[8]

λ_{max}^{EtOH} 240 nm; $E_{1cm}^{1\%}$ 387 ± 1%

2.4 Mass Spectrum

The low resolution mass spectrum, shown in figure 3, was obtained on an AEI MS-902 spectrometer equipped with a frequency modulated analog tape recorder.[6]

It demonstrates the expected M^+ of m/e 428. The m/e 385 ion represents the loss of the C-17 acetyl group or C_3H_7 from the acylate, probably the former is the predominant pathway. The loss of the C-17 acylate gives rise to the m/e 330 ion and the m/e 312, 313 ions. Prominent ions at m/e 71 and m/e 99 correspond to the acylate portion of the molecule. The m/e 287 ion is shown below. The m/e 269 ion can be envisoned as the dehydrated ion of the m/e 287 ion. A series of fragment ions occur from the progressive loss of D-ring, C-ring and B-ring carbons but,

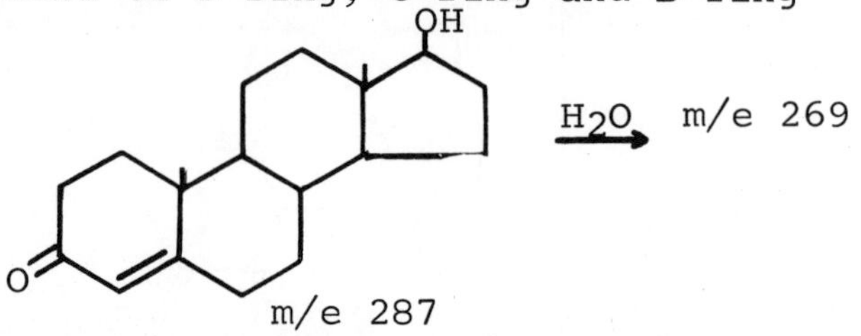

OH

O

$\xrightarrow{H_2O}$ m/e 269

m/e 287

while present, are not particularly intense enough to indicate the presence of the Δ^4-3-keto group. However, the ions at m/e 121-124 are present and have the compositions expected for the A-ring group. The assignment of some of the diagnostic ions is depicted below.[6]

Fragmentation Pattern of Hydroxyprogesterone Caproate

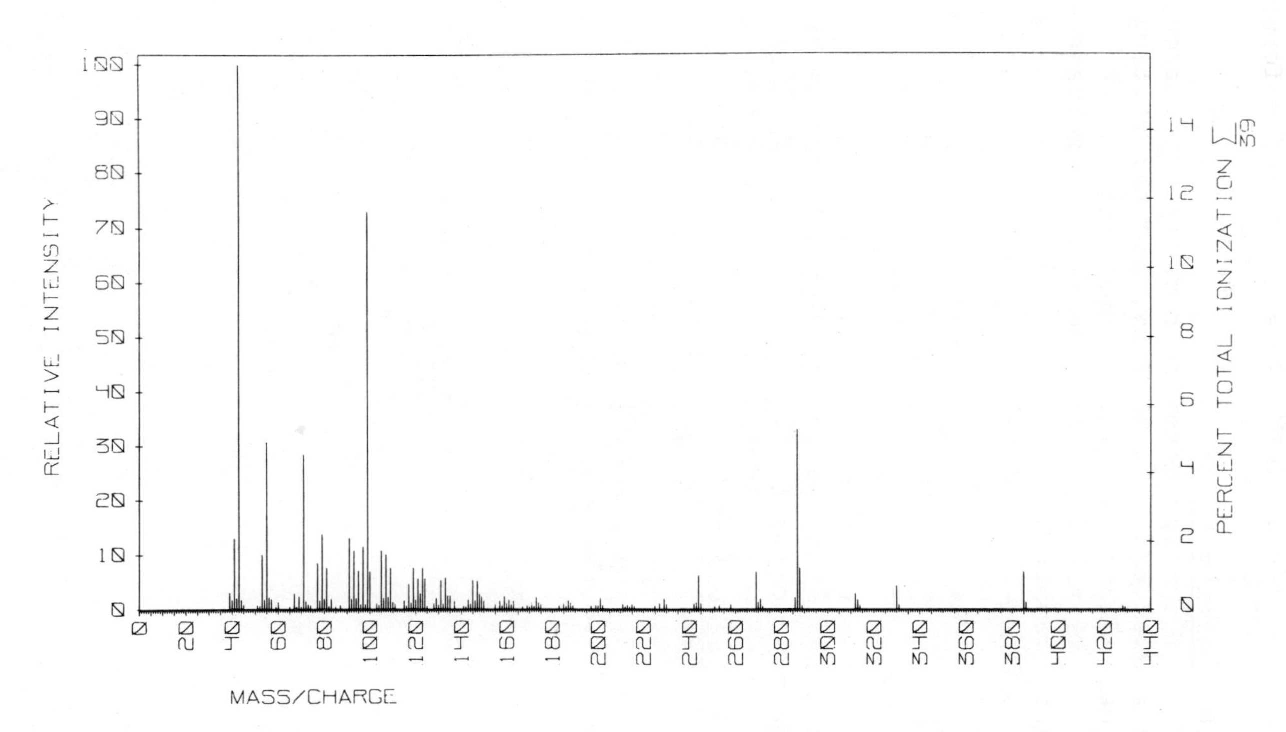

Figure 3. Low Resolution Mass Spectrum of Hydroxyprogesterone Caproate. Instrument: AEl MS-902.

2.5 Rotation

$[\alpha]^{20}_{D} = +\ 61^{o}$ (c = 1, in chloroform)[2]

2.6 Melting Range

Like many steroids, hydroxyprogesterone caproate does not melt sharply, and the melting point depends on the rate of heating[7].

The following melting range temperatures (^{o}C) were reported.

117-122[8]
120.0-121.0[9]
115-118[3]
119-121[2]
120.1-121.9[10]

The thermomicroscopic[8] and fusion properties[11] of hydroxyprogesterone caproate have also been described.

2.7 Solubility

Insoluble in water[10]

50 mg/ml in ethyl ether[10]
1.2 mg/ml in benzene[10]
25-29 mg/ml in sesame oil[2]
350-400 mg/ml in levulinic acid butyl ester[2]
Soluble in a mixture of benzyl benzoate and sesame or castor oil[10].

2.8 Crystal Properties

Dense needles from isopropanol[2]. The powder X-ray diffraction pattern is presented in Table I.

TABLE I
Powder X-ray Diffraction Pattern of Hydroxyprogesterone Caproate[23]

d(Å)*	I/I_1**	d(Å)*	I/I_1**
12.2	0.26	4.64	0.21
9.1	0.11	4.53	0.33
7.15	0.31	4.05	0.09
6.90	0.10	4.00	0.11
6.63	0.13	3.88	0.10
6.35	0.37	3.80	0.11
5.80	1.00	3.70	0.15
5.47	0.42	3.62	0.11
5.28	0.21	3.30	0.08
5.08	0.28	2.88	0.09

*d = interplanar distance $\frac{n\lambda}{2 \sin\Theta}$

λ = 1.539Å; Radiation; K_1 and K_2 Copper

**Relative intensity based on highest intensity of 1.00.

3. Synthesis

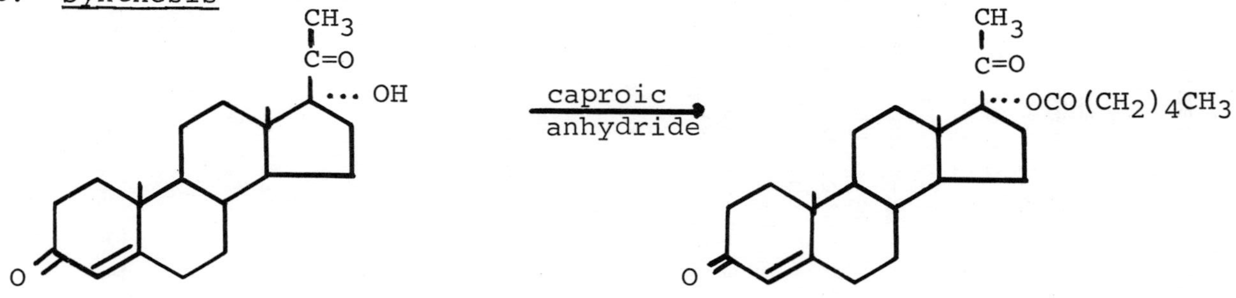

CH_3, C=O, OH, O

caproic anhydride →

CH_3, C=O, $OCO(CH_2)_4CH_3$, O

Hydroxyprogesterone caproate is synthesized by acylation of 17α-hydroxyprogesterone (cf.4) with caproic acid in the presence of p-toluene sulfonic acid[3]. Variations of this basic method have also been reported[5]. It can also be prepared by oxidation of the corresponding 5-prengnen-17-ol-3-one caproate[2].

4. Stability - Degradation

Hydroxyprogesterone caproate is very stable as a solid. The 17α-esterlinkeage is sterically hindered and therefore quite stable, but can be saponified under forcing conditions. In solution photolytic degradation of the A-ring is possible, when exposed to ultraviolet light or ordinary fluorescent laboratory lighting (cf. 12).

5. Drug Metabolic Products

Excretion and tissue distribution of labeled hydroxyprogesterone caproate was studied in rats[13], steers[14] and pregnant women[16]. Langecker[15] was not able to recover unchanged hydroxyprogesterone caproate in amounts exceeding 0.1% after administration to man. Plotz[16] incubated rat liver homogenates with 4-^{14}C-labeled hydroxyprogesterone caproate under anaerobic conditions and recovered allopregnane-3β, 17α-diol-20-one-17α-caproate in 25% yield and pregnane-3β, 17α-diol-20-one-17α-caproate in 2% yield, together with some unchanged starting material. That the caproic acid ester was not removed attest to the stability of the bond (see section 4) and there is evidence that metabolites in urine also still carry the ester group. Plotz also found that in homogenate of human placenta the 20-keto group was not reduced under conditions which reduce the 20-keto group of progesterone.

6. Methods of Analysis

6.1 Elemental Analysis

	Calc. %
C	75.66
H	9.41
O	14.93

6.2 Spectrophotometric Analysis

The compendial assay of hydroxyprogesterone caproate is based on the U.V. absorption maximum at 240 nm[17].

6.3 Spectrofluorometric Analysis

Fluorescence was used for detection in biological fluids[15].

6.4 Polarimetric Analysis

Rotation has been used to determine hydroxyprogesterone caproate in oil vehicles[18].

6.5 Polarographic Analysis

The half wave potential (E-1/2 versus standard calomel electrode) was determined as -1.65 volts in 1M lithium chloride in 95% methanol due to the reduction of the α, β unsaturated ketone[22]. The diffusion current constant (Id) was found to be 5.27 $\pm$ 3%.

6.6 Chromatographic Analysis

6.61 Paper

The quantitative determination and separation from the free steroid of hydroxyprogesterone caproate by paper chromatography in oily vehicles has been described by Roberts and Florey[19]. After spotting, the paper strips were impregnated with 30% diethylene glycol monoethyl ether (Carbitol)in chloroform and developed for 4 hours with methylcyclohexane saturated with diethylene glycol monoethyl ether. After locating of the steroid spots with a fluorescent paddle, the spots are eluted and reacted with isonicotinic acid hydrazide in methanol containing also hydrochloric acid. The absorbance of the yellow hydrazone is read at 415 nm.

6.62 Thin-Layer

Thin-layer chromatographic systems have been compiled in Table II.

Table II

Absorbent	Solvent System	Rf	Ref.
Magnesium silicate	Benzene-Ethanol (98:2)	0.35	20
Magnesium silicate	Benzene-Ethanol (9:1)	0.61	20
Magnesium silicate	Chloroform	0.44	20
Magnesium silicate	Chloroform-Ethanol (98:2)	0.53	20
Magnesium silicate	Chloroform-Ethanol (96:4)	0.60	20
Magnesium silicate	Benzene-Dioxane (2:1)	0.71	20
Magnesium silicate	Ether-Ethanol (98:2)	0.83	20
Silica gel	Benzene-Acetone (4:1)	0.67	21
Silica gel	Benzene-Methanol (9:1)	0.61	21
Silica gel	Pet. ether, Benzene, Acetic acid, Water (67:33:85:15)	0.44	21
Alumina	Benzene-Acetone (4:1)	0.61	21

Detection system: U.V. light; sulfuric acid (dark brown); sulfuric-acetic acid (yellow red); vanillin-sulfuric acid (dark yellow)

7. References

1. B. Toeplitz, The Squibb Institute, Personal Communication.
2. E. Kaspar, K. H. Pawlowski, Karl Junkmann and Martin Schenk, U.S.Patent 2,753,360 (1956).
3. V. M. Bakshi and Y. K. Hamied, Ind. J. Chem. 2,294(1964).
4. L. F. Fieser and M. Fieser, "Steroids" Reinhold Publishing Corp. 1959.
5. Brit. Patent 848,881 (Chem.Abstr. 57,2348g); Japan Patent 3574(62) (C.A. 58,91896); Belg. Patent 661,975(C.A. 65,5510); U.S.S.R. Patent 210,858 (C.A. 69,96,998e).
6. A. I. Cohen, The Squibb Institute, Personal Communication.
7. I. Gyenes and A. Laszlo, Magyar Kem. Folyoirat 67, 360(1961) (C.A. 56,149506, 1962).
8. M. Kuhnert-Brandstatter, E. Junger and A. Kofler, Microchem. J. 9,105(1965).
9. K. Junkmann, Arch.exper.Path.u.Pharmakol. 223,244(1954).
10. G. A. Brewer, The Squibb Institute, Personal Communication.
11. J. P.Crisler, N. F. Witt and M. H.Crisler, Microchimica Acta 1962,317.
12. D. R. Barton and W. C. Taylor, J.Am.Chem.Soc. 80,244(1958);J.Chem.Soc.1958,2500.
13. H. Kiesling and A. Elmquist, Acta.Endocrinol. 28,502(1958).
14. F.X. Gassner, R. P. Martin, W. Shimoda and J. W. Algeo, Fertility and Sterility 11,49 (1960).
15. H. Langecker, A. Harwart and K. Junkmann, Arch. exptl.Pathol.Pharmacol. 225,309(1955).
16. M. Wiener, C. I. Lupu and E. J. Plotz, Acta. Endocrinol. 36,511(1961).

17. U.S.P. XVIII.
18. V. Gerosa and M. Melandri, Ann.Chim.(Rome) 47,1388,(C.A. 52,7620(1958)).
19. H. R. Roberts and K. Florey, J.Pharm.Sci. 51,794(1962).
20. V. Schwarz, Pharmazie 18,122(1963).
21. S. Hara and K. Mibe, Chem.Pharm.Bull. 15,1036 (1967).
22. N. H. Coy and C. M. Fairchild, The Squibb Institute, Personal Communication.
23. Q. Ochs, The Squibb Institute, Personal Communication.

Literature surveyed through December 1972.

The help of H. Gonda and A. Mohr in the preparation of this profile is gratefully acknowledged.

ISOSORBIDE DINITRATE

Luciano A. Silvieri and Nicholas J. DeAngelis

CONTENTS

1. Description

1.1 Name, Formula, Molecular Weight

Isosorbide dinitrate is designated by the following chemical names: 1,4:3,6-dianhydro-d-glucitol dinitrate[1]; 1,4:3,6-dianhydrosorbitol 2,5-dinitrate[1]; dinitrosorbide[1]. In Chemical Abstracts the compound is listed under the heading (Glucitol:1,4:3,6-dianhydro, "dinitrate, D"). Some of the commonly used trade or trivial names are: Isordil, Isorbid, Vascardin, and Carvanil.

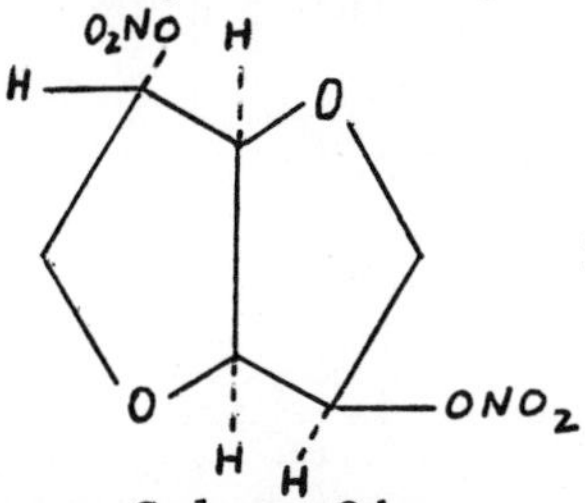

O_2NO H H O O H H ONO_2

M.W. = 236.14

1.2 Appearance, Color, Odor

Isosorbide dinitrate is a white, odorless, crystalline powder. However, diluted isosorbide dinitrate, which is a dry mixture of isosorbide dinitrate with lactose, mannitol, or other inert excipients to permit safe handling, occurs as an ivory-white powder. The mixture usually contains about 25% of isosorbide dinitrate.

2. Physical Properties

2.1 Infrared Spectra

The infrared (I.R.) spectrum of isosorbide dinitrate[46] is given in Figure 1.[2] The I.R. spectrum was obtained in a KBr pellet and is identical to that published by Sammul et.al.[3] Assignment of some of the significant absorption bands is given in Table I.

Table I

Infrared Spectral Assignments of Isosorbide Dinitrate

Frequency ($cm.^{-1}$)	Vibration Mode	Reference
2950-2850	aliphatic C-H stretching	48
1665 and 1635	asymmetric NO_2 stretching	47
1460	methylene scissoring vibration	48
1285-1270	symmetrical NO_2 stretching	47
~1100	asymmetrical C-O-C stretching	48
865	O-NO_2 stretching	47

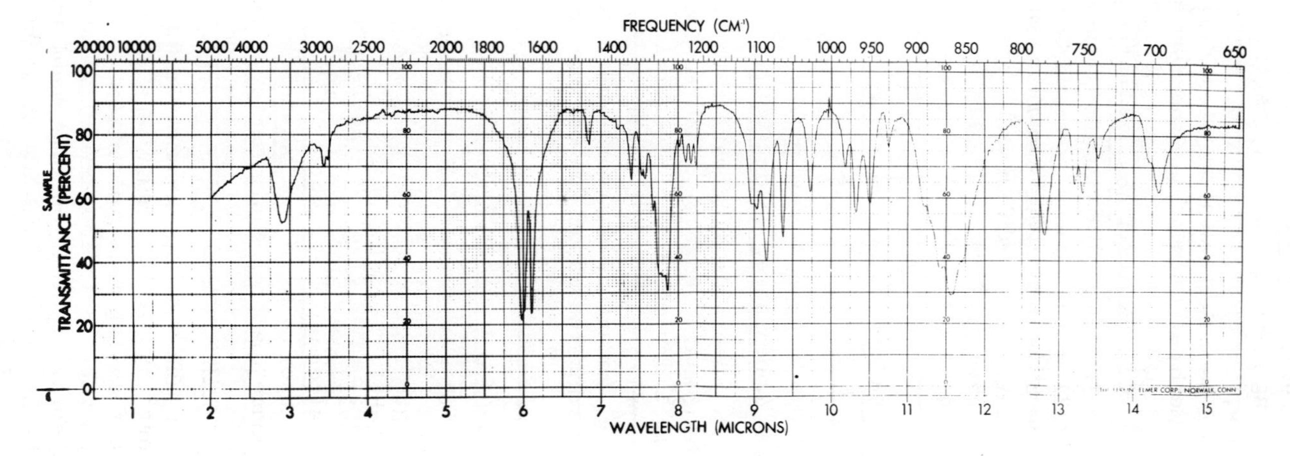

Figure 1 - I.R. Spectrum of Isosorbide Dinitrate, Lot No. 5123 V 6/1. 0.25% KBr Pellet - Instrument: Perkin Elmer Model 21.

Hayward et.al.[4] reported the nitrato and hydroxyl stretching bands of nitrate esters of 1,4:3,6-dianhydrohexitols, including isosorbide dinitrate, in dilute solutions of benzene, acetonitrile, chloroform, isopropyl ether, methylal, pyridine, and carbon tetrachloride.

2.2 Nuclear Magnetic Resonance Spectra

The nuclear magnetic resonance (N.M.R.) spectrum (Figure 2) was obtained by preparing a saturated solution of isosorbide dinitrate[46] in deutero chloroform containing tetramethylsilane as internal reference.[5] There are no exchangeable protons. The NMR proton spectral assignments are given in Table II.

Table II

NMR Spectral Assignments of Isosorbide Dinitrate

Chemical Shift (ppm.)	Proton
3.9-4.2	-O-$\underline{CH_2}$
4.4-5.1	$>C(\underline{H})-O-$
5.2-5.5	$-C(\underline{H})-ONO_2$

Hopton and Thomas[6] have reported on NMR studies of isosorbide which elucidate the conformation of this molecule. Their findings indicate the conformation is a composite of the envelope and half-chair forms.

2.3 Ultraviolet Spectrum

Isosorbide dinitrate gives no peak maxima in the wavelength region of 400 nm. to 220 nm.[8], but does absorb U.V. light. The absorption begins at about 290 nm. and continually increases as the wavelength decreases. In U.S.P. alcohol at a concentration of approximately 0.1 mg./ml. an absorbance of 0.1 is obtained at 250 nm., and an absorbance of 1.5 is obtained at 220 nm.[7]

2.4 Mass Spectra

The mass spectra of isosorbide dinitrate[46] were obtained[9] by direct insertion of the sample into an MS-902 double focusing mass spectrometer modified with a SRI dual CIEI source. Both low resolution electron empact (EI) and chemical ionization (CI) mass spectra were obtained. For the EI spectrum the ionizing electron beam energy was 70 eV and for the CI spectrum the beam energy was 500 eV.

The mass spectrum assignments[9] of the prominent

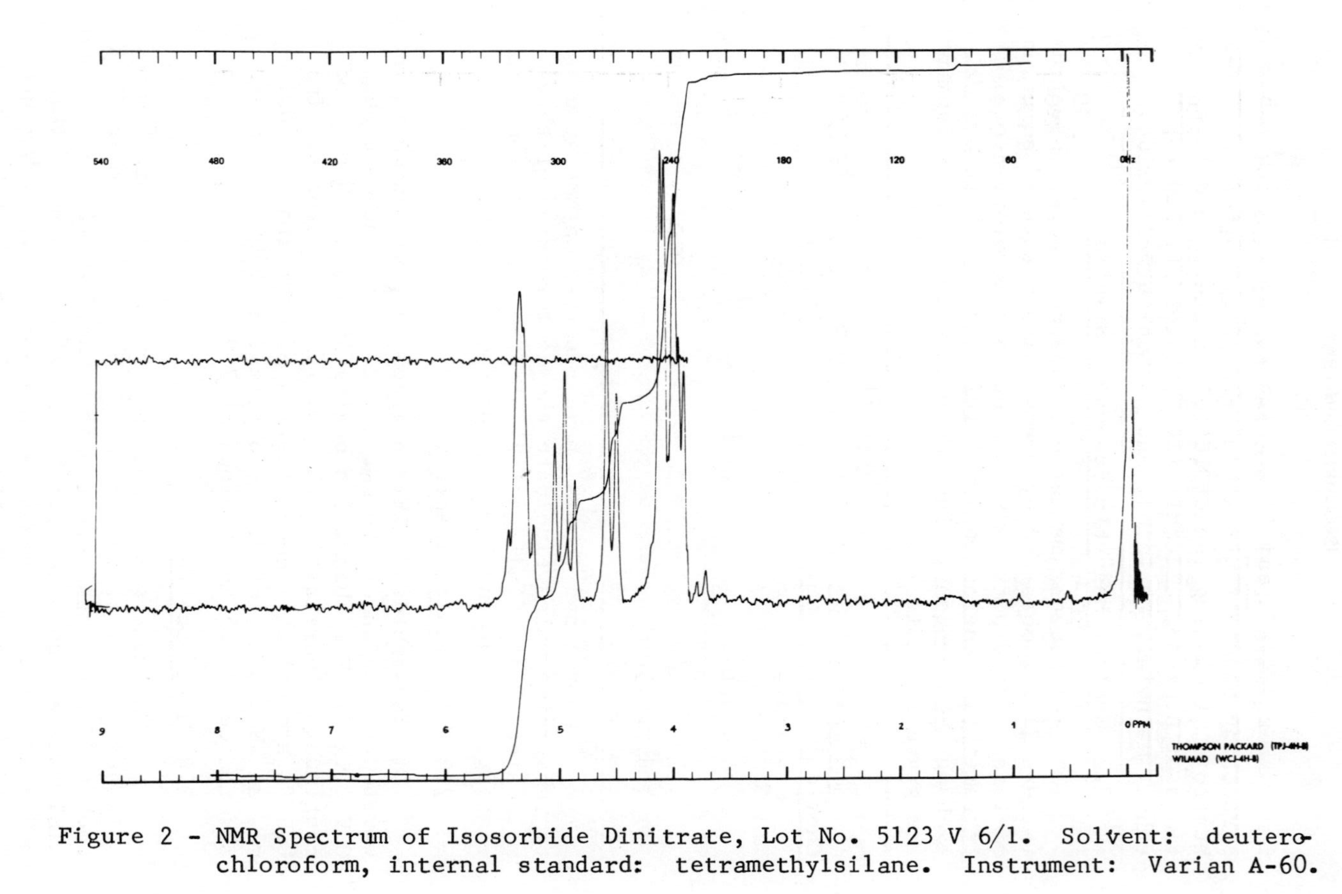

Figure 2 - NMR Spectrum of Isosorbide Dinitrate, Lot No. 5123 V 6/1. Solvent: deuterochloroform, internal standard: tetramethylsilane. Instrument: Varian A-60.

ions under both EI and CI conditions are given in Table III.

Table III

Mass Spectral Assignments of Isosorbide Dinitrate

Chemical Ionization			Electron Impact		
m/e	R.I.(%)	Species	m/e	R.I.(%)	Species
237	15	MH^+	236	2.5	M^+
190	24	MH^+-HNO_2	143	75	M-HNO_2 , NO_2
144	60	MH^+-HNO_2 , NO_2	126	100	M-HNO_3 , HNO_2
127	100	MH^+-HNO_3 , HNO_2	114	27	M-HNO_2 , $CHONO_2$
			100	25	M-HNO_2 , CH_2CHONO_2

2.5 Optical Rotation

The specific rotation $[\alpha]^D_{25}$ of isosorbide dinitrate[46] was determined[7] to be +137° in U.S.P. alcohol at a concentration of 3 mg./ml. The Merck Index[1] reports an $[\alpha]^D_{25}$ of +135°. Jackson and Hayward[10] found the $[\alpha]^D_{20}$ to be +141° and Goldberg[11] +134°.

2.6 Melting Range

There are two ranges reported in the literature for the melt of isosorbide dinitrate. The Merck Index[1], Wiggins[12], Goldberg[11], and Hayward et.al.[8], report values of 70°C.-71°C. Jackson and Hayward[10], Forman et.al.[13], and Hayward et.al.[8] report 50.5°C.-52°C. We found a value of 70°C.-71.5°C., Class Ia[14], for isosorbide dinitrate.[46] This difference in melting point may be a result of polymorphism. The melting temperature range does not change significantly with variations in heating rate of from 1 to 5°C./min.

2.7 Differential Thermal Analysis

The differential thermal analysis (DTA) curve of isosorbide dinitrate[46] run from room temperature to the melting point exhibits no endotherms or exotherms other than that associated with the melt. The DTA curve[7] run on a DuPont 900 DTA using a micro cell and a heating rate of 5°C./min. is shown in Figure 3.

2.8 Solubility

The following solubilities have been determined for isosorbide dinitrate at room temperature.

Solvent	Authors[7]	Merck Index[1]	U.S.P.[15]
water	$<$ 0.5 mg./ml.	sparingly soluble	very slightly soluble
acetone	$>$ 1 g./ml.	freely soluble	very soluble
alcohol	24 mg./ml.	freely soluble	sparingly soluble

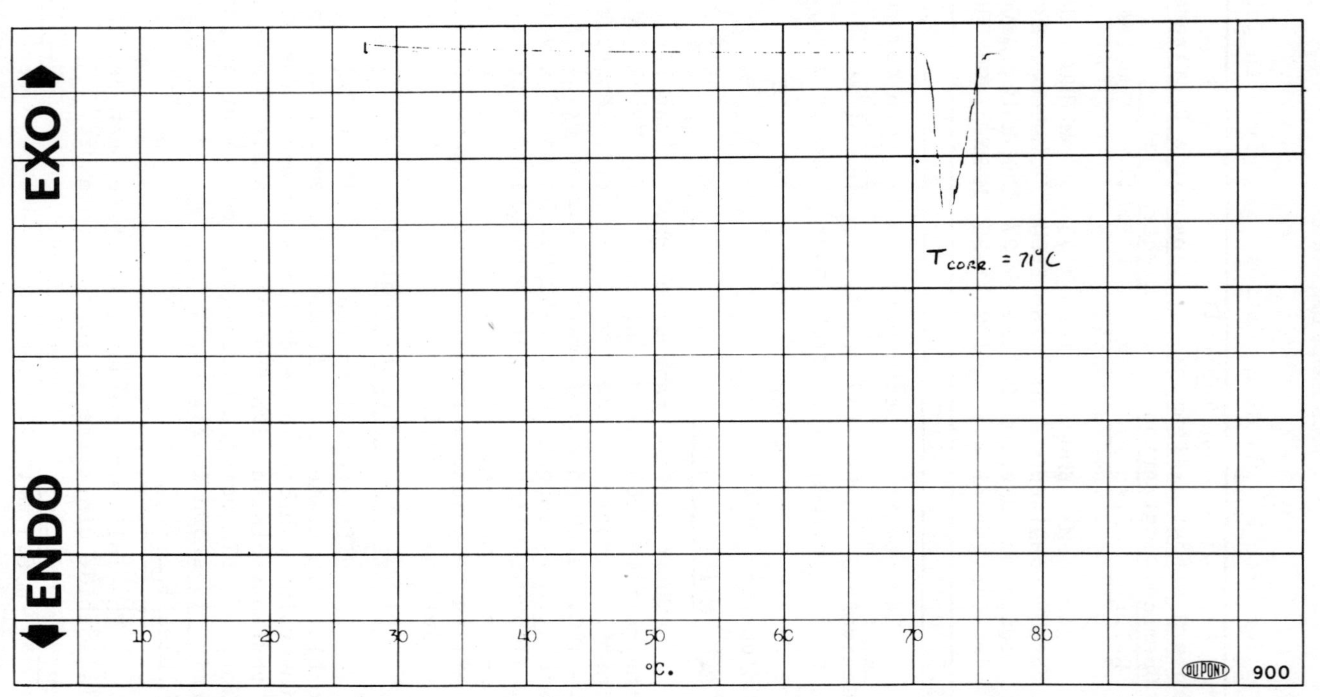

Figure 3 - DTA Spectrum of Isosorbide Dinitrate, Lot No. 5123 V 6/1, heating rate 5°/min.
Instrument: DuPont 900 DTA.

Solvent	Authors[7]	Merck Index[1]	U.S.P.[15]
ether	43 mg./ml.	freely soluble	
chloroform	330 mg./ml.		freely soluble
hexane	<1 mg./ml.		

2.9 Crystal Properties

The X-ray powder diffraction pattern of isosorbide dinitrate[46] obtained[7] with a Philips diffractometer[16] using CuK$_\alpha$ radiation is shown in Figure 4. The calculated d spacings[7] for the diffraction pattern shown are given in Table IV.

Table IV

X-Ray Powder Diffraction Pattern for Isosorbide Dinitrate

2θ (degrees)	d(Å)	I/Io
10.0	8.85	40
16.1	5.50	25
17.3	5.13	100
19.5	4.55	3
20.2	4.40	95
21.1	4.21	1
23.2	3.83	2
23.8	3.74	9
25.1	3.55	32
25.6	3.48	20
27.2	3.28	2
28.9	3.09	3
29.3	3.05	2
31.4	2.849	5
32.6	2.747	5
33.2	2.698	2
34.4	2.607	9
35.8	2.508	3
41.1	2.196	12

d(interplanar distance) $= \frac{n\lambda}{2\sin\theta}$

I/Io = relative intensity (based on highest intensity = 100)

3. Synthesis

Isosorbide dinitrate has been prepared by routes utilizing both L-sorbose and D-glucitol (sorbitol).

In one synthetic route,[10] isosorbide was prepared from L-sorbose by the method of Wiggins,[17] This was accomplished by hydrogenation of L-sorbose followed by dehydration to obtain isosorbide. The nitrate ester of isosorbide was then prepared by the method of Forman et.al.[18] in a yield of about 50%; the yield was increased to 85-90% by nitration in

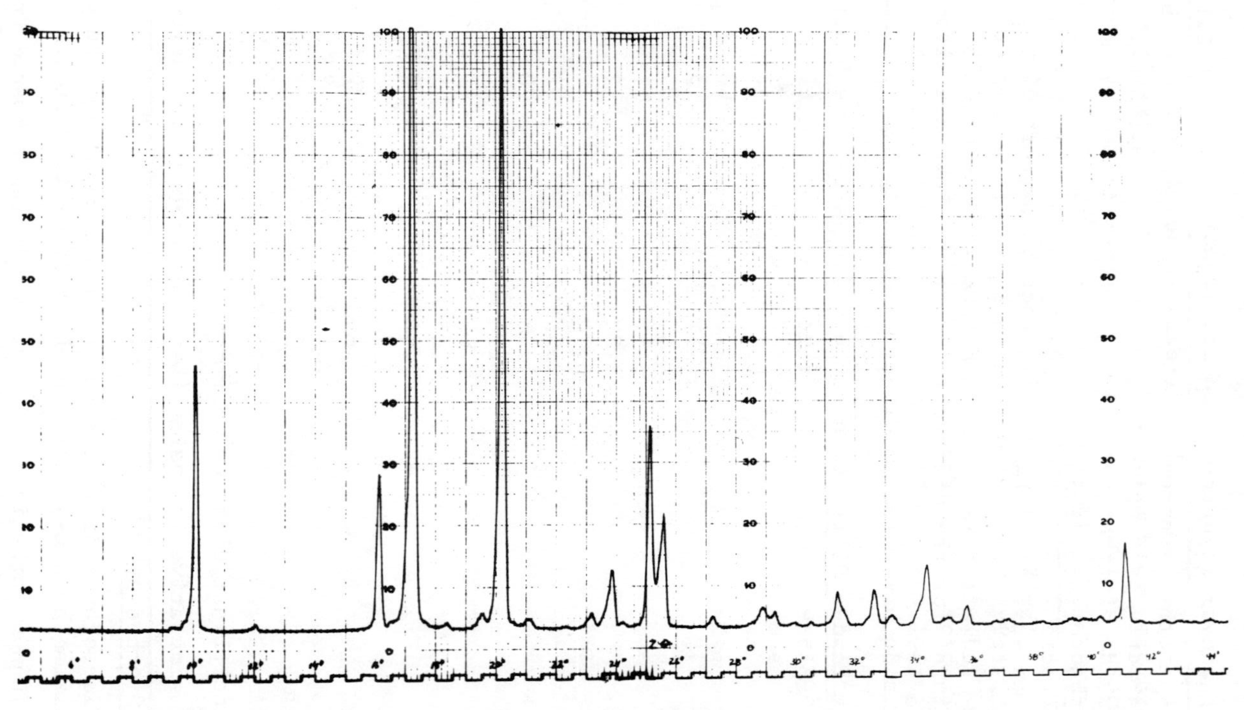

Figure 4 - X-Ray Diffraction Pattern of Isosorbide Dinitrate, Lot No. 5123 V 6/1.
Radiation: CuK$_{\alpha}$ Instrument: Norelco Philips Diffractometer.

acetic anhydride-nitric acid-acetic acid mixture.

In another synthesis,[19] shown in Figure 5, isosorbide was prepared by dehydration of D-glucitol (sorbitol). Isosorbide dinitrate was subsequently obtained by treating isosorbide with different nitrating agents. Concentrated nitric acid (98%) gave a yield of 83%, a mixture of nitric acid and sulfuric acid (30 and 60% respectively) gave a yield of 65%.

Isosorbide dinitrate-C^{14} was prepared by anhydrization of sorbitol using p-toluene sulfonic acid.[20] The resulting isosorbide was nitrated using 97% nitric acid.

4. Stability and Degradation

The stability of isosorbide dinitrate, in the solid form, has been studied at various temperatures. It was found to be stable at a temperature of 45°C. for a period of 12 months and at room temperature for a period of 60 months.[21].

In acidic medium, under vigorous hydrolytic conditions, isosorbide dinitrate degrades in a stepwise manner forming 2-mononitrate and 5-mononitrate as intermediates with the final products being isosorbide and inorganic nitrate.[21] For example, 25% decomposition[22] results from heating isosorbide dinitrate in 1 N hydrochloric acid at 100°C. for one hour. Decomposition in base is somewhat more rapid, i.e., a 1 N sodium hydroxide solution of isosorbide dinitrate when heated at 100°C. for one hour, decomposes about 45%[22].

Jackson and Hayward[10] studied the decomposition of isosorbide dinitrate in pyridine, and found that a slow decomposition to a polymer, nitrogen oxides, and pyridinium nitrate took place when solutions were heated above 50°C.

5. Metabolism

Dietz[23] reasoned that isosorbide dinitrate is completely metabolized in man and in the dog since no unchanged drug was detected in urine. Less than 1% of the dose was recovered as 5-isosorbide mononitrate and 2-isosorbide mononitrate and therefore he postulated that the removal of the nitrate groups is a stepwise process ending with the completely denitrated isosorbide as the major metabolite.

Sherber et.al.[24] demonstrated that 86% of isosorbide dinitrate, intravenously administered, is cleared from

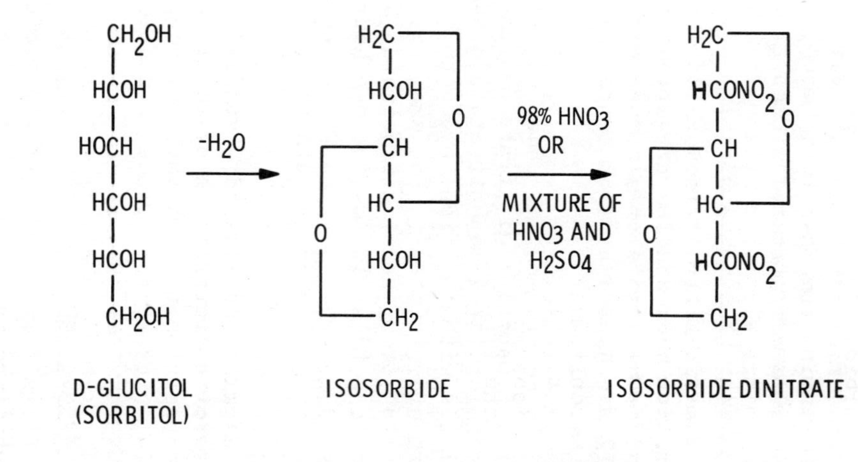

Figure 5 - A Synthetic Route for Isosorbide Dinitrate

rabbit blood within 90 seconds.

In a later study, Reed et.al.[20] found that after oral doses of isosorbide dinitrate-C^{14} nearly all of the drug, with carbon skeleton intact, was excreted in the urine of dogs during the first 24 hours. Some 20 to 30% of the carbon skeleton of isosorbide dinitrate was excreted as neutral material, principally as isosorbide, 2-isosorbide mononitrate and 5-isosorbide mononitrate and the ether monoglucuronide of isosorbide.

Sisenwine and Ruelius[25] made a complete investigation of the plasma concentration and urinary excretion of isosorbide dinitrate and its metabolites in dogs. They found that the initial biotransformation occurs by denitration of the two isomeric mononitrates, 2-isosorbide mononitrate and 5-isosorbide mononitrate. As the mononitrate esters disappear a small amount of isosorbide appears, probably stemming more from hydrolysis of the more labile nitrate group in 2-isosorbide mononitrate than from the sterically hindered nitrate group in 5-isosorbide mononitrate. Further transformation of the isosorbide molecule does not take place as demonstrated by the finding that only unchanged isosorbide is excreted after administration of isosorbide-C^{14} to dogs. The disappearance of 5-isosorbide mononitrate from plasma is probably caused by transformation of the molecule to glucuronide and other conjugates. The isosorbide glucuronide found in urine is a metabolite formed from the 5-isosorbide mononitrate glucuronide and not from isosorbide, since the latter does not undergo transformation or conjugation. A postulated biotransformation scheme is illustrated in Figure 6. These investigators did not find any 2-isosorbide mononitrate in dog urine contrary to the findings of Dietz[23] and Reed et. al.[20]

6. Identification

Isosorbide dinitrate can be identified by virtue of its characteristic IR and X-ray spectra (see 2.1 and 2.9).

7. Assay Methods

7.1 Elemental Analysis

Element	% Theory	Determined[5,46]
C	30.52	30.39
H	3.41	3.24
N	11.86	11.77

7.2 Colorimetric Analysis

The most common procedures for colorimetric analysis

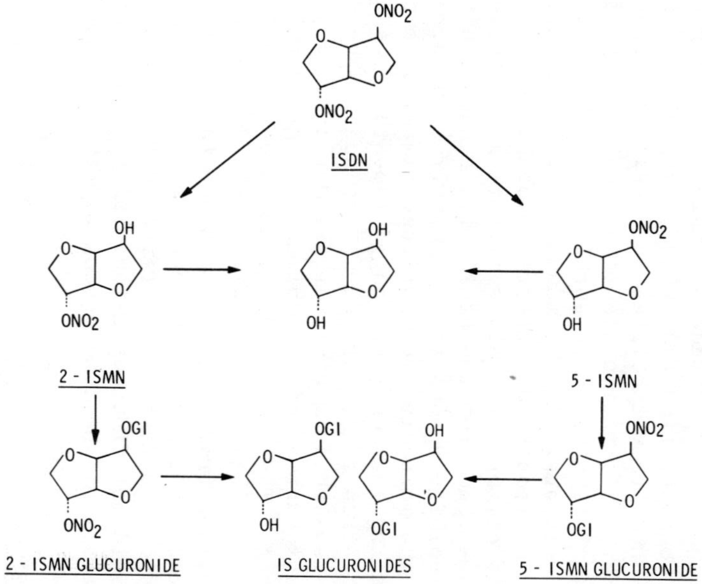
ONO2
ONO2
ISDN
OH
ONO2
OH
OH
ONO2
OH
2 - ISMN
5 - ISMN
OGI
ONO2
OGI
OH
OH
OGI
ONO2
OGI
2 - ISMN GLUCURONIDE
IS GLUCURONIDES
5 - ISMN GLUCURONIDE

ISDN = ISOSORBIDE DINITRATE
ISMN = ISOSORBIDE MONONITRATE
IS = ISOSORBIDE

Figure 6 - Metabolic Pathways of Isosorbide Dinitrate

of nitrate esters involves alkaline hydrolysis and colorimetric measurement of liberated nitrite. This basic procedure has been used by investigators for at least the last 40 years.[26] For isosorbide dinitrate more direct methods have been used. Most of these involve reaction with the nitrate group.

Reed <u>et.al.</u>[20] measured isosorbide dinitrate by reacting it with 1% 2,6-xylenol in glacial acetic acid in the presence of 80% (w/v) sulfuric acid for one-half hour to form nitroxylenol. The nitroxylenol was extracted from the aqueous reaction mixture using chloroform and then re-extracted into sodium hydroxide to form the yellow sodium salt whose absorbance was measured at 420 nm. This method is extremely sensitive.

Jackson and Hayward[10] reacted isosorbide dinitrate with 70% (v/v) sulfuric acid and a 1% solution of 3,4-dimethylphenol for 30-60 minutes at 30°C. Water was then added and the solution distilled collecting the distillate in a flask containing a solution of 2% sodium hydroxide. The absorbance of the resulting solution was then measured at 435 nm.

Nagese <u>et.al.</u>[27] determined isosorbide dinitrate in pharmaceuticals by two colorimetric methods. One was based on the Griess-Romijin reagent and the other on the phenoldisulfonic acid reagent.

The phenoldisulfonic acid colorimetric method for determining mannitol hexanitrate[28-31] was adapted[32] for measuring isosorbide dinitrate in tablets and injectables. In this method a sample containing 10 mg. of isosorbide dinitrate is extracted from an acidic solution with a total of 100 ml. of ether or chloroform. An aliquot (4 ml.) of the extract is then evaporated to dryness and 1 ml. of glacial acetic acid and 2.0 ml. of phen[illegible]
are added. After 15 minutes th[illegible]
with ammonium hydroxide and its [illegible]
nm. In this procedure it is not necessary to use isosorbide dinitrate as a standard, any of the inorganic nitrates, such as sodium or potassium nitrate, can be utilized.

A method used for determining nitrate[33] has been adopted[34] to study solutions of isosorbide dinitrate. This method involves hydrolyzing isosorbide dinitrate with sodium hydroxide to form nitrite ions. The nitrite is then diazotized with p-nitroaniline to form a diazonium ion which when coupled with azulene in the presence of perchloric acid exhibits an absorbance maximum at 515 nm.

7.3 Infrared Analysis

Infrared spectrophotometric procedures used for the estimation of nitrate esters[35,36] have been adopted[32] for determining isosorbide dinitrate in tablets and injectables. The method consists of simply extracting isosorbide dinitrate from its dosage form with chloroform and reading the absorbance at 1650 $cm.^{-1}$ in a sodium chloride cell. Carbon tetrachloride can be substituted for chloroform in this procedure.[37]

7.4 Gas Liquid Chromatographic Analysis

The colorimetric and IR methods mentioned above are virtually non-specific. This is a disadvantage especially in cases where mixtures of esters are present due to partial denitration of the parent ester. Gas liquid chromatographic methods overcome this difficulty, provide adequate separation and, at the same time, maintain sensitivity of detection.

Sherber et.al.[24] determined isosorbide dinitrate in rabbit blood by extracting it in ethyl acetate and injecting the extract without further purification in a 6 ft. column packed with 3.8% SE-30 on Gas Chrom P. The system was operated isothermally under the following conditions: oven temperature, 110°C., flash heater, 130°C., detector, 190°C.; nitrogen flow rate, 55 ml./min. These same investigators also used a 3% XE-60 on Gas Chrom Q column with an oven temperature of 150°C.; flash heater at 160°C.; flame ionization detector temperature at 180°C.; and nitrogen flow rate at 55 ml./min. The internal standard used was m-dinitrobenzene. The sensitivity of this method is such that as little as 0.01 µg. can be detected with a flame ionization detector.

Davidson et.al.[38] used a gas chromatographic method for separation and quantitation of isosorbide dinitrate in the presence of other organic nitrate esters of common therapeutic use. They used two columns: one packed with 1% SE-30 on trimethylchlorosilylated Chromosorb P and the other packed with 1% Dexsil 300 GC on dimethylchlorosilylated Chromosorb W. The injection port and detectors were maintained at 200°C. with nitrogen as the carrier gas. The columns were programmed at a rate of 10°/min. with an initial setting of 65°C.

A mixture of isosorbide dinitrate and its two mononitrates was assayed by gas chromatography using columns packed with either 3% XE-60 or 3.5% QF-1 on Gas Chrom Q.[39]

Both columns were operated isothermally, the QF-1 at 110°C. and the XE-60 at 150°C. The injection port was maintained at 160°C. Good separation of the different nitrates was obtained on the QF-1 column. The XE-60 column was less satisfactory as 5-isosorbide mononitrate overlapped with isosorbide dinitrate. The minimum amount of isosorbide dinitrate that could be determined was 0.5 µg. with a flame ionization detector and 8 ng. with an electron capture detector. These same investigators[40] have subsequently determined isosorbide dinitrate in human plasma on the QF-1 column. Isoidide dinitrate was used as the internal standard.

Another gas chromatographic method for the determination of isosorbide dinitrate in tablets[22] uses a 5 ft. x 1/8 in. stainless steel column packed with 1% OV-17 on Diatoport S, 80/100 mesh operated at 155°C. The isosorbide dinitrate is extracted from an acidic solution with chloroform, and ethyl pentadecanoate is used as the internal standard. This method will separate isosorbide dinitrate from isosorbide and the two mononitrates.

7.5 Polarographic Analysis

A polarographic method used for determining glyceryl trinitrate[41] has been adopted[42] for assaying isosorbide dinitrate in injectables and tablets. This method involves the reduction of the nitrate ester at the dropping mercury electrode in a 0.5 N tetramethylammonium hydroxide aqueous solution buffered at pH 8.7 with an ammonium buffer (0.5 N in NH_4Cl and NH_4OH). The reduction wave has a half-wave potential of -0.78V. Nitrate and nitrite ions do not interfere.

7.6 Automated Analysis

The phenoldisulfonic acid reaction (7.2) has been utilized for determining isosorbide dinitrate in tablets using an automated procedure.[43] Glacial acetic acid is used as the extraction solvent.

7.7 Radiochemical Analysis

Isosorbide dinitrate-C^{14} was analyzed[20,25] in a liquid scintillation spectrometer under conditions appropriate for counting C^{14}.

7.8 Thin Layer Chromatographic Analysis

Several TLC systems for the separation of isosorbide dinitrate from its metabolites and similar structure compounds have been reported in the literature. In one method[23] the sample was spotted on chromatographic plates coated with silica gel (.25 mm) and were developed by ascending techniques with two solvent systems, one a benzene: ethyl acetate (1:1) mixture, and the second a 2-propanol: ammonium hydroxide (4:1) mixture. For visualization the plates were sprayed with 1% diphenylamine in methanol solution and exposed to ultraviolet light for 5 minutes. Diphenylbenzadium in sulfuric acid has also been used as a spray reagent.[24] The latter reagent was found to be 10 times more sensitive than diphenylamine for nitrate esters. The R_f for isosorbide dinitrate is .68 in the first solvent system and .77 in the second. Both of these systems separated isosorbide dinitrate from its two mononitrates. Other investigators using these same solvent systems visualized isosorbide with a metaperiodate-permanganate spray.[25]

Another TLC method[44] uses as an adsorbent a mixture of silicic acid and plaster of Paris (70:30 w/w). Chromatograms were developed with mixtures of anhydrous solvents in the following volume ratio: benzene-petroleum ether (1:1); ether-petroleum ether (3:39); ether-benzene (1:19). A 1% solution of diphenylamine in 95% ethanol, followed by exposure to a shortwave ultraviolet light for 10 minutes was used for detection.

Isosorbide dinitrate[21] was separated from isosorbide and its two mononitrates on plates coated with Silica Gel G by using a mixture of carbon tetrachloride and acetone (8:2) as the solvent.

7.9 Paper Chromatographic Analysis

Jackson and Hayward[21] developed several methanol-hydrocarbon solvent systems useful for paper chromatography of isosorbide dinitrate. For visualization the paper was sprayed with 1% alcoholic diphenylamine and exposed to UV light for 5 to 10 minutes.

8. References

1. Merck Index, 8th Ed., Merck and Co., Inc.(1968).
2. Benjamin K. Ayi, Wyeth Laboratories, Personal Communication.
3. O. R. Sammul, W. L. Brannon, and A. L. Hayden, J. Ass. Off. Agr. Chem., 47 (5), 918(1964).
4. L. D. Hayward, D. J. Livingstone, M. Jackson, and V. M. Csizmadia, Can. J. Chem., 45, 2191(1967).
5. Bruce Hoffmann, Wyeth Laboratories, Personal Communication.
6. F. J. Hopton and G. H. S. Thomas, Can. J. Chem., 47, 2395(1969).
7. N. J. DeAngelis, Wyeth Laboratories, unpublished results.
8. L. D. Hayward, R. A. Kitchen, and D. J. Livingstone, Can. J. Chem., 40, 434(1962).
9. Charles Kuhlman, Wyeth Laboratories, Personal Communication.
10. M. Jackson and L. D. Hayward, Can. J. Chem., 38 496(1960).
11. L. Goldberg, Acta Physiol. Scand., 15, 173(1948); C.A. 42, 5564d.
12. L. F. Wiggins, Advances in Carbohydrate Chem., 5, 206 (1950).
13. S. E. Forman, C. J. Carr, and J. C. Krantz, J. Am. Pharm. Assoc., 30, 132(1941).
14. USP XVIII, Mack Printing Co., P. 935(1970).
15. USP XIX, in print.
16. North American Philips Company, Mount Vernon, New York.
17. L. F. Wiggins, J. Chem. Soc., 4, (1945).
18. S. E. Forman, C. J. Carr, and J. C. Krantz, J. Am. Pharm. Assoc., 30, 132(1941).
19. P. M. Kochergin and R. M. Titkova, Med. Prom. S.S.S.R. 13, No. 8, 18-20(1959).
20. D. E. Reed, J. F. May, L. G. Hart, and D. H. McCurdy, Arch. Int. Pharmacodyn., 191, 318-336(1971).
21. J. Rutgers, Wyeth Laboratories, unpublished results.
22. K. Dilloway, Wyeth Laboratories, unpublished results.
23. A. J. Dietz Jr., Biochem. Pharmacol., 16 (12), 2447-8(1967).
24. D. A. Sherber, M. Marcus, and S. Kleinberg, Biochem. Pharmacol., 19, (2), 607-12(1970).
25. S. F. Sisenwine and H. W. Ruelius, J. Pharmacol. Exp. Ther., 176(2), 296-301(1971).
26. L. A. Crandall, C. D. Leake, A. S. Leovenhart, and C. W. Muehlberger, J. Pharmacol. Exp. Ther., 37, 283

(1929).
27. Y. Nagase, Y. Kanoya, A. Sugiyama, and H. Haruhiko, Yakaguaku Zasshi, 85(2), 119-25(1965).
28. E. Sornoff, J. Ass. Off. Agr. Chem., 38, 637(1955).
29. Ibid., 39, 630(1956).
30. Ibid., 40, 815(1957).
31. "Official Methods of Analysis of the Association of Official Agricultural Chemists", 10th Ed., 605(1965).
32. D. Gutekunst, Wyeth Laboratories, unpublished results.
33. E. E. Garcia, Anal. Chem., 39, 1605(1967).
34. J. Poole, Wyeth Laboratories, unpublished results.
35. Jonas Carol, J. Ass. Off. Agr. Chem., 42, 468(1959).
36. Ibid., 43, 259(1960).
37. D. Woo, J. K. C. Yen, and P. Sofronas, Anal. Chem., 45(12), 2144-45(1973).
38. I. W. F. Davidson, F. J. Dicarlo, and E. I. Szabo, J. Chromatogr., 57, 345-52(1071).
39. M. T. Rossell and M. G. Bogaert, J. Chromatogr., 64, 364-7(1972).
40. M. T. Rossell and M. G. Bogaert, J. Pharm. Sci., 62 (5), 754(1973).
41. B. C. Flann, J. Pharm. Sci., 58(11), 122(1969).
42. J. Bathish, Wyeth Laboratories, unpublished results.
43. C. Davis, Wyeth Laboratories, unpublished results.
44. L. D. Hayward, R. A. Kitchen, and D. J. Livingstone, Can. J. Chem., 40, 434-40(1962).
45. M. Jackson and L. D. Hayward, J. Chromatogr., 5, 166-69(1961).
46. Lot No. 5123 V 6/1, Wyeth Laboratories.
47. R. D. Guthrie and H. Spedding, J. Chem. Soc., 953 (1960).
48. L. Bellamy, "The Infrared Spectra of Complex Molecules," 2nd Ed., J. Wiley and Sons, Inc., New York, N. Y., 1964.

METHAQUALONE

Dahyabhai M. Patel, Anthony J. Visalli, Jerome J. Zalipsky and Nelson H. Reavey-Cantwell

CONTENTS

Analytical Profile - Methaqualone

1. Description

1.1 Name, Formula, Molecular Weight

Methaqualone is 2-methyl-3-o-tolyl-4(3H)-quinazolinone

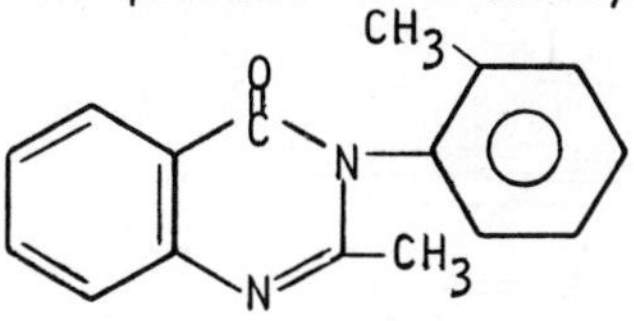

$C_{16}H_{14}N_2O$ Molecular Weight: 250.30

1.2 Appearance, Color, Odor

Methaqualone occurs as a white, odorless, crystalline powder.

2. Physical Properties

2.1 Infrared Spectrum (IR)

The infrared spectrum of methaqualone is presented in Figure 1 (1). The spectrum was obtained from a KBr pellet consisting of a 0.5% dispersion of methaqualone in KBr. The assigned principal absorption bands are listed in Table I (1).

Table I

Principal Bands in the IR Spectrum of Methaqualone

Wavelength (cm^{-1})	Assignment
3049, 2994 and 2907	CH stretching
1672	Amido carbonyl stretching
1603, 1595 and 1471	Aromatic skeletal vibration
1340 and 1326	Aromatic C-N stretching
778, 760, 720 and 703	CH deformation of aromatic ring

Figure 1

Infrared Spectrum of Methaqualone

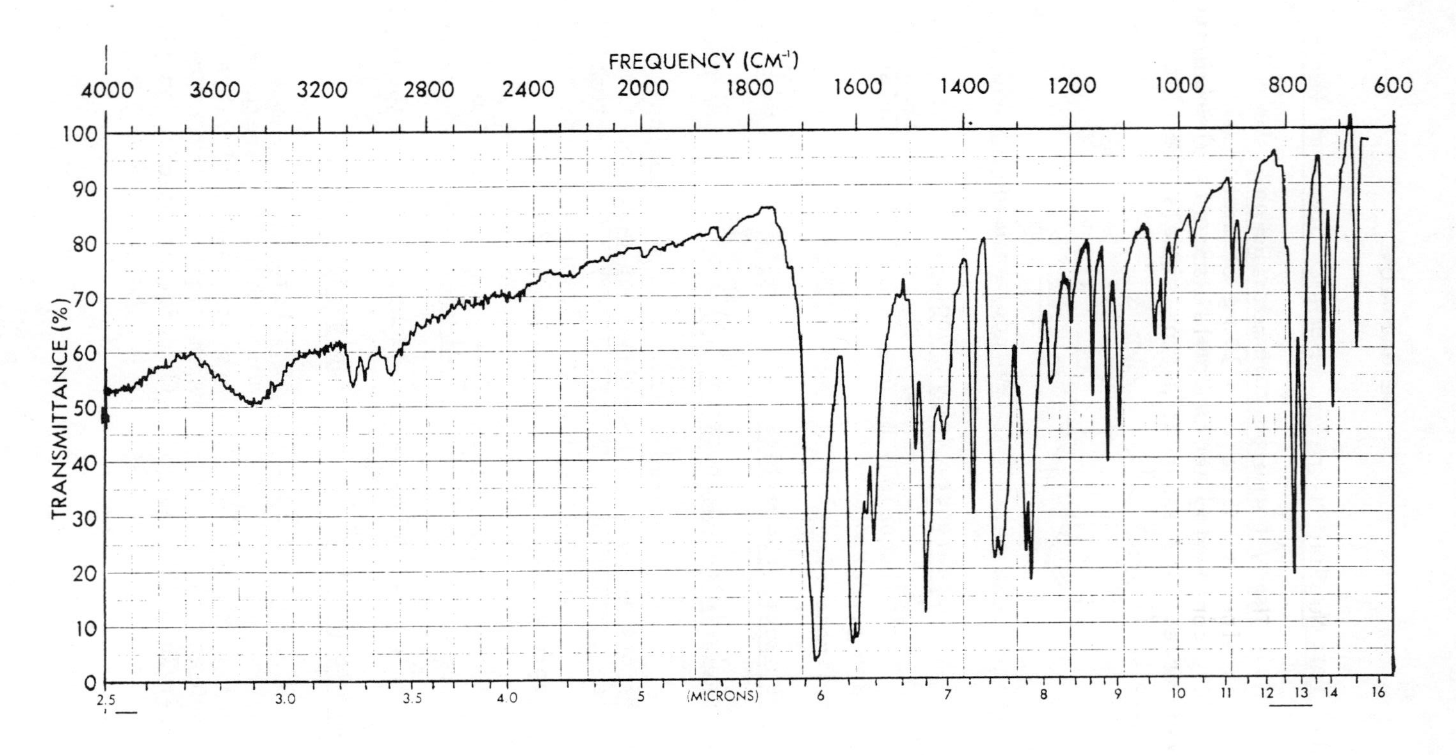

2.2 Nuclear Magnetic Resonance Spectrum (NMR)

The NMR spectrum of methaqualone shown in Figure 2 was obtained by dissolving 30 mg of methaqualone in 0.5 ml of CCl_4 containing tetramethylsilane as the internal reference. The assigned proton signals are listed in Table II (1).

Table II

NMR Absorption Signals of Methaqualone

Group	No. of Protons	Chemical Shift (ppm)	Signal Appearance
CH_3 (Quinazolinone ring)	3	2.06	singlet
CH_3 (Tolyl)	3	2.15	singlet
Aromatic Protons	8	7-8.4	multiplet

2.3 Ultraviolet Absorption Spectra (UV)

Methaqualone exhibits characteristic ultraviolet absorption properties in different solvents. These characteristic properties were established using a Cary Model 14 Spectrophotometer in 1 cm cells. The UV Spectra are presented in Figure 3 and the relevant UV absorption data are listed in Table III (1).

Table III

UV Absorption Characteristics of Methaqualone

Solvent	λ max. nm	λ min. nm	$a_{1\,cm}^{0.1\%}$ at λ max. (ε)
Methanol	225.5	214.0	140.9 (35.26×10^3)
	265	249.0	38.0 (9.51×10^3)
Ethanol (95%)	225.5	214.0	139.6 (34.94×10^3)

Figure 2

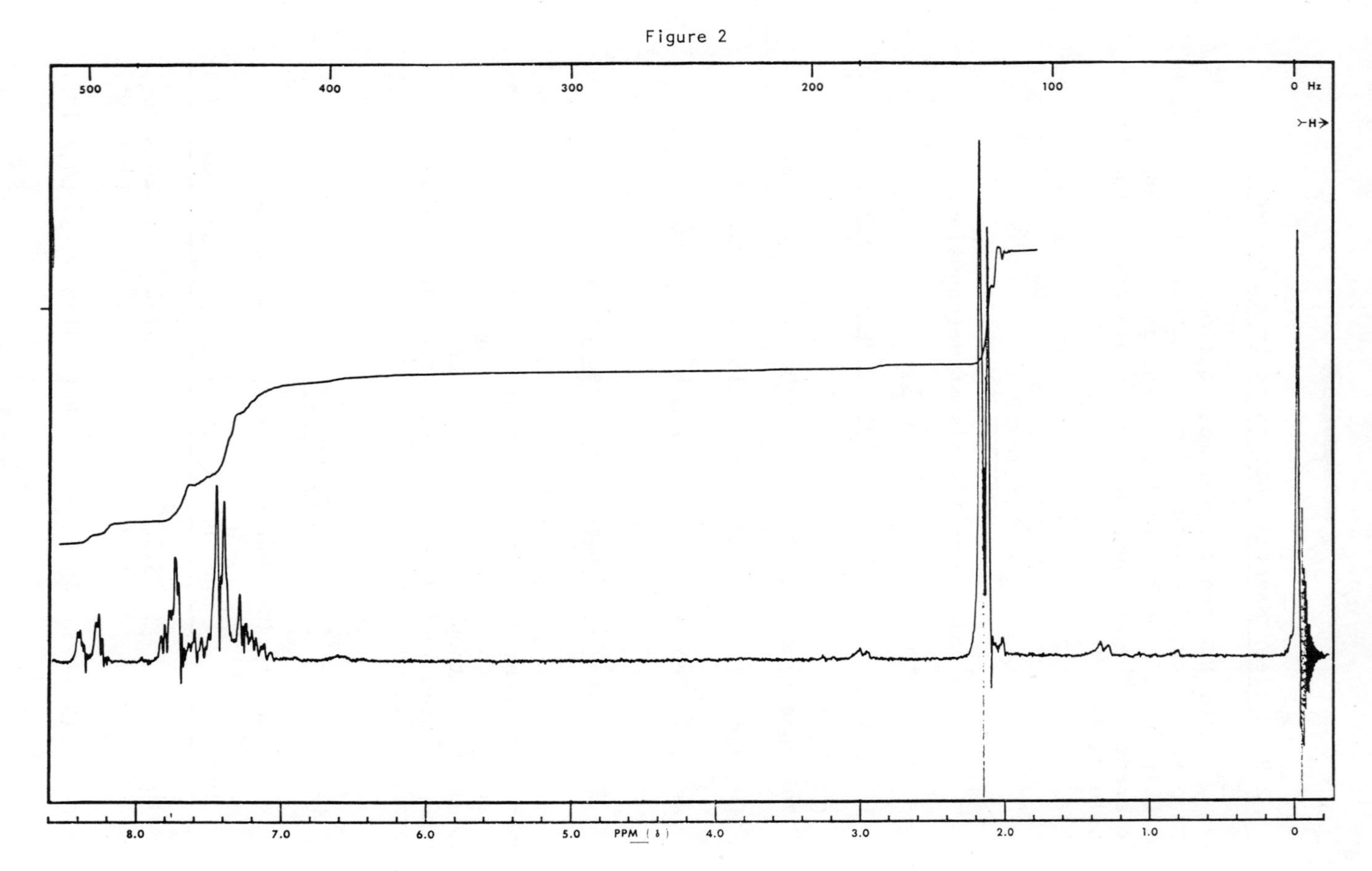

NMR Spectrum of Methaqualone

Figure 3

Ultraviolet Absorption Spectra of Methaqualone

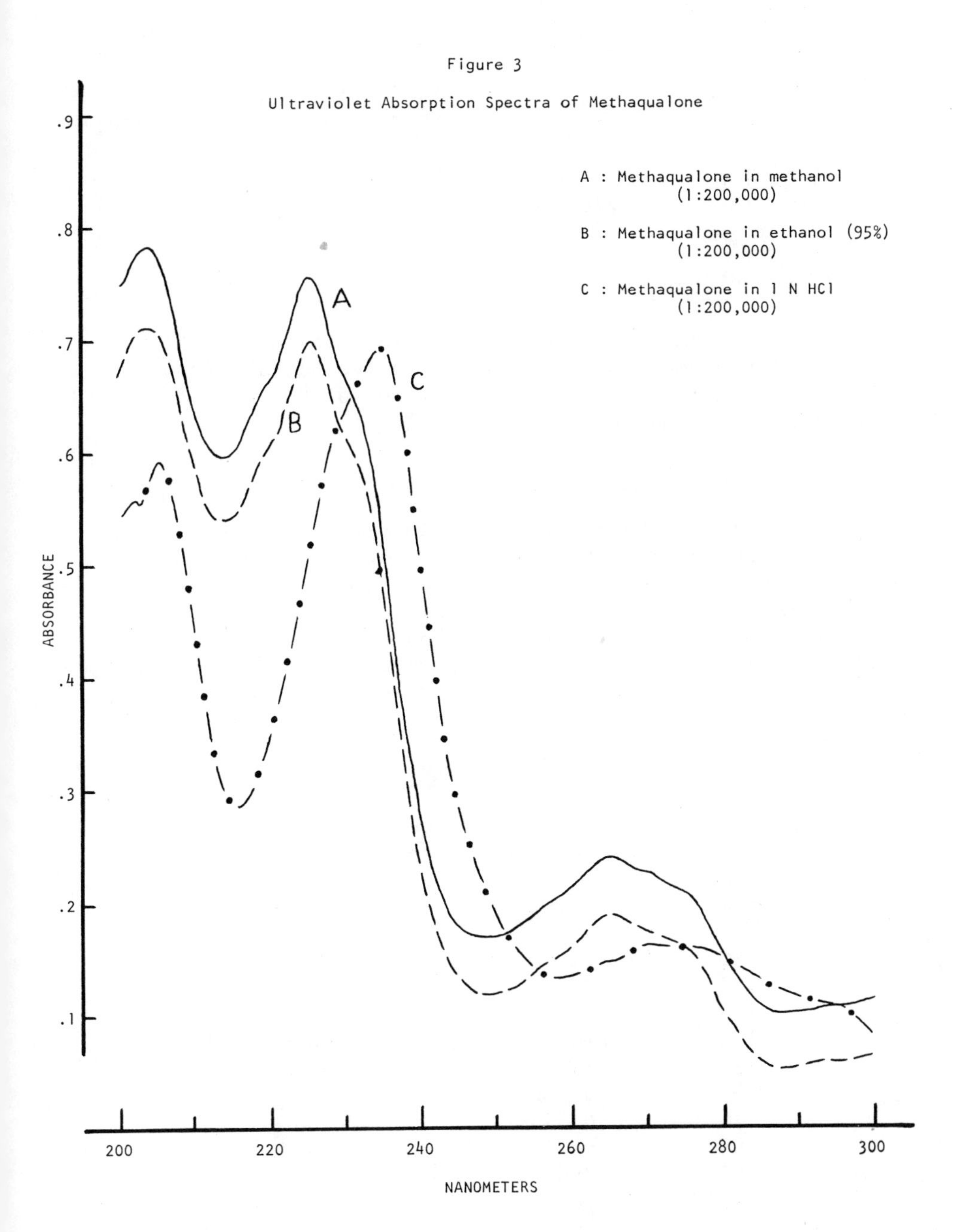

	265.0	249.0	37.7 (9.43 X 10^3)
HCl (1 N)	235.0	215.5	133.0 (33.42 X 10^3)
	~270	258.5	30.8 (7.71 X 10^3)
HCl (0.1 N)	234.5	215.5	133.4 (33.39 X 10^3)
	~270	257.5	31.5 (7.88 X 10^3)

2.4 Fluorescence Spectrum

Methaqualone does not exhibit any native fluorescence. However, the reduced derivative tetrahydroquinazolinone exhibits an emission maximum at 450 nm with excitation at 345 nm in methanol (2).

2.5 Mass Spectrum

The mass spectrum of methaqualone in the form of a bar graph is presented in Figure 4. The spectrum was obtained on a Hitachi-Perkin Elmer Model RMU-6D Mass Spectrometer using the direct inlet probe with sample temperature at 60°C. The electron energy was 70 ev and the accelerating voltage was 1700 volts. The characteristic fragmentation pattern is presented in Table IV (1).

Table IV

Fragmentation Pattern Produced in Mass Spectral Examination of Methaqualone

Mass (m/e)	Intensity	Fragment Ion
250	Strong	Molecular ion (M)
235	Very Strong	$M-CH_3$
145	Weak	$M-C_7H_7N$
144	Weak	$M-[C_7H_7+ CH_3]$
132	Medium	$M-C_7H_4NO$
91	Strong	$M-C_9H_7N_2O$
76	Medium	$C_7H_7-CH_3$
65	Medium	$C_7H_7-C_2H_2$
50	Medium	$C_6H_4-C_2H_2$
41	Very Weak	$C_9H_{10}N-C_7H_7$

Figure 4

Mass Spectrum of Methaqualone

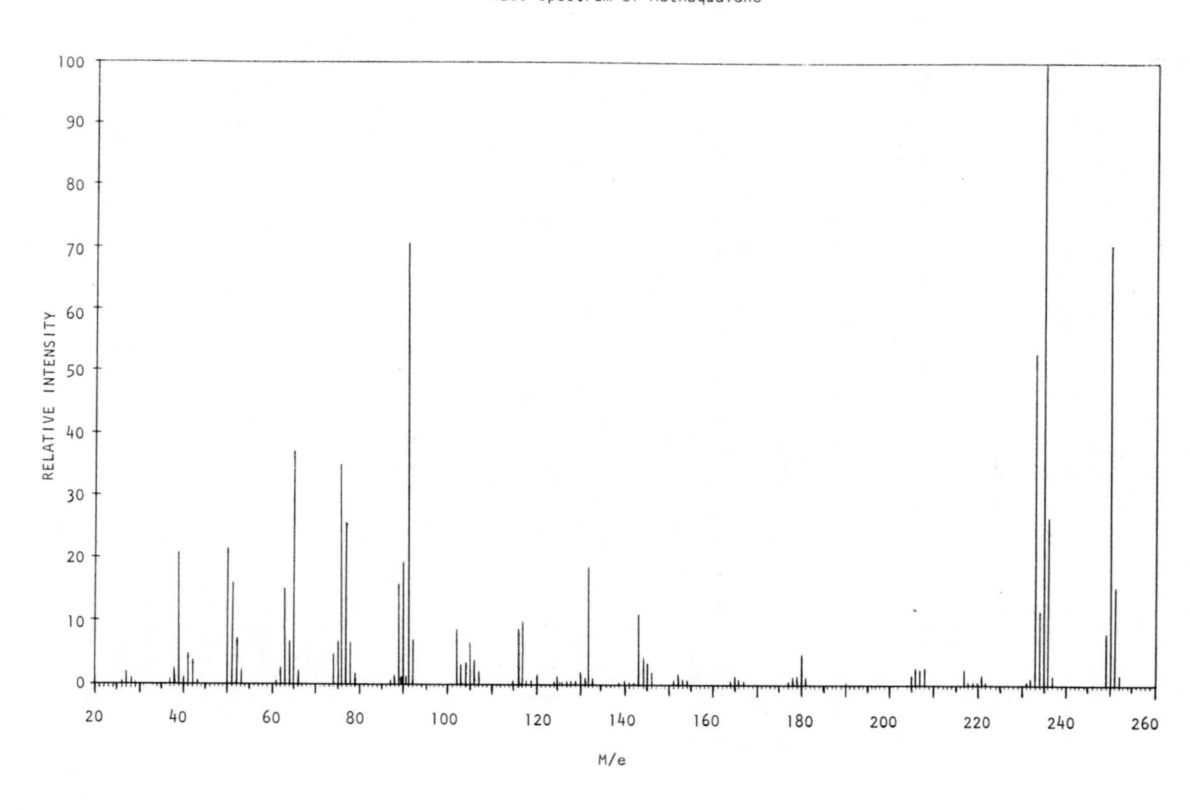

39	Medium	$C_2H_3N-H_2$

2.6 Melting Range

The melting range of methaqualone is dependent on the rate of heating. Using the U.S.P. XVIII Class 1 procedure, methaqualone melts between 114°-117°C (1).

2.7 Optical Rotation

Methaqualone exhibits no optical activity when analyzed at a series of mercury and sodium spectral lines (1).

2.8 Differential Scanning Calorimetry (DSC)

The DSC thermogram of methaqualone is shown in Figure 5. A melting endotherm is observed at 385°K(112°C) with the temperature programing at 10°C/minute. The ΔH_f was found to be 5.55 KCal/mole (1).

2.9 Solubility

The equilibrium solubility data obtained on methaqualone at 23°C is listed in Table V (1).

Table V

Methaqualone Solubility

Solvent	Solubility (g/100 ml)
Water	0.03
Methanol	16.58
Ethanol	12.53
Isopropanol	4.76
Ether	3.71
Chloroform	>40
Benzene	28.00
Acetonitrile	>40
Petroleum Ether (B.P. 30-60°)	0.33

Figure 5

DSC Thermogram of Methaqualone

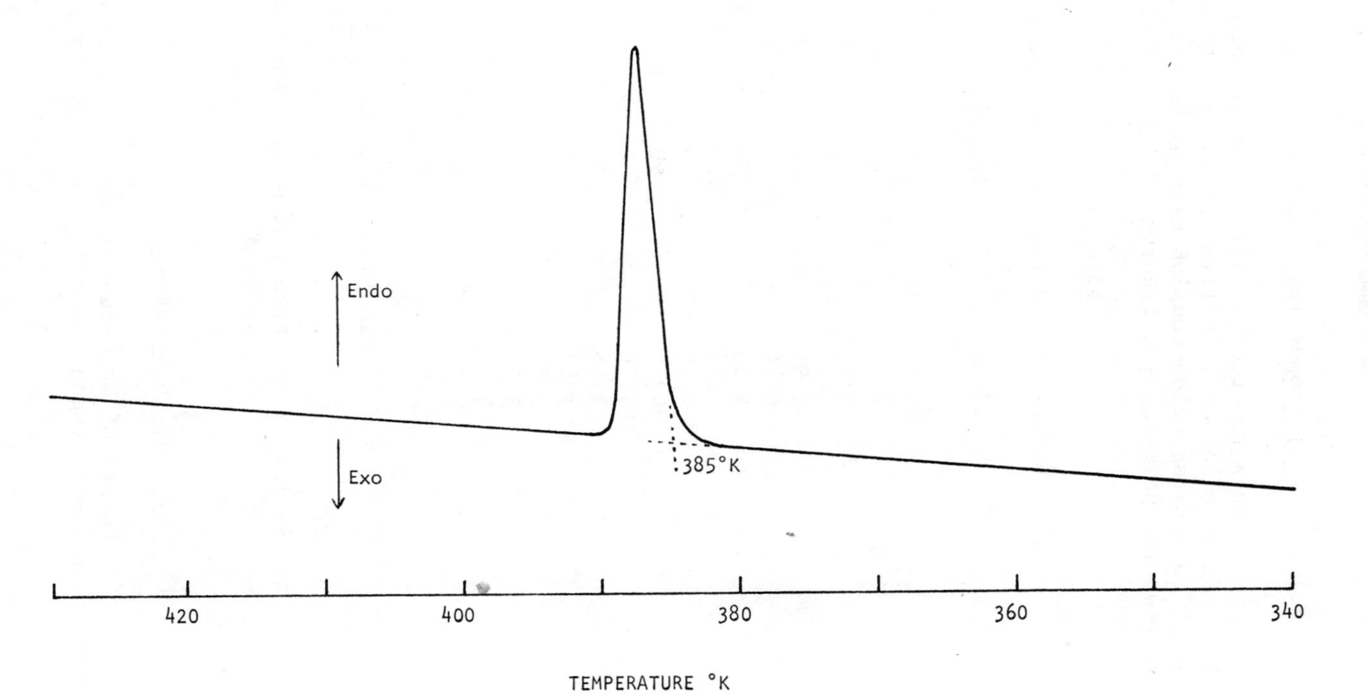

2.10 Crystal Properties

The X-ray powder diffraction pattern of methaqualone was obtained on a Phillips scanning diffractometer using copper radiation coupled with Nickel filter. The diffraction pattern is presented in Table VI (1).

Table VI

X-ray Diffraction Pattern of Methaqualone

d(A[1])*	I/I_o**
9.4	15
8.2	25
7.76	25
7.30	50
6.95	65
6.10	25
5.53	75
4.76	100
4.13	30
4.00	20
3.75	30
3.66	30
3.53	15
3.39	60
3.24	15
3.17	15
3.12	30
3.05	25

* d = (interplanar distance) $\frac{n\lambda}{2 \sin \theta}$

** I/I_o = relative intensity (based on highest intensity of 1.00)

2.11 Dissociation Constant

The pKa of methaqualone by the spectrophotometric method at ionic strength $\mu = 0.1$ was found to be 2.54 (1).

3. Synthesis

Methaqualone is synthesized by the reaction scheme shown in Figure 6. Anthranilic acid is reacted with acetyl chloride to yield n-acetylanthranilic acid. This intermediate product is condensed with o-toluidine in the presence of PCl_3 to yield methaqualone (3).

4. Stability-Degradation

On refluxing methaqualone with 1.0, 0.1 N NaOH and 1.2 N HCl, the principal degradation products found are anthranilic acid, o-toluidine and acetic acid. In addition, 2-aminobenzo-o-toluidide was found on refluxing methaqualone with 1.0 and 0.1 N NaOH. Also, n-acetylanthranilic acid was found as a degradation product on refluxing methaqualone with 0.1 N NaOH. Under these stringent conditions, a low yield of degradation products was obtained; thereby indicating good stability of the drug under normal conditions (4). The degradation scheme is presented in Figure 7.

5. Metabolism

Methaqualone (MTQ) and its hydrochloride salt is rapidly absorbed ($K_a = 0.82\ hr^{-1}$) from the stomach and duodenum following oral administration of tablet or capsule dosage forms. MTQ exists in blood primarily in the plasma phase bound to serum albumin. Peak serum levels of 3-4 μg/ml are seen 1-1 1/2 hours after a 300 mg dose. The serum elimination curve is biexponential with a distributional phase ($\alpha = 0.97\ hr^{-1}$) and an elimination phase ($\beta = 0.036\ hr^{-1}$). The principle tissues of distribution are lipid tissue, liver and kidney. MTQ is metabolized by hepatic microsomal oxidoreductases. Hydroxylation of 2 and 2' methyl substituents, tolyl sidechain and quinazolene nucleus, gives rise to the principle monohydroxy metabolites. O-glucuronide conjugation and O-methylation occur prior to excretion of metabolites in bile or urine. Ring opening involving formation of N-1-oxide and 2-nitrobenzo-o-toluidide is a minor metabolitic path. Biliary excretion and entero-

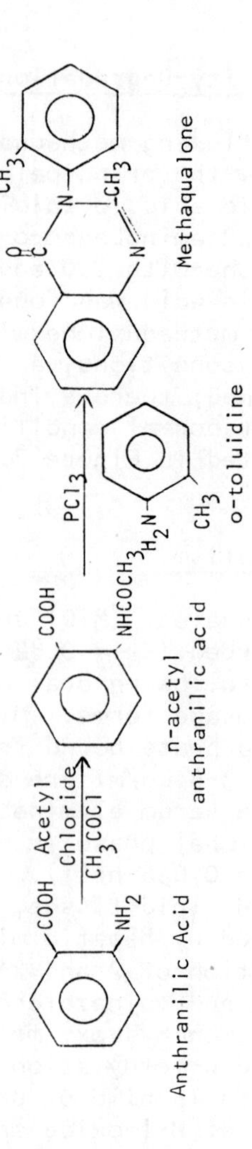

Figure 6

Synthesis of Methaqualone

Figure 7

Degradation Profile of Methaqualone

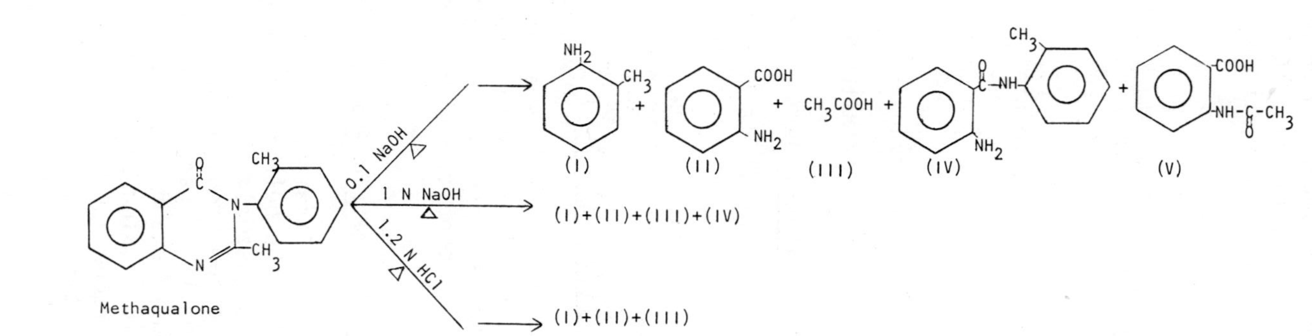

(I) = o-toluidine

(II) = Anthranilic acid

(III) = Acetic Acid

(IV) = 2-aminobenzo-o-toluidide

(V) = n-acetylanthranilic acid

hepatic recirculation of metabolites occur. At a therapeutic dose of 75 to 300 mg in man, MTQ is completely biotransformed and little, if any, unchanged drug appears in urine (5).

6. Methods of Analysis

6.1 Elemental Analysis

The results from the elemental analysis are listed in Table VII (1).

Table VII

Elemental Analysis of Methaqualone

Element	% Theory	% Found
C	76.77	76.50
H	5.64	5.74
N	11.19	10.89

6.2 Phase Solubility Analysis

Phase solubility analysis was carried out using n-hexane-absolute alcohol (145:5 v/v) as a solvent system. The results of the analysis are shown in Figure 8 along with the listings of the conditions under which the analysis was performed (1).

6.3 Nonaqueous Titrimetric Analysis

The nonaqueous titration is the official method of analysis in B.P. for the bulk methaqualone. An accurately weighed 0.5 g sample is dissolved in 80 ml of glacial acetic acid, a few drops of crystal violet T.S. is added and the solution is titrated to an emerald green end point against previously standardized 0.1 N $HClO_4$ in glacial acetic acid (6). King and Perry (7) reported the application of the nonaqueous titrimetric method for assaying methaqualone tablets.

Figure 8

Phase Solubility Analysis of Methaqualone

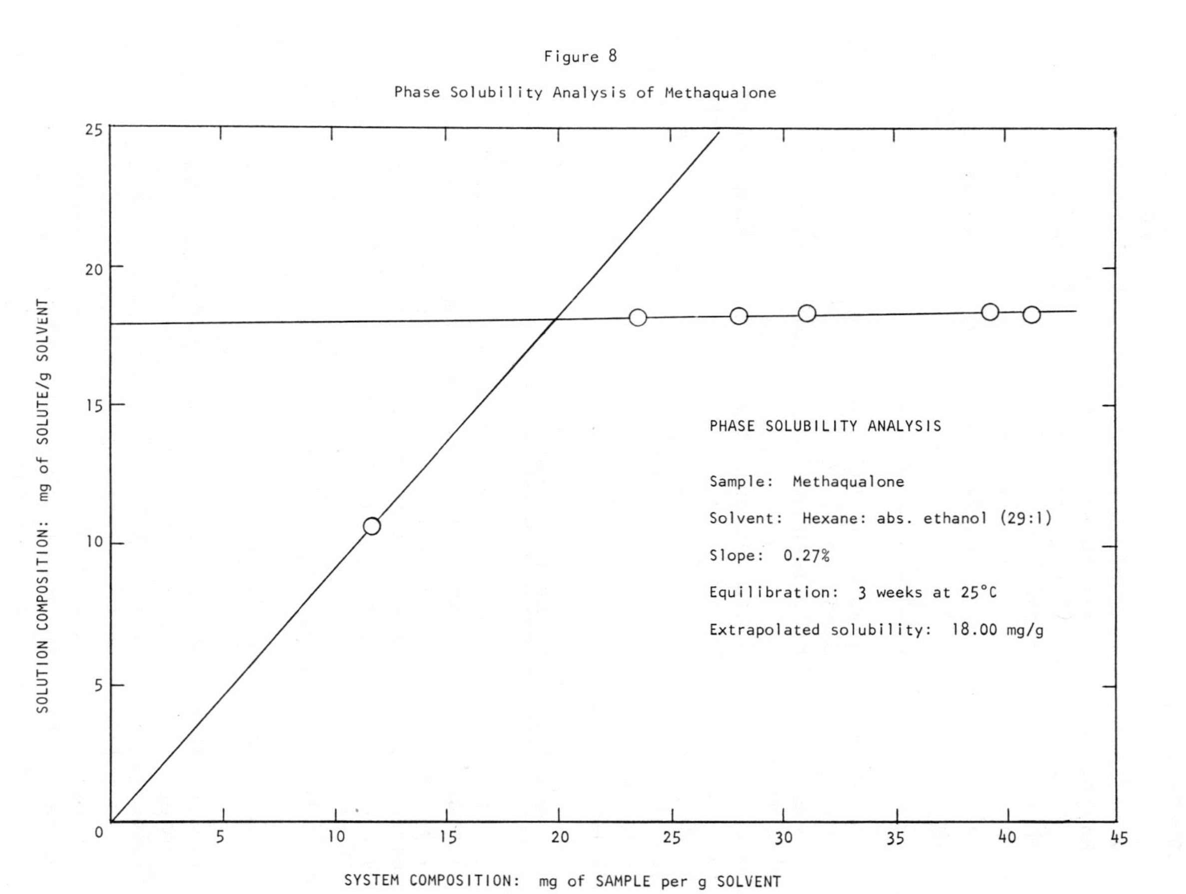

6.4 Colorimetric Analysis

Nakano et al. (8) reported a colorimetric assay method for methaqualone in microgram quantities based upon diazotization of the diazotizable amines formed by acid hydrolysis of methaqualone. Pirl et al. (9) developed assay methods for methaqualone in biological specimens as low as 0.3 μg/ml. The method entails isolation of methaqualone from the biological specimen and condensing with p-dimethylaminobenzaldehyde in alkaline media. The resulting chromophore is measured at 503 nm in weak acid.

6.5 Spectrophotometric Analysis

Akagi Masuo et al. (10) described two methods for assaying methaqualone in biological fluids and tissues. In the first method, methaqualone is extracted from alkalinized biological material with hexane, the solvent is evaporated and the residue after dissolving in dil HCl is measured at 234 nm. The second method eliminates the normal biological blank by geometric treatment of the absorbances at three wavelengths. These methods are specific for methaqualone in that they do not include degradation products of methaqualone.

6.6 Spectrofluorometric Analysis

A fluorimetric method of assaying therapeutic plasma levels of methaqualone has been reported by Brown, S. S. et al. (2). The method is based on reducing methaqualone to its dihydro-derivative using lithium borohydride as an effective reducing agent. The resulting dihydro-methaqualone derivative exhibits emission maximum at 450 nm at the activation maximum of 345 nm.

6.7 Gas Chromatographic Analysis

There are several reports regarding the gas chromatographic analysis of methaqualone in biological fluids (1, 10, 11, 12, 13). These methods can be applied for formulation analysis, evaluation of quality, and characterization of methaqualone. Table VIII summarizes the gas chromatographic systems with references.

Table VIII

Gas Chromatographic Systems Used for Methaqualone Analysis

Note: All systems listed used flame ionization detectors.

Reference	Column	Carrier Gas	Column Temp.°C	Internal Standard	*RR_T
(1)	6 ft. X 4 mm., I.D., glass tubing, 10% SE-30 on Chromosorb WAW(DMCS), 100-120 mesh	N_2 45 ml/min	240°	Diphenhydramine	2.17
(1)	4 ft. X 2 mm., I.D., glass tubing, 3% OV-17 on Chromosorb W(H.P.), 80-100 mesh	N_2 30 ml/min	240°	Codeine	0.53
(1)	4 ft. X 4 mm., I.D., glass tubing, 3.8% UC-W98 on Chromosorb W(H.P.), 80-100 mesh	N_2 45 ml/min	230°	Codeine	0.56
(10)	2 ft. X 4 mm., I.D., glass tubing, 3% XE-60 on Gas Chrom Q, 100-120 mesh	He 50 ml/min	180°	Butyl Stearate	2.81
(11)	7 ft. X 4 mm., I.D., glass tubing, 3% Cyclohexanedimethanol succinate 85-100 mesh	N_2 55 ml/min	200°	Butobarbitone	2.14
(12)	2 ft. X 4 mm., I.D., glass tubing, 2.5% SE-30 on Chromosorb G, 80-100 mesh	N_2 40 ml/min	200°	Codeine	0.42
(13)	2-m. X 3 mm., I.D., glass tubing, 4% SE-30 and 6% QF-1 on	N_2 30 ml/min	190°	Butobarbital	0.24

Chromosorb WAW (DMCS), 80-100 mesh	Pento-barbital	0.30
	Pheno-barbital	0.69

* Retention time relative to internal standard.

6.8 Thin Layer Chromatographic Analysis

There are several TLC systems reported in the literature which are listed in Table IX. These systems are used for detecting the associated impurities in methaqualone, identification of methaqualone or characterization of methaqualone and its metabolic products.

Table IX

Thin Layer Chromatographic Systems for Methaqualone Analysis

Reference	Support	Solvent System	Application
6	Kieselghur G	Anaesthetic Ether saturated with water	Detection of associated impurities
14	Silica Gel G	Chloroform-acetone	Identification
14	Silica Gel G	Isopropanol-chloroform-25% ammonium hydroxide (45:45:10)	Identification
14	Silica Gel G	Petroleum ether (B.P. 50-75°) Pyridine (75:15)	Identification
15	Silica Gel G	Chloroform-acetone -25% ammonium hydroxide (50:50:2)	Separation of methaqualone from the metabolites

15	Silica Gel G	Cyclohexane-chloro-form-diethylamine (7:2:1)	Separation of methaqualone from the metabolites
15	Silica Gel G	Propanol-ethyl acetate-25% ammonium hydroxide (6:2:2)	Separation of methaqualone from the metabolites

7. References

1. Analytical and Physical Chemistry Department contribution, Research Division*, William H. Rorer, Inc., Fort Washington, Pa.

2. Brown, S. S. and Smart, G. A., J. Pharm. Pharmacol., 21, 466-468 (1969).

3. Herbert, M. and Pichat, L., Synthese De La Methyl-2, o-Tolyl-3, Quinazolone-4 ^{14}C-2, Report CEA No. 1302, Centre D'Etudes Nucleaires De Saclay, Service De Documentation, Boite postale n°2-Gif-sur-Yvette, S.-et-O (1959).

4. Zalipsky, J., Patel, D. M. and Reavey-Cantwell, N. H., Hydrolytic Degradation of Methaqualone, submitted to J. Pharm. Sci. for publication.

5. Smyth, R. D., Research Division, William H. Rorer, Inc., Personal Communication.

6. British Pharmacopoeia, 296 (1972).

7. King, E. E. and Perry, A. R., J. Ass. Pub. Anal. 7, 55-59 (1969).

8. Nakano, M., Yakuzaigaku, 22, 267-269 (1962).

9. Pirl, Joerg, N., et al. Anal. Chem. 44, 1675-1676 (1972).

10. Douglas, J. F. and Shahinian, S., J. Pharm. Sci., 62, 835-836 (1973).

11. Berry, D. J., J. Chromatog., 42, 39-44 (1969).

12. Finkle, B. S., Cherry, E. J. and Taylor, D. M., J. Chromatog. Sci., 9, 393-414 (1971).

13. Hatch, R. C., Am. J. Vet. Res., 33, 203-207 (1972).

14. Stahl, E., Thin Layer Chromatography, 2nd Edition, Spring-Verlag Berlin-Heideberg, New York, 539.

15. Hirtz, Jean L., Analytical Metabolic Chemistry of Drugs, Marcel Dekker, Inc., New York, 242-243.

*Mass spectrum and X-ray diffraction analysis was performed by Morgan Schaffer, Inc. and Walter C. McCrone Associates, respectively.

8. Acknowledgements

The authors wish to thank all members of the Analytical and Physical Chemistry Department, Research Division, William H. Rorer, Inc. for their help and cooperation. Also, special thanks are due to Mrs. Sandy Landis for her valuable secretarial help in preparing this monograph.

NORETHINDRONE

Arvin P. Shroff and Ernest S. Moyer

NORETHINDRONE

1. Description

1.1 Name, Formula, Molecular Weight

Norethindrone is 17α-ethinyl-17β-hydroxy-4-estren-3-one. It is also known by the following chemical names.

a. 17-Hydroxy-19-nor-17α-pregn-4-en-20-yn-3-one
b. 19-Nor-17α-ethynyltestosterone
c. 17α-Ethinyl-19-nortestosterone
d. 19-Nor-17α-ethynyl-17β-hydroxy-4-androsten-3-one
e. 19-Nor-17α-ethynylandrosten-17β-ol-3-one
f. 17α-Ethynyl-19-nortestosterone
g. Anhydrohydroxynorprogesterone
h. 19-Norethisterone
i. Norpregneninolone

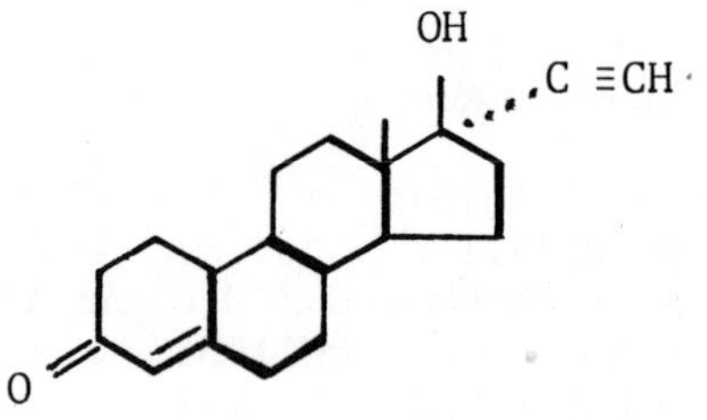

OH
C ≡CH
O

$C_{20}H_{26}O_2$ Molecular Wt. 298.41

1.2 Appearance, Color, Odor

White to creamy white, odorless, nonhygroscopic, crystalline powder.

2. Physical Properties

2.1 Infrared Spectrum

The infrared absorption spectrum of norethindrone is presented in Figure 1. The spectrum was taken in a KBr pellet with a Beckman IR-8 Spectrophotometer. Some of the absorption assignments are given in Table I.[1]

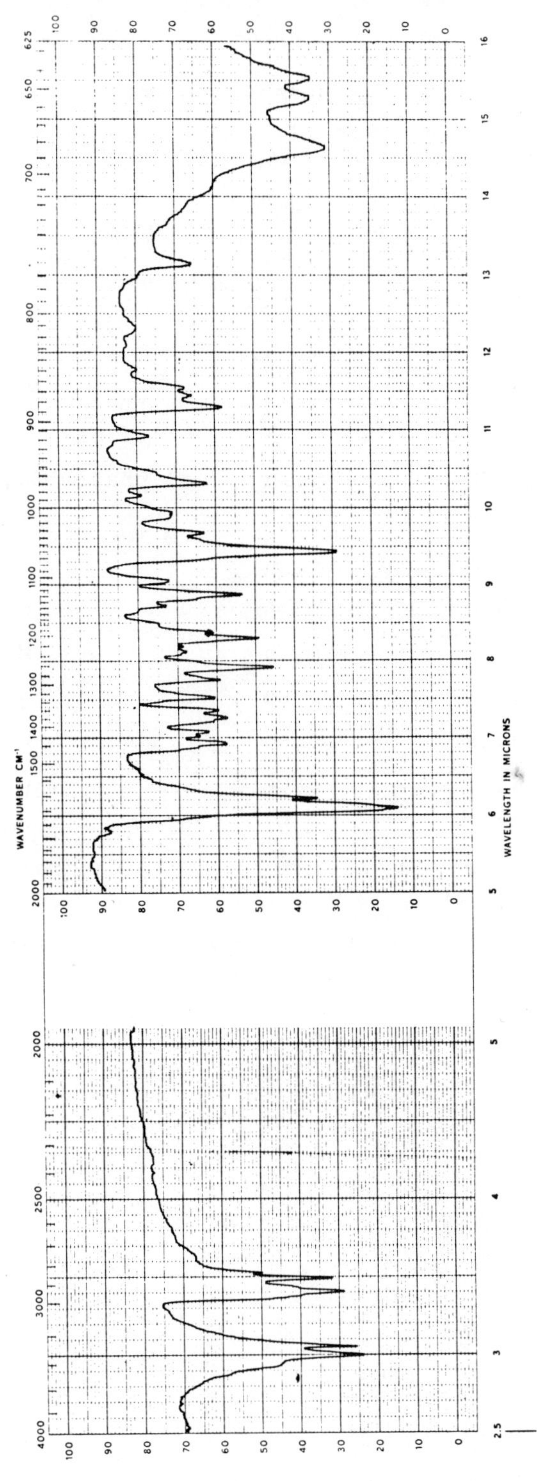

Figure 1. Infrared spectrum of Norethindrone

TABLE I

INFRARED ASSIGNMENTS FOR NORETHINDRONE

λ	$\bar{\nu}$	Intensity*	Vibrational Assignments
2.95	3400	v,sh,b	OH stretching
2.98	3300	s,sh	C ≡CH stretching
3.05	3220	m,sh	Olefinic C-H stretching
3.4(3.5)	2970(2840)	m,sh	Aliphatic C-H stretching
6.1	1650	s,sh	C=O stretching
6.3	1590	m,sh	C=C stretching
6.7(7.5)	1490(1340)	m,sh	C-H deformation
6.92	1445	m,sh	$C-H_2$ bending
7.1	1390	m,sh	CH_3 deformation
8.7	1135	m,sh	OH bending
9.42	1060	s,sh	C-O deformation
11.29	885	m,b	OH wagging
14.65	680	m,b	Olefinic C-H wagging

*s = strong, m = medium, v = variable, b = broad, sh = shoulder

Mesley[2] assigned the following bands (cm^{-1}) to norethindrone.

a. Absorption associated with 17β-hydroxyl group: 1065, 1132.

b. Characteristic absorption due to 4-ene-3-keto function: 1416, 1337, 1269, 1211, 1199, 1018, 960, 890, 764, 685.

These assignments as well as those made by Djerassi[3] and others[4] essentially agree with absorption peaks and shoulders presented in the spectrum, Figure I.

2.2 Nuclear Magnetic Resonance Spectrum

Cross, Landis, and Murphy[5] reported the chemical shift of the C_{18} angular methyl group of norethindrone in various solvents. They observed the chemical shifts at 54.5 cps in d-$CHCl_3$, 54.5 cps in $CHCl_3$, 47.5 cps in d_6-DMSO and 53.5 cps in d_7-DMF.

A reference NMR spectrum of norethindrone recorded on a Varian A-60 Spectrometer is presented in Figure II. An 8% solution in deuterated chloroform containing tetramethylsilane as an internal reference was used. Characteristic chemical shifts are observed at 5.87 ppm (C_4 Vinyl); 2.37 ppm (C-17 OH); 2.57 ppm (C-17 C ≡ CH); and 0.90 ppm (C-18 CH_3).

2.3 Mass Spectrum

A low resolution mass spectrum of norethindrone on a magnetic instrument (Atlas CH-4) was reported by De Jongh and co-workers.[6] Major fragments were observed at m/e 110, 215, and 231 representing the following fragmentation pattern.

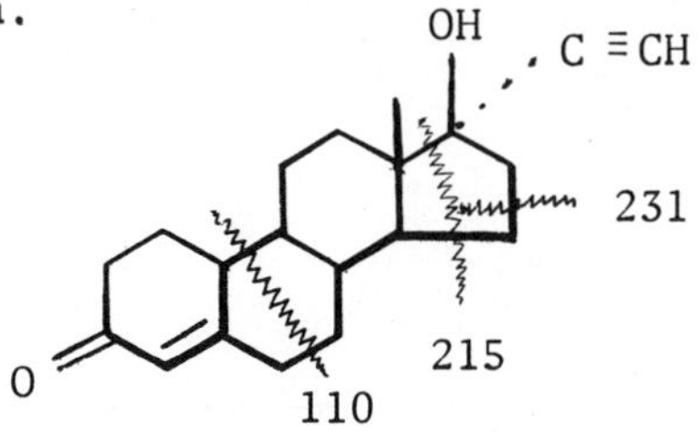

OH
C ≡ CH
231
215
O
110

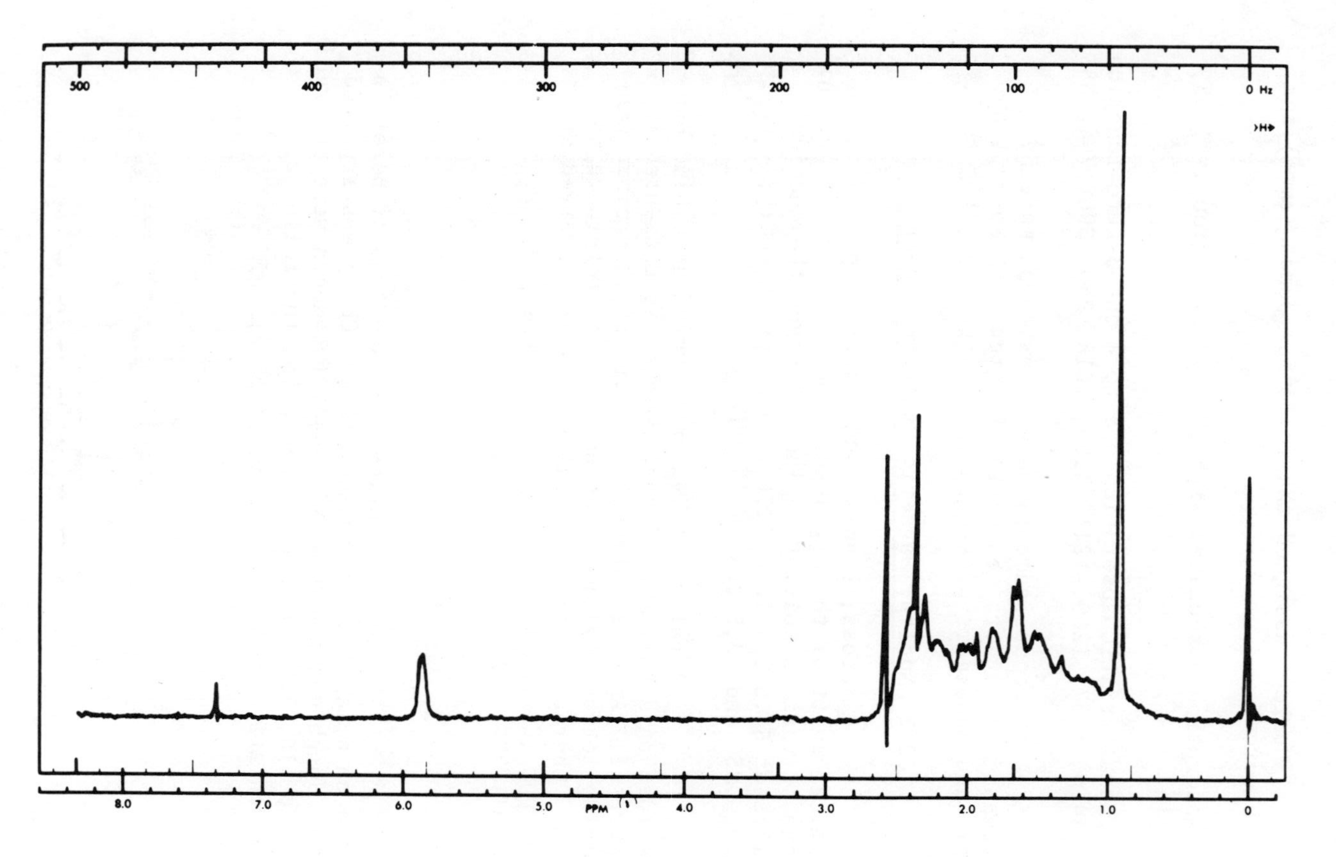

Figure II. NMR Spectrum of Norethindrone

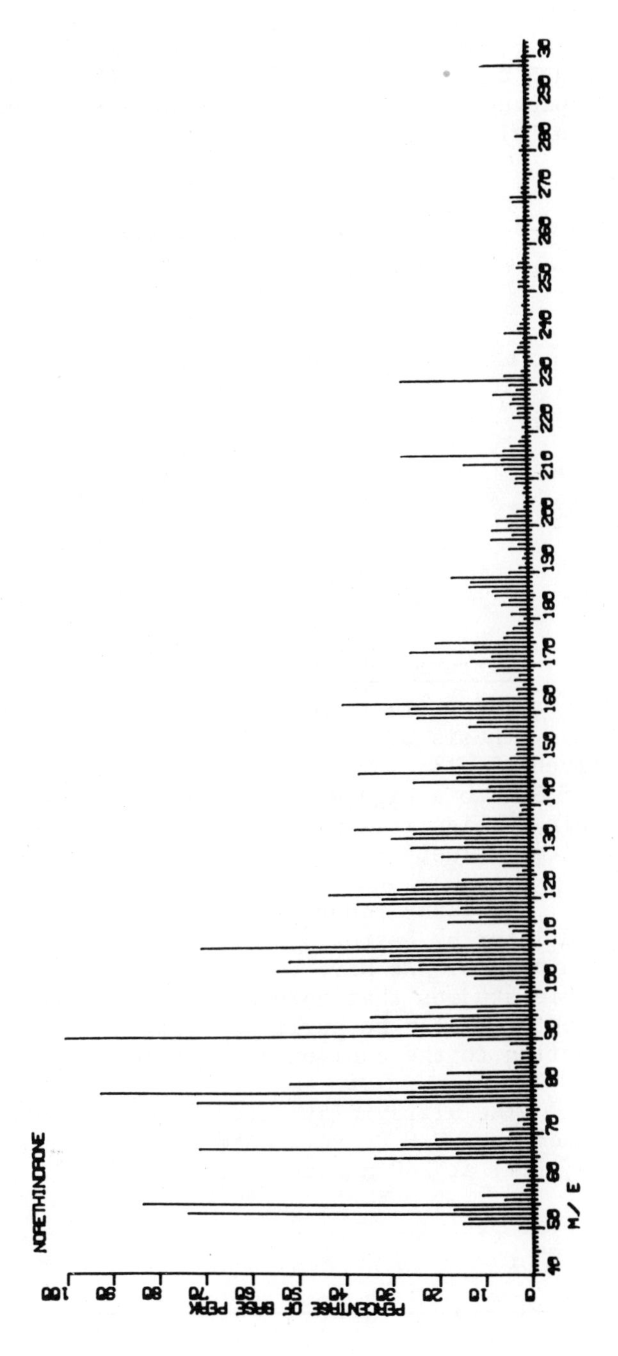

Figure III. Mass Spectrum of Norethindrone

In Figure III, the mass spectrum of norethindrone obtained on a quadrupole instrument (Finnigan Model 1015) is depicted. The relative intensities of the major ions are as follows:

m/e	%I
55	76.63
67	74.94
79	95.36
91	100.00
105	59.15
110	75.78
215	44.77
231	48.35
298	44.63

2.4 Ultraviolet Absorption Spectrum

The E (1%, 1cm) value at 240 nm reported with the first synthesis of norethindrone[3] was later corrected.[7] Different E (1%, 1cm) values have been suggested from a low of 525 to a high of 590. Many of these preparations contain foreign related steroids which affect the E (1%, 1cm) value.[8]

Nielsen[9] reported a value of 560 which was confirmed later by Bastow[10] who isomerized norethynodrel to norethindrone. We have determined[11] by thin-layer chromatographic studies that norethindrone, no matter how it is manufactured, can be purified chemically by first converting to the enol ether[12] followed by hydrolysis.

The ultraviolet curve shown in Figure IV was obtained on a U.S.P. reference standard material. Its E (1%, 1cm) at 240 nm was 575.

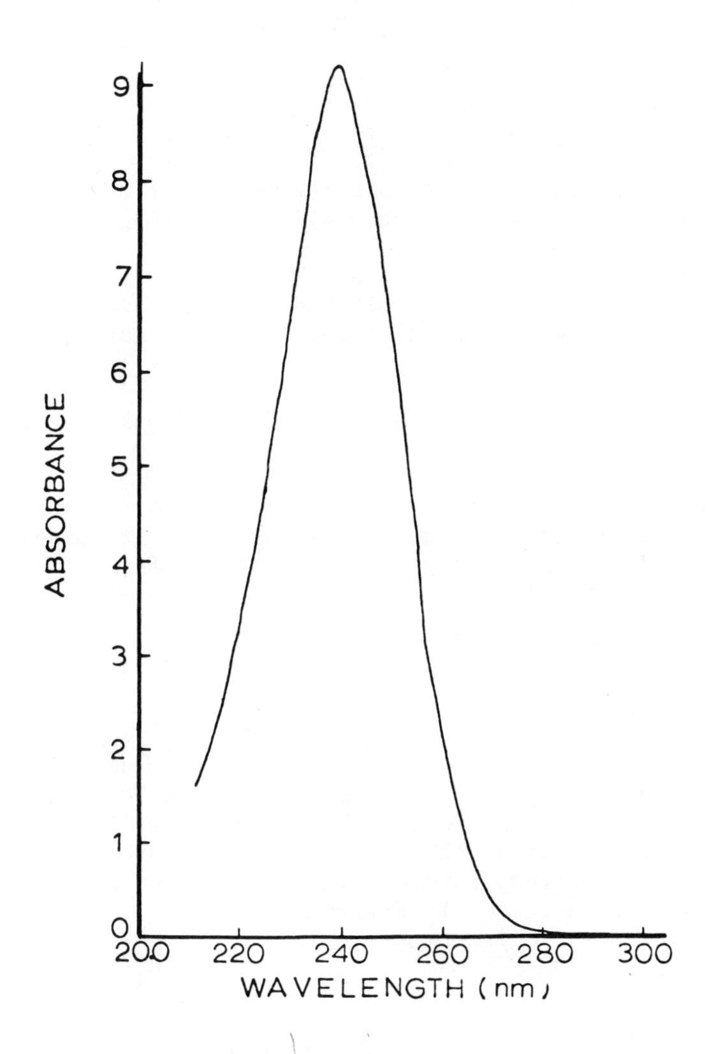

Figure IV Ultraviolet Spectrum of Norethindrone

2.5 Melting Range

The melting point also varied according to different investigators. For example, Djerassi *et al*[3] reported a melting point of 203-204°C, whereas Nielsen[9] found a range of 199-208°C with decomposition. Kofler *et al*[13] on the other hand, suggested a melting point range of 202-207°C. In our experience, most of the batches had a melting range of 3°C and all melted between 202-208°C (Class IA U.S.P.).[14]

2.6 Differential Scanning Calorimetry

Figure V shows the DSC thermogram of norethindrone. A Perkin-Elmer model DSC-1B employing a heating rate of 10°C/minute was used.

The difference between the "extrapolated onset" and the "peak" for reference standard U.S.P. norethindrone was 2.5°C.

The "extrapolated onset" and the "peak" are defined as follows:

a. "Extrapolated Onset": The temperature corresponding to the intersection of the extrapolation of the baseline and the longest, sharpest line section of the low temperature side of the peak.

b. "Peak": The temperature of reversal.

2.7 Solubility

Nielsen[9] reports the solubilities of norethindrone to be: almost insoluble in water, 1 part in 80 parts ethanol, 1 part in 400 parts ether, and 1 part in 15 parts chloroform. A plot of the solubilities of norethindrone as a function of temperature for eleven common organic solvents is illustrated in Figure VI.[8]

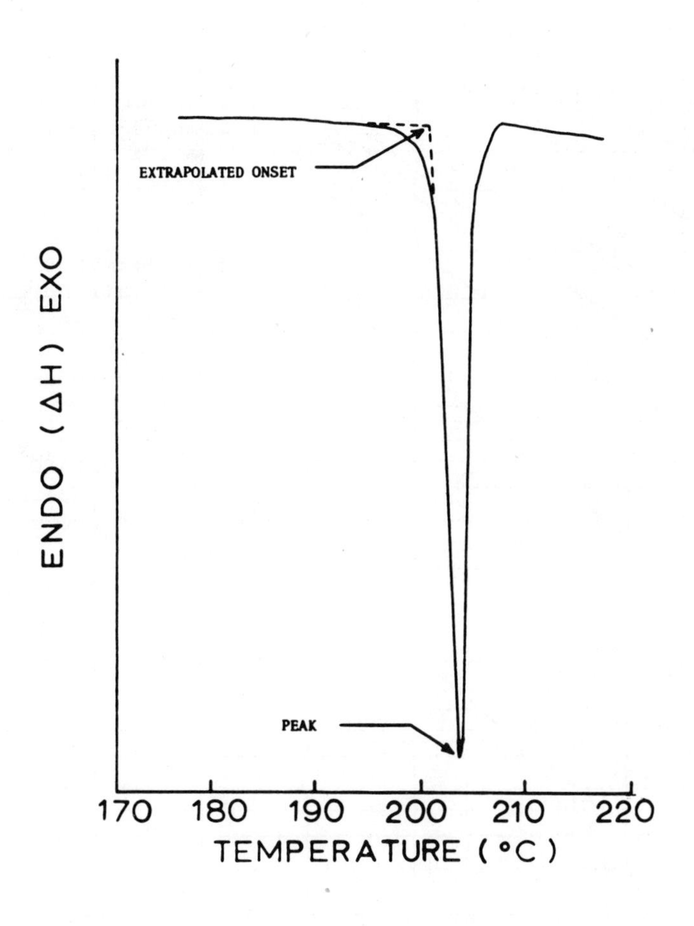

Figure V DSC Thermogram of Norethindrone

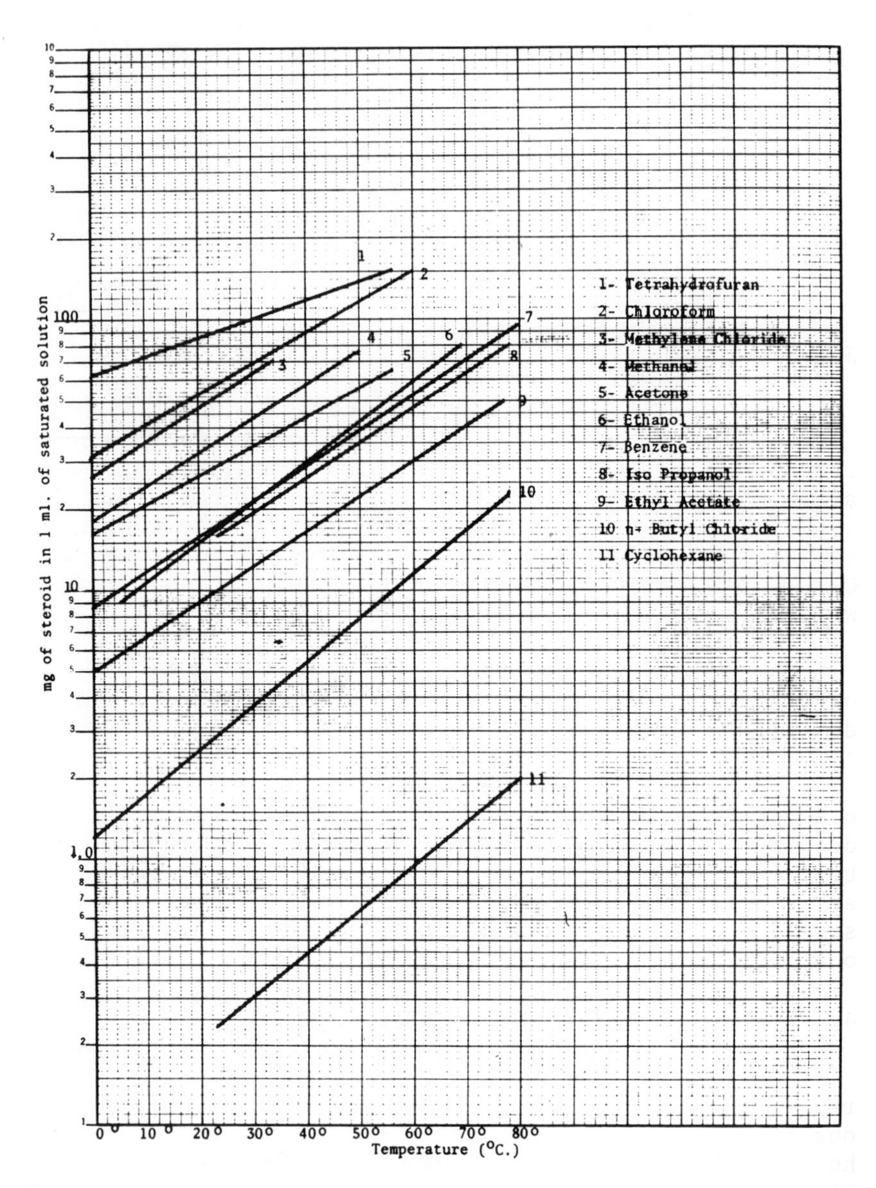

Figure VI Solubility of Norethindrone in Organic Solvents

2.8 Optical Rotation

The following rotations have been reported:

$[\alpha]_D$	-25°	(Chloroform)[3]
$[\alpha]_D$	-24°	(Chloroform)[4]
$[\alpha]_D$	-31.7°	(Chloroform)[15]
$[\alpha]_E$	-33°	(Chloroform)[4]

3. Synthesis

Djerassi, Miramontes, Rosenkranz and Sondheimer[3] first reported the synthesis (Scheme I) of norethindrone from 19-nortestosterone I. Chromium trioxide oxidation afforded estr-4-ene-3,17-dione II. Reaction of the dione with ethyl orthoformate under carefully controlled conditions resulted in selective enol ether formation at C-3. Ethynylation followed by cleavage of the enol ether with aqueous mineral acid gave norethindrone.

Alternate methods using estradiol 3-methyl ether or estrone tetrahydropyranyl ether have also been described.[16,17]

In a recent patent[18] the preparation of 17α-ethynyl-4-androstene-17β,19β-diol-3-one was reported in detail. It was mentioned that the 19-hydroxy compound can be converted to the 19-nor derivative with a strong base as reported by Barber et al[19] and Meyer.[20] The general scheme of preparing norethindrone by this method is outlined in Scheme II.

4. Metabolism

Tissue distribution of tritiated norethindrone in rats was studied by Watanabe and co-workers.[21] They observed no selective uptake by any tissue up to four hours after administration. Matsuyoshi[22] incubated norethindrone with rat liver and kidney homogenate and obtained dihydro and tetrahydro derivatives of norethindrone.

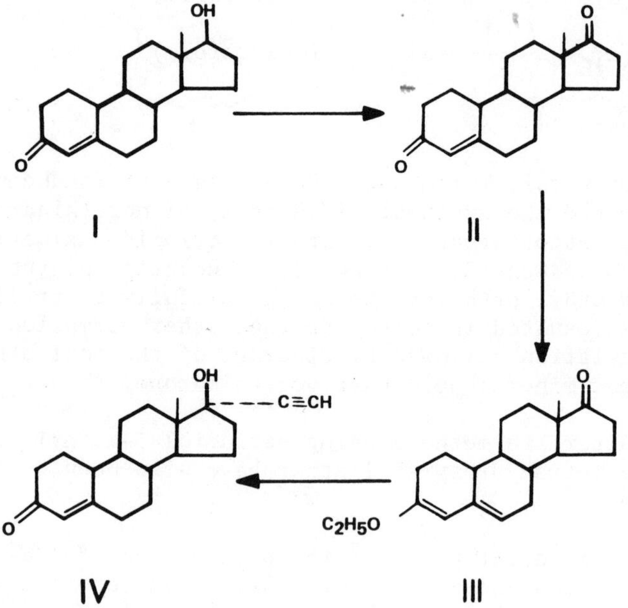

OH O

O O

I II

OH O

---C≡CH

O C_2H_5O

IV III

Scheme 1 Synthetic Pathway to Norethindrone

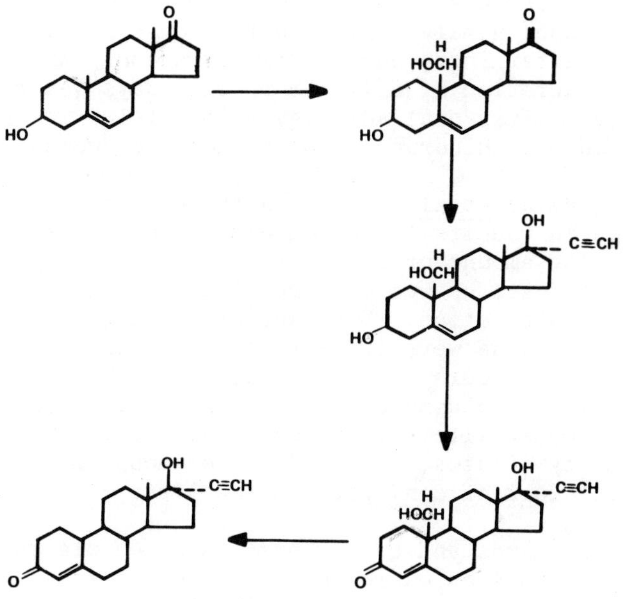

Scheme II Alternate Synthetic Pathway to Norethindrone

The metabolic fate of norethindrone in humans has been studied by several investigators.[23-39]

Kamyab and co-workers[31] administered ^{14}C-norethindrone intravenously to seven women and found the radioactivity excreted mainly in the urine. In the first twenty-four hours 32.1%, and in five days 53.9% of the activity given appeared in the urine. Layne and co-workers[34] found a similar excretion pattern with tritiated norethindrone; namely, 50.4% in five days after oral dose and 70.2% after intravenous administration. A radioactivity excretion of only 33% in five days was reported by Murata[35] after oral administration of 100 mg of ^{14}C or ^{3}H labeled norethindrone to postmenopausal women.

Kamyab _et al_[30] observed the urinary radioactivity excretion in rabbits to be comparable to that found in humans. The amount of radioactivity in the conjugated and unconjugated states was measured[31] in plasma after intravenous administration. Substantial fractions of the administered dose were present in the circulation for up to forty-eight hours after injection. The unconjugated activity in the plasma decreased rapidly to less than 0.4% of the dose per liter of plasma after twelve hours whereas after thirty minutes, a considerable amount of activity was in the form of conjugated metabolites.

Gerhards and co-workers[27] studied plasma radioactivity as well as excretion in urine and feces after p.o. administration of 20 mg of ^{14}C-norethindrone of 0.25 mg ^{3}H-norethindrone to males. The radioactive substances in plasma were characterized as: 17α-ethynyl-5β-estrane-3α,17β-diol; 17α-ethynyl-5β-estrane-3β,17β-diol and 17α-ethynyl-5α-estrane-3α,17β-diol.

Earlier, a number of investigators[23,25,29,31,32,36] had reported excretion of 17α-ethynyl-17β-estradiol after oral administration of norethindrone. These observations, however, were not supported by _in vitro_ investigations. As an explanation of this discrepancy, Breuer[24] recently suggested that the increased estrogen content in _in vivo_ experiments probably is due to artifacts produced during work-up of the urine. Recent work of Stillwell and co-workers[39] substantiated this observation. They found no ethynyl estradiol in urine after administration of norethindrone to a female.

All norethindrone metabolites characterized from humans and rabbits possess the 17α-ethynyl group. One exception to this was the *in vitro* experiments of Palmer.[37] He identified 4-estrene-3,17-dione as a metabolite from norethindrone. The overall metabolic pathway of norethindrone is illustrated in Scheme III.

5. Methods of Analysis

5.1 Elemental Analyses

The elemental analyses obtained on USP reference standard norethindrone and that reported by Djerassi and co-workers[3] are as follows:

Element	% Theory	Ref. Std.	Reported[3]
C	80.49	80.69	80.83
H	8.79	8.64	8.80

5.2 Phase Solubility

The phase solubility analysis of norethindrone was carried out using ethyl acetate as the solvent. Calculations as per USP XVIII were performed on a time-sharing computer using an in-house program. The results are illustrated in Figure VII.[40]

5.3 Thin-Layer Chromatographic Analysis

A TLC procedure for separating norethindrone from possible impurities is described in the USP (XVIII).[41] One hundred micrograms of the norethindrone sample and one microgram each of four possible impurities (19-norandrostendione, norethynodrel, 10β-hydroxy-norethynyltestosterone and 10β-hydroxy-norandrostendione) are spotted in individual lanes on a silica gel plate. The TLC plate is developed to a height of 17.5 cm by ascending chromatography employing a solvent system of 95:5 chloroform-methanol. After air drying the plate, it is sprayed with a mixture of 1:3 H_2SO_4 and dehydrated alcohol. The plate is heated in an oven at 105°C for five minutes and the intensities of the various foreign related

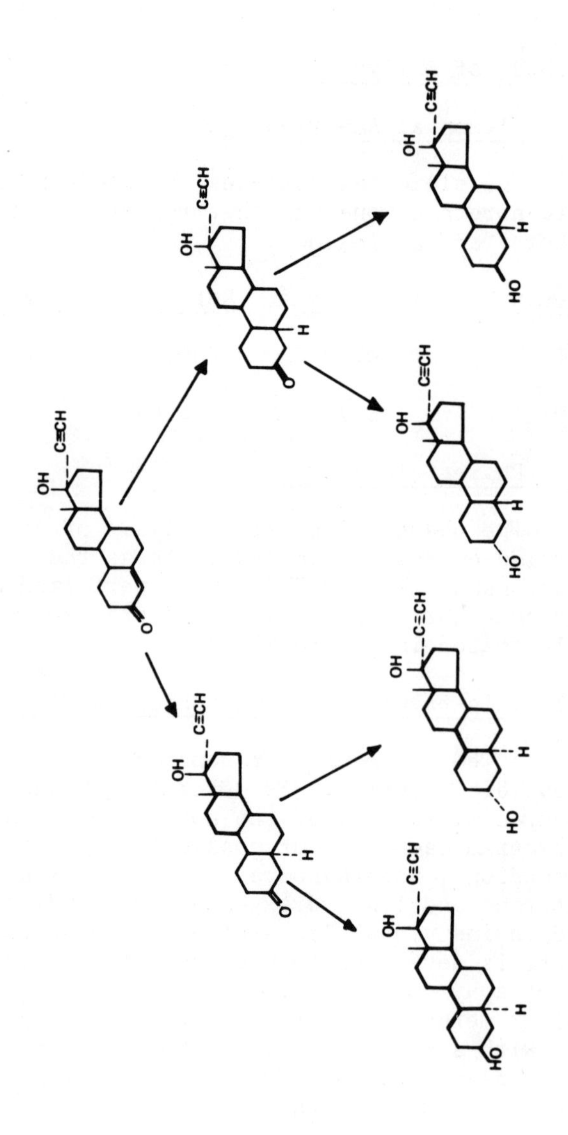

Scheme III. Norethindrone and its major metabolites

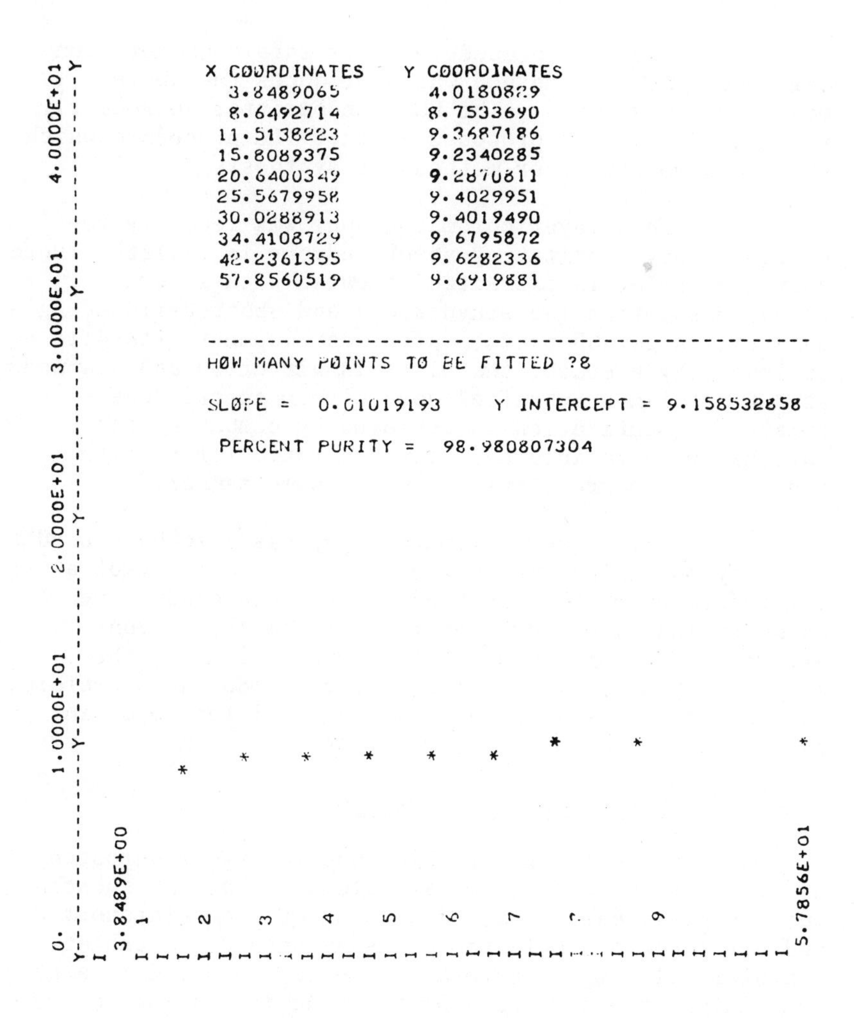

Figure VII Phase Solubility of Norethindrone

steroid spots are compared visually under long-wavelength ultraviolet light with the intensities produced by the corresponding steroids in the norethindrone sample. No more than two impurities totaling a maximum of two percent are allowed.

This TLC procedure takes into consideration only one synthetic method which produces the above mentioned impurities. The USP Subcommittee #6 recently recognized this and proposed another TLC procedure which will accommodate other methods of synthesis.[42]

Thin-layer chromatography was recently reported[43] for quantitating norethindrone in tablets. Here a single tablet is extracted, treated with internal standard solution (norethynodrel) and spotted (10-20 μg) on a silica gel GF plate. After development with 4:1 benzene-ethylacetate, the plate is air dried and scanned at 250 nm with a Schoeffel double-beam spectrodensitometer. Quantitation is achieved by comparing the norethindrone to internal standard ratio (area) with that of a standard obtained in the same manner.

Thin-layer chromatography has also been used to identify norethindrone in tablets.[44] An aliquot of a chloroform extract is spotted and the plate developed using cyclohexane-ethylacetate (1:1) as the solvent system. After drying at 100°C for ten minutes, the plate is sprayed with antimony trichloride in chloroform. Norethindrone has an R_f of 0.47 and a violet color in daylight and red color under ultraviolet light.

5.4 Vapor Phase Chromatography

Oxidation of norethindrone to 19-norandrostendione was observed[45] on metal columns. Lodge[46] substituted a glass column packed with 5% QF-1 on Diatoport S and developed a single tablet assay method. A tablet containing 1-5 mg of norethindrone is dissolved in acetone containing pregnenolone acetate as an internal standard and an aliquot is injected directly into the glass column operated isothermally at 200°C.

5.5 Spectrophotometric Analysis

Keay[44] reported an ultraviolet method for the quantitative analysis of norethindrone in tablets. Here, powdered tablets equivalent to 6 mg of norethindrone were extracted with 100 ml of chloroform and a 20 ml aliquot evaporated to dryness. The residue is redissolved in 100 ml methanol, its absorbance measured at 240 nm and the norethindrone content calculated. The accuracy of the method has been claimed to be better than 5%. A modification of the above procedure has been included in the NF XIII[47] for determining content uniformity of norethindrone tablets.

5.6 Colorimetric Analysis

A number of colorimetric methods have been used to detect and determine the quantity of norethindrone.

A. Blue Tetrazolium

The reaction of blue tetrazolium with various steroids has been studied thoroughly by Meyer and Lindberg.[48] They noted that Δ^4-3-ketosteroids (devoid of an α-ketol function) react with blue tetrazolium and reduce it to form a reddish-purple colored formazan. This reactivity is considerably enhanced in the case of 19-norsteroids and may result from a complex oxidative mechanism resulting in aromatization of the A-ring of these steroids. Smith and co-workers[49] extended this concept for the analysis of norethindrone in tablets. They extracted the progestin with a chlorinated solvent mixture and developed the color by adding blue tetrazolium and tetramethylammonium hydroxide. The reaction is allowed to proceed in absence of light for sixty minutes at room temperature. Quantitation is achieved by measuring the color intensities of the standard and sample preparations at 530 nm.

B. Isonicotinic Acid Hydrazide

Reaction of Δ^4-3-ketosteroids with isonicotinic acid hydrazide (isoniazid) produces a yellow hydrazone which can be used for quantitation.[50,51,52] Barth[14] used this reaction in automating the colorimetric quantitation of norethindrone in tablets, and Wu[53,54] used the manual approach for the AOAC collaborative study on norethindrone. Recently,[55] the Food and Drug Administration Laboratory published an automatic method which is capable of performing single tablet assays.

5.7 Titrimetric Analysis

The British Pharmacopeia[56] recommends a potentiometric titration method for the assay of norethindrone. Here a tetrahydrofuran solution of norethindrone, which also contains silver nitrate, is titrated with sodium hydroxide solution and the end point is determined potentiometrically.

Rizk and co-workers[57] modified the method so that 30 mg of norethindrone can be easily titrated as compared to 200 mg required for the Britith Pharmacopeia method.

6. References

1. C. J. Shaw, Ortho Research Foundation, personal communication.
2. R. J. Mesley, Spectrochimica Acta 22, 889 (1966).
3. C. Djerassi, L. Miramontes, G. Rosenkranz, and F. Sondheimer, J. Am. Chem. Soc. 76, 4092 (1954).
4. W. Neudert and H. Röpke, "Atlas of Steroid Spectra", Springer-Verlag, Inc., New York, 1965 Spectrum #770.
5. A. D. Cross, P. W. Landis and J. W. Murphy, Steroids 5, 655 (1965).
6. D. C. DeJongh, J. D. Hribar, P. Littleton, K. Fotherby, R. W. A. Rees, S. Shrader, T. J. Foell, and H. Smith, Steroids 11, 649 (1968).
7. A. Zaffaroni, personal communication, 1961.
8. A. D. Mebane, Ortho Research Foundation, personal communication.
9. L. S. Nielsen, Arch. Pharm. Chemi. 74, 78 (1967).
10. R. A. Bastow, J. Pharm. Pharmac. 19, 41 (1967).
11. A. P. Shroff and G. Karmas, unpublished data.
12. H. J. Ringold, C. Djerassi, and A. Bowers, U.S. Patent 3,138,589.
13. M. Kuhnert-Brandstatter, E. Junger and A. Kofler, Microchem. J. 9, 105 (1965).
14. H. Barth, Ortho Research Foundation, personal communication.
15. The Merck Index 8th Edition, Merck and Co., Inc., Rahway, N.J. 1968, p.748
16. F. B. Colton, U. S. Patents 2,655,518 (1952), 2,691,028 (1953), and 2,725,378 (1955).
17. C. Gandolfi and P. deRuggieri, Gazz. Chim. Ital. 94, 675 (1964).
18. A. Bowers, Canadian Patent 746,499.
19. G. W. Barber and M. Ehrenstein, J. Org. Chem. 20, 1253 (1955).
20. A. S. Meyer, Experientia 11, 99 (1955).
21. H. Watanabe, N. N. Saha and D. S. Layne, Steroids 11, 97 (1968).
22. K. Matsuyoshi, Folia Endocr. Jap. 43, 91 (1967).
23. H. Breuer, U. Dardenne, and W. Nocke, Acta. Endocr. 33, 10 (1960.
24. H. Breuer, Lancet II, 615 (1970).

25. J. B. Brown and H. A. F. Blair, Proc. Roy. Soc. Med. 53, 433 (1960).
26. K. Fotherby, S. Kamyab, P. Littleton, and K. J. Dennis, Acta Endocr. Suppl. 119, 136 (1966).
27. E. Gerhards, W. Hecker, H. Hitze, B. Nieuweboer, and O. Bellmann, Acta Endocr. 68, 219 (1971).
28. S. Ishihara, Folia Endocr. Jap. 42, 55 (1966).
29. R. Kaiser and H. Stecher, Arch. Gynäk. 194, 146 (1960).
30. S. Kamyab, P. Littleton, and K. Fotherby, J. Endocr. 39, 423 (1967).
31. S. Kamyab, K. Fotherby, and A. Klopper, J. Endocr. 41, 263 (1968).
32. H. Langecker, Acta Endocr. 37, 14 (1961).
33. C. Lauritzen and W. D. Lehmann, Arch. Gynäk. 204, 212 (1967).
34. D. S. Layne, T. Golab, K. Arai, and G. Pincus, Biochem. Pharmac. 12, 905 (1963).
35. S. Murata, Folia Endocr. Jap. 43, 1083 (1967).
36. H. Okada, M. Amatsu, S. Ishihara, and G. Tokuda, Acta Endocr. 46, 31 (1964).
37. K. H. Palmer, J. F. Feierabend, B. Baggett, and M. E. Wall, J. Pharmac. Exp. Ther. 167, 217 (1969).
38. C. A. Paulsen, Metabolism 14, 313 (1965).
39. W. G. Stillwell, E. C. Horning, M. G. Horning, R. N. Stillwell, and Z. Zlatkis, J. Ster. Biochem. 3, 699 (1972).
40. R. E. Huettemann, Ortho Research Foundation, personal communication.
41. U. S. Pharmacopeia, 18th Rev., Mack Publishing Co., Easton, Pa. 1970, p.453.
42. K. Florey, U.S.P. Sub-Committee #6, personal communication.
43. A. P. Shroff and C. J. Shaw, J. Chromatog. Sci. 10, 509 (1972).
44. G. R. Keay, Analyst 93, 28 (1968).
45. Facts and Methods, F & M Scientific Research 5, (3) 1964.
46. B. A. Lodge, Can. J. Pharm. Sci. 5, 74 (1970).
47. National Formulary, 13th Edition, Mack Publishing Co., Easton, Pa., 1970, p.488.
48. A. S. Meyer and M. C. Lindberg, Anal. Chem. 27, 813 (1955).

49. R. V. Smith, T. H. Hassall, and S. C. Liu, J. Ass. Off. Anal. Chem. 53, 1089 (1970).
50. L. L. Smith and T. Foell, Anal. Chem. 31, 102 (1959).
51. E. J. Umberger, Anal. Chem. 27, 768 (1955).
52. A. Ercoli, L. deGiuseppe and P. deRuggiere, Farm. Sci. Tec. (Pavia) 7, 170 (1952).
53. J. Wu, J. Ass. Off. Anal. Chem. 53, 831 (1970).
54. J. Wu, J. Ass. Off. Anal. Chem. 54, 617 (1971).
55. L. K. Thornton, Public Health Service (FDA) Drug Auto Analysis Manual, Method #22, April 1972.
56. British Pharmacopoeia, 1973, University Printing House, Cambridge 1973, p.323.
57. M. Rizk, J. J. Vallon, and A. Badinand, Anal. Chim. Acta 65, 220 (1973).

NORGESTREL

Andrew M. Sopirak and Leo F. Cullen

CONTENTS

1. Description

1.1 Name, Formula, Molecular Weight

Norgestrel is designated as dl-13-ethyl-17α-ethynyl-17-hydroxygon-4-en-3-one and is listed in the subject indices of Chemical Abstracts under the heading: dl-13-ethyl-17-hydroxy-18,19-dinor-17α-pregn-4-en-20-yn-3-one.

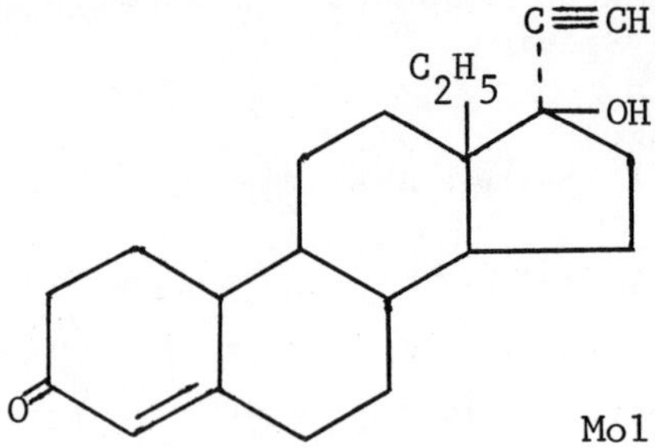

$C_{21}H_{28}O_2$ Mol. Wt.: 312.46

1.2 Appearance, Color, Odor

Norgestrel is a white to off-white, practically odorless, crystalline powder.

2. Physical Properties

2.1 Infrared Spectra

An infrared absorption spectrum of a potassium bromide dispersion of norgestrel (Wyeth Reference Standard material, Lot #C-10484) is presented in Figure 1. This spectrum agrees with a published spectrum[1]. The spectral band assignments are listed in Table I.

Table I

Infrared Spectral Assignments of Norgestrel

Frequency ($cm.^{-1}$)	Vibration Mode	Reference
3350	OH stretching	2
3270	Acetylenic C-H stretching	1
2940 and 2865	Aliphatic C-H stretching	3
1655	Conjugated C=O stretching	1
1615	C=C stretching	3
1448	Methyl C-H asymmetrical bending	3
1420 and 1335	OH in plane bending	3
1360	Methyl C-H symmetrical bending	3
1065	Alcoholic C-O stretching	1
700 - 670	Acetylenic C-H bending	3

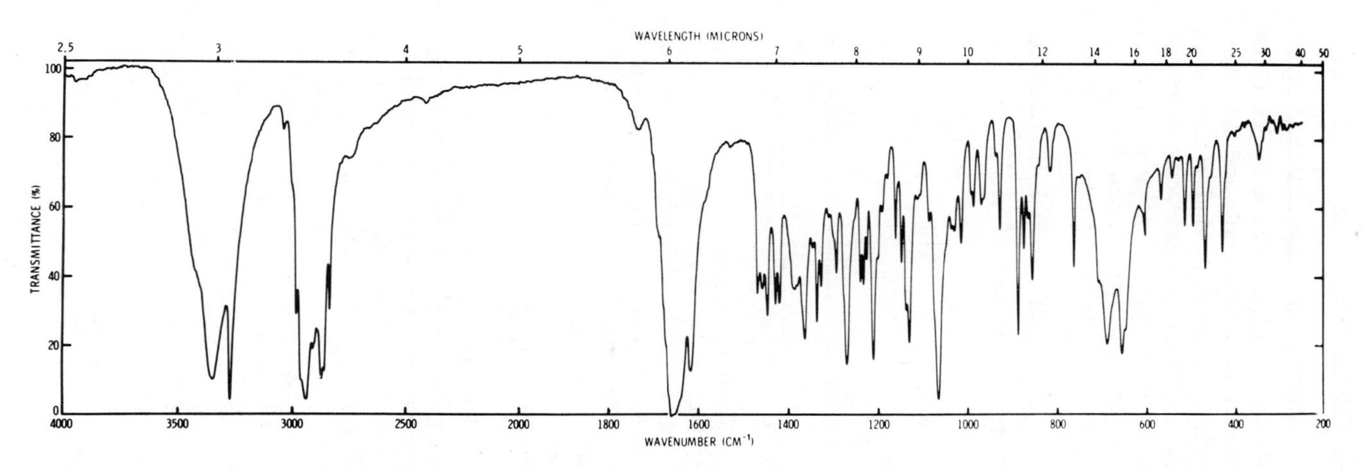

Figure 1 - I.R. Spectrum of Norgestrel (Wyeth Reference Standard Material, Lot #C-10484) - 1% KBr Pellet

2.2 Nuclear Magnetic Resonance

The 60 MH_z nuclear magnetic resonance (NMR) spectrum, shown in Figure 2, was obtained by dissolving norgestrel (Wyeth Reference Standard material, Lot #C-10484) in deuterochloroform containing tetramethylsilane as an internal reference. The only exchangeable proton is the hydrogen associated with oxygen at position 17. The NMR proton spectral assignments are listed in Table II.

Table II

NMR Spectral Assignments of Norgestrel

Chemical Shift (ppm.)	Proton	Splitting
5.86	-C(=O)-C<u>H</u>=C<	singlet
2.61	-C≡C<u>H</u>	singlet
2.31	O<u>H</u>	singlet
1.02	$-CH_2-\underline{CH_3}$	triplet

2.3 Ultraviolet Spectra

Fernandez and Noceda[4] reported a λmax. of 240 nm. for norgestrel in absolute methanol (a = 56.0). Norgestrel in 95% ethanol (Wyeth Reference Standard material (Lot #C-10484) when scanned between 360 nm. and 200 nm. exhibited a λmax. at 240 nm. (a = 54.6). This spectrum in 95% ethanol is shown in Figure 3.

2.4 Mass Spectra

The mass spectrum of norgestrel (Wyeth Reference Standard material, Lot #C-10484) was obtained by direct insertion of the sample into an MS-902 double focusing, high resolution mass spectrometer. The sample was run at 200°C. and 1.0×10^{-6} torr with the ionization electron beam energy at 70 eV. The high resolution data were compiled and tabulated with the aid of an on-line PDP-8 Digital Computer. Results are presented as a bar graph in Figure 4 and the high resolution mass spectrum assignments of prominent ions are given in Table III.[5] This spectrum is in agreement with that presented by DeJongh *et.al.*[6]

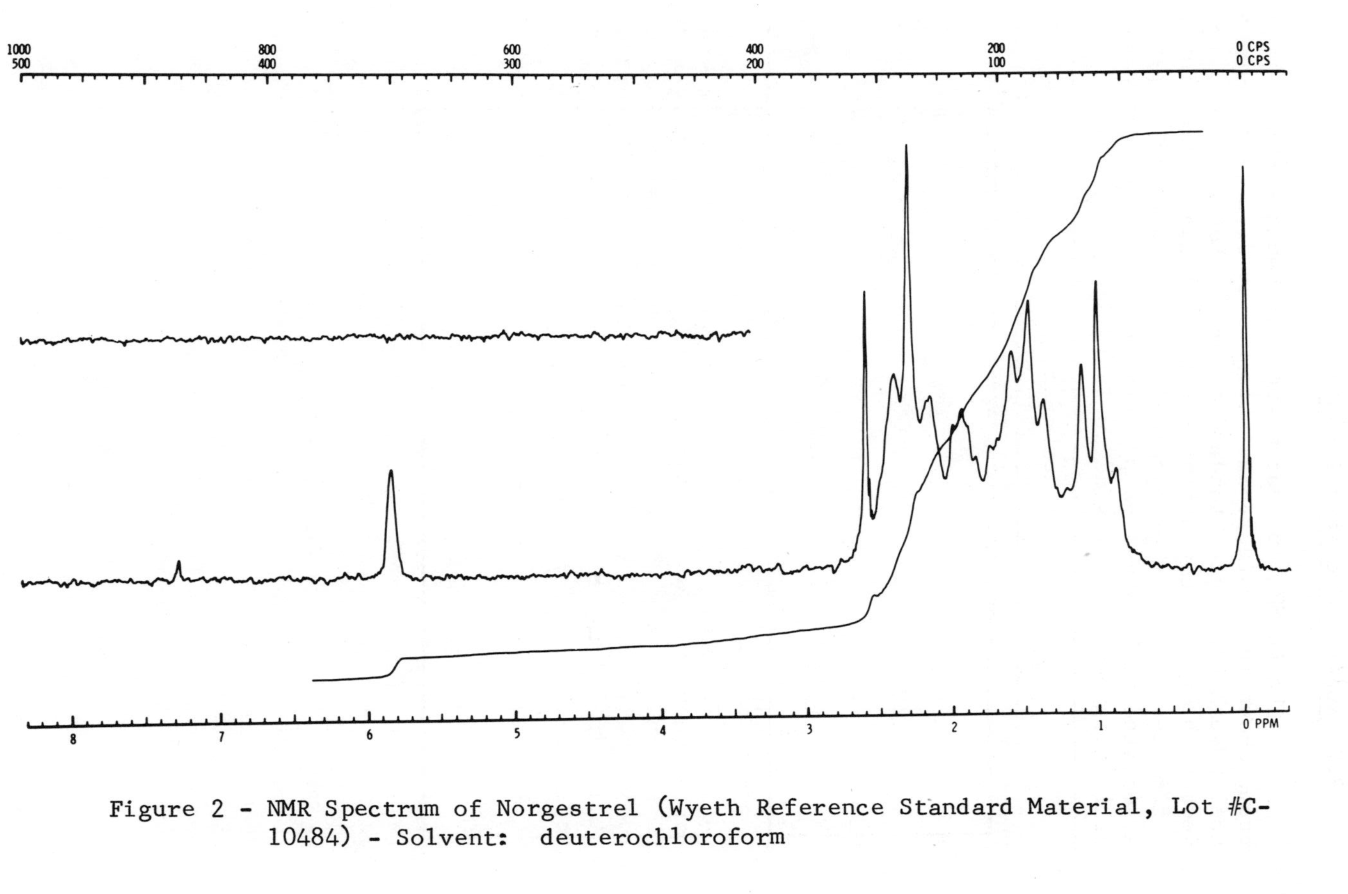

Figure 2 - NMR Spectrum of Norgestrel (Wyeth Reference Standard Material, Lot #C-10484) - Solvent: deuterochloroform

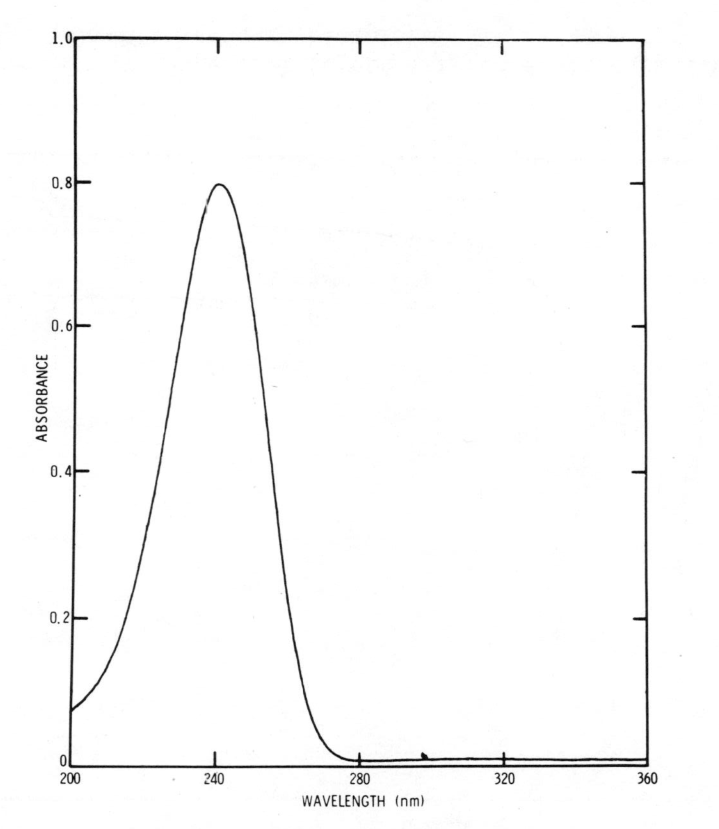

Figure 3 - Ultraviolet Spectrum of Norgestrel (Wyeth Reference Standard Material, Lot #C-10484) - Solvent: 95% ethanol

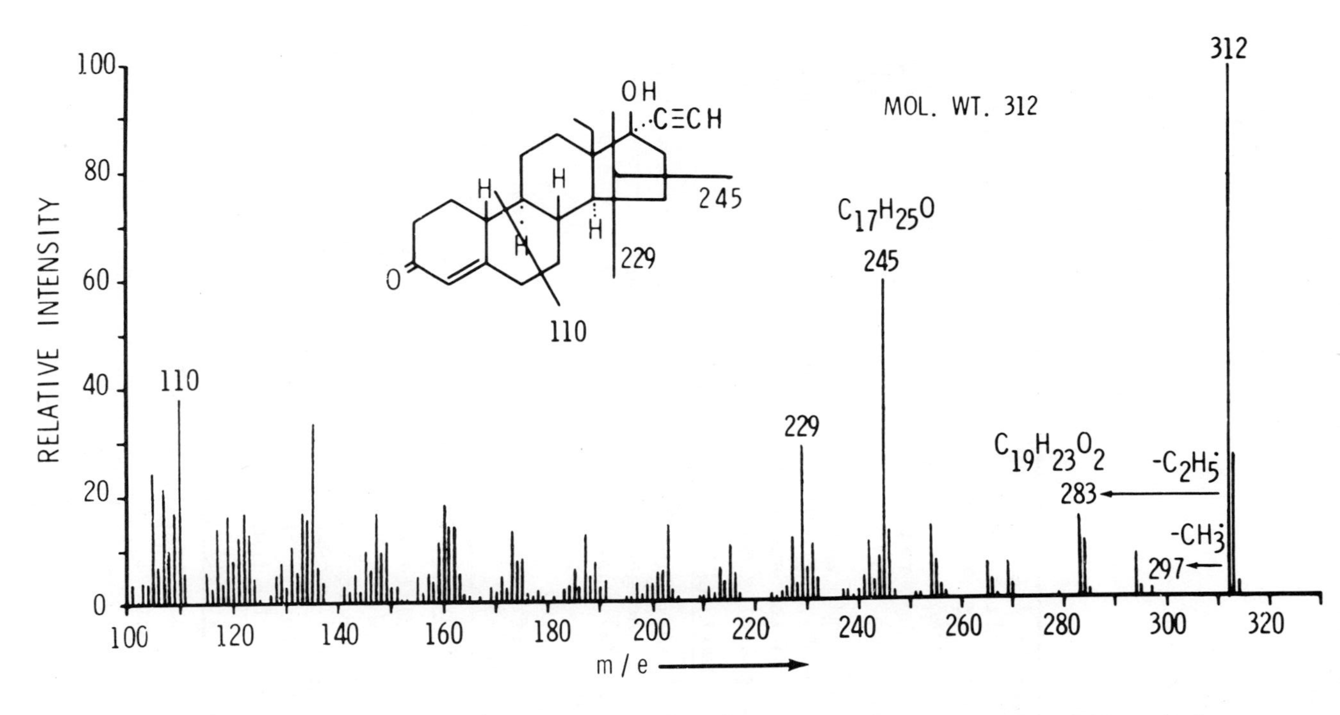

Figure 4 - Mass Spectrum of Norgestrel (Wyeth Reference Standard Material, Lot #C-10484)

Table III

High Resolution Mass Spectral Assignments of Norgestrel

Measured Mass	Calculated Mass	Formula
312.2082	312.2088	$C_{21}H_{28}O_2$
284.1812	284.1775	$C_{19}H_{24}O_2$
283.1698	283.1698	$C_{19}H_{23}O_2$
245.1902	245.1905	$C_{17}H_{25}O$
229.1596	229.1592	$C_{16}H_{22}O$
110.0730	110.0731	$C_7H_{10}O$

Norgestrel gives a molecular ion at m/e 312 as its base peak. The first prominent fragment occurs at m/e 297 which corresponds to the loss of the methyl radical, followed by the more intense peak at m/e 283 which represents cleavage of the 13-ethyl radical ($M^+ - C_2H_5\cdot$). A peak at m/e 284 is observed from the loss of C_2H_4.

Ions of m/e 245 and m/e 229 are explained by the cleavage of the D-ring with hydrogen transfer.[6] The intense fragment in the spectrum at m/e 110 is indicative of the remaining A-ring molecular fragment with the formula $C_7H_{10}O$.

2.5 Melting Range

The following melting temperature range has been observed on norgestrel (Wyeth Reference Standard material, Lot #C-10484) employing the U.S.P. XVIII, Class I conditions:[7] 207-210°C.[8] Short et.al. reports a melting range temperature of 206-207°C.[1] The melting temperature range does not change significantly with variations in heating rates from 1 to 5°C./min.

2.6 Differential Thermal Analysis

The differential thermal analysis (DTA) curve of norgestrel (Wyeth Reference Standard material, Lot #C-10484) obtained from room temperature to the melting point at a heating rate of 20°C./min. is presented in Figure 5. A sharp endothermic change observed at 209°C. corresponds to the melt of the drug.

2.7 Solubility

The following solubility data were obtained at room temperature:[9]

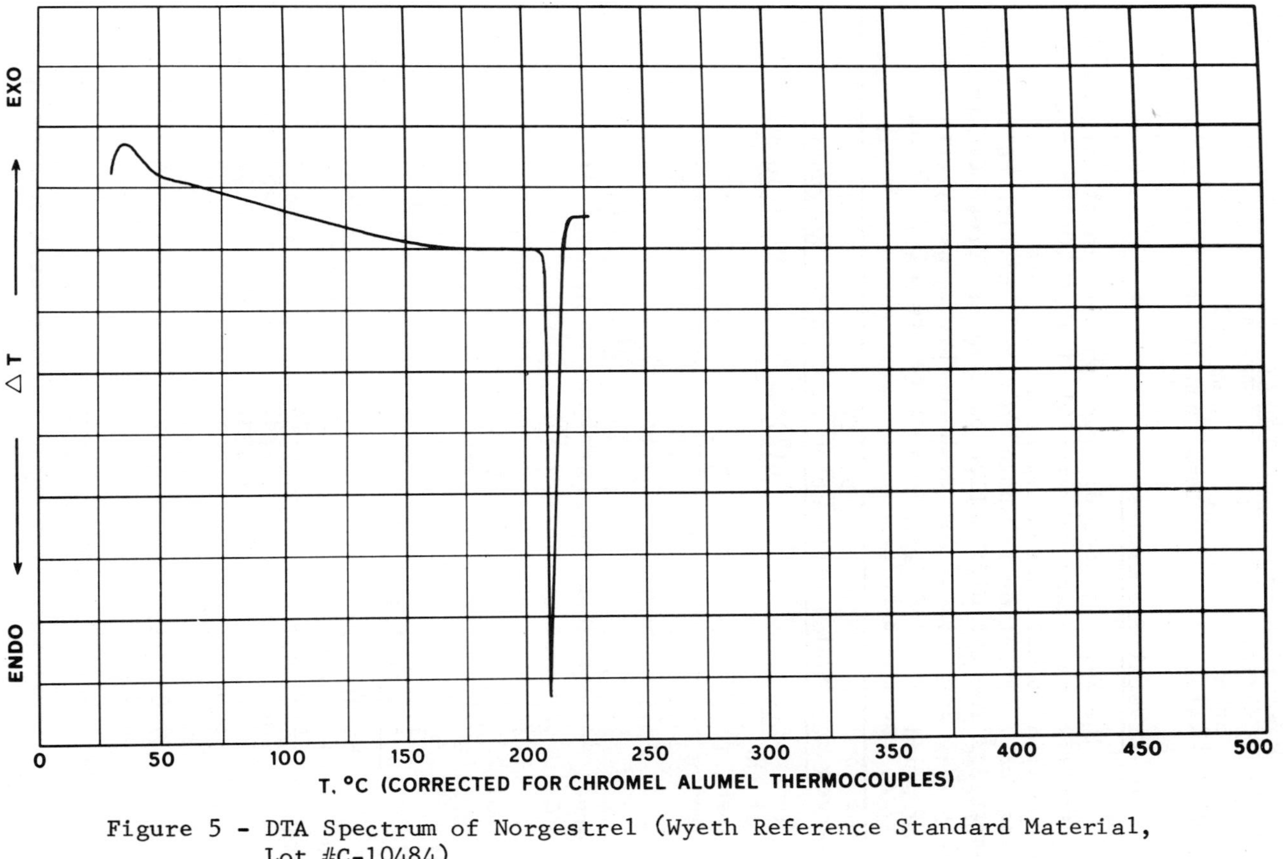

Figure 5 - DTA Spectrum of Norgestrel (Wyeth Reference Standard Material, Lot #C-10484)

chloroform	111 mg./ml.
acetone	15 mg./ml.
95% ethanol	13 mg./ml.
benzene	7 mg./ml.
ether	2 mg./ml.
water	less than .01 mg./ml.

2.8 Crystal Properties

The X-ray powder diffraction pattern of norgestrel (Wyeth Reference Standard material, Lot #C-10484) obtained with a Philips diffractometer at a scan rate of 1°/min. using CuK$_\alpha$ radiation is shown in Figure 6. The calculated d-spacings for the diffraction pattern are presented in Table IV. The crystal has orthorhombic symmetry with space group $P2_12_12_1$. Unit cell parameters are a = 20.673Å, b = 12.979Å, and c = 6.567Å.[8]

Table IV
X-Ray Powder Diffraction Pattern for Norgestrel

2θ	d(Å)	I/Io
10.95	8.08	0.02
13.9	6.37	1.00
14.2	6.24	0.53
14.5	6.11	0.81
15.2	5.83	0.20
15.8	5.61	0.72
17.4	5.09	0.17
18.0	4.79	0.06
18.9	4.69	0.06
19.8	4.48	0.42
21.2	4.19	0.04
21.8	4.075	0.05
23.3	3.817	0.19
24.9	3.575	0.24
25.3	3.520	0.12
26.0	3.428	0.09
26.3	3.389	0.09
27.6	3.232	0.07
29.05	3.073	0.07
30.4	2.940	0.10
31.8	2.814	0.10
34.7	2.585	0.07

d = (interplanar distance) $\frac{n\lambda}{2 \sin\theta}$

I/Io = relative intensity (based on highest intensity of 1.00)

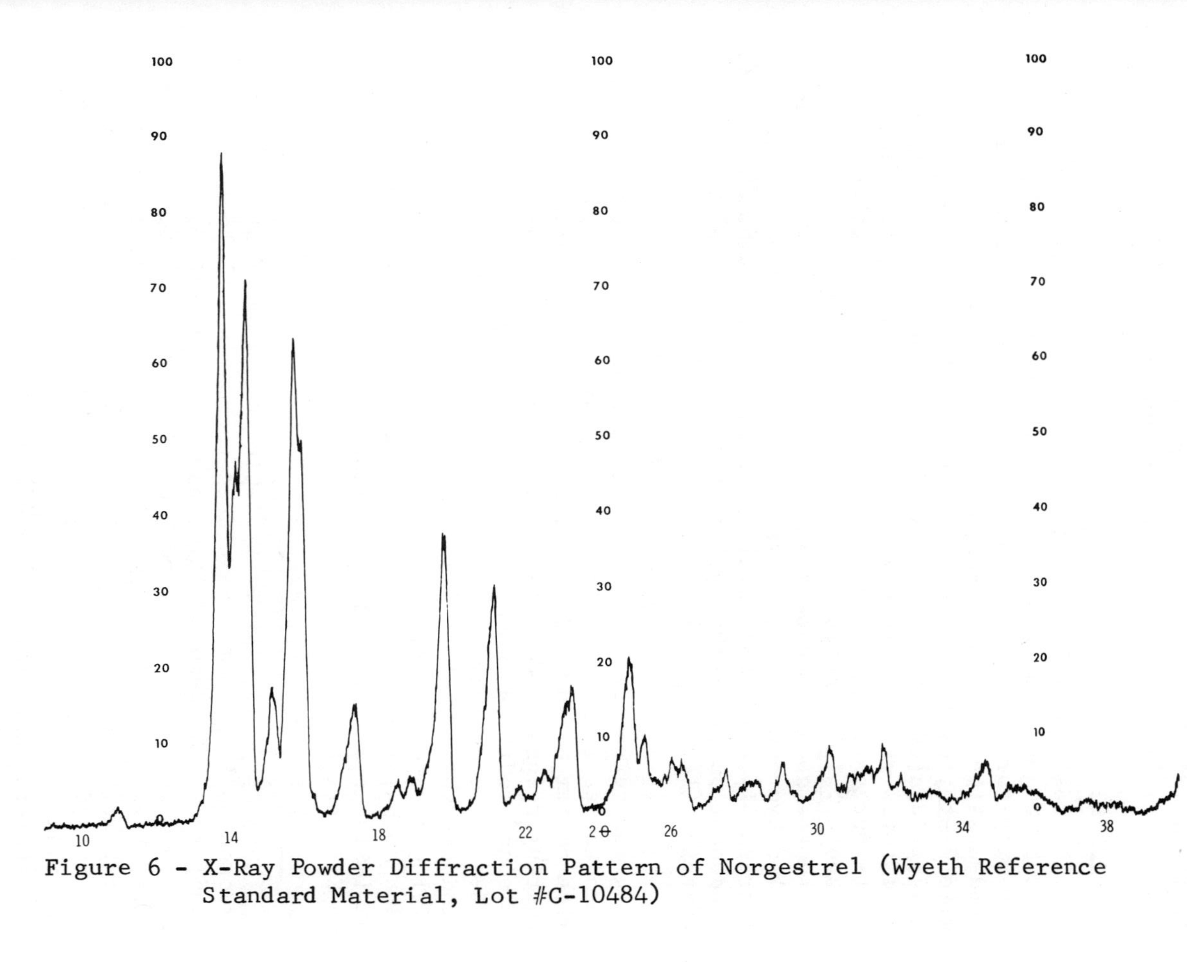

Figure 6 - X-Ray Powder Diffraction Pattern of Norgestrel (Wyeth Reference Standard Material, Lot #C-10484)

2.9 Optical Rotation

dl-norgestrel is a racemate and therefore exhibits no optical rotation. A solution of 500 mg. of norgestrel in 10 ml. of chloroform has a reading of 0° ± 0.05° in a 1 decimeter cell at 25°C. with sodium D light.[9]

3. Synthesis

The basic synthetic route for norgestrel has been developed by Hughes and Smith[10,11] and is presented in Figure 7. The ring formation is accomplished by the base catalyzed condensation of 6-(m-methoxyphenyl)-hex-1-ene-3-one (I) with 2-ethylcyclopentan-1,3-dione (II) to yield the trione (III). Cyclodehydration of (III) with p-toluenesulfonic acid forms gonapentaene-17-one (IV). An alternate synthetic route for this ring structure (IV) has been described by Hughes et.al.[12] and is also presented in Figure 7. In this route the gonapentaene-17-one (IV) is prepared by the methanolic hydrochloric acid cyclization of the seco-oestratetraene-14,17-dione (IIIA) prepared from the base catalyzed condensation of 2-ethylcyclopentan-1,3-dione (IIA) with tetralol (IA).

Catalytic hydrogenation of (IV) yields the corresponding gonatetraene-17-one (V) and subseqnently converted into the estradiol derivative (VII) by reduction initially with sodium borohydride followed by potassium-ammonia-tetrahydrofuran. Conversion of (VII) by the Wilds and Nelson[13] modification of the Birch reduction[14] gives the 1,4-dihydro derivative (VIII). An Oppenauer oxidation[15] produces the 17-ketone (IX), which is then ethynylated, giving the enol-ether intermediate (X). Treatment of (X) with methanolic hydrochloric acid yields norgestrel (XI). A similar synthetic route is presented in a review article by Klimstra.[16]

4. Stability and Degradation

Norgestrel is extremely stable in the solid state when exposed to long term accelerated thermal and photochemical conditions. No decomposition has been observed for this compound when stored for periods of 5 years at room temperature, 1 year at 45°C., 1 month at 75°C. and 1 month under direct UV light.[17] Decomposition has been reported under severe hydrolytic and more accelerated photolytic and thermal conditions.[18]

Studying steroids structurally related to norgestrel, Savard et.al.[19], demonstrated that exposure of Δ^4-3-keto-steroids to ultraviolet light produced saturation of the 4,5 double bond and/or migration of the double bond from that

continued . . . .

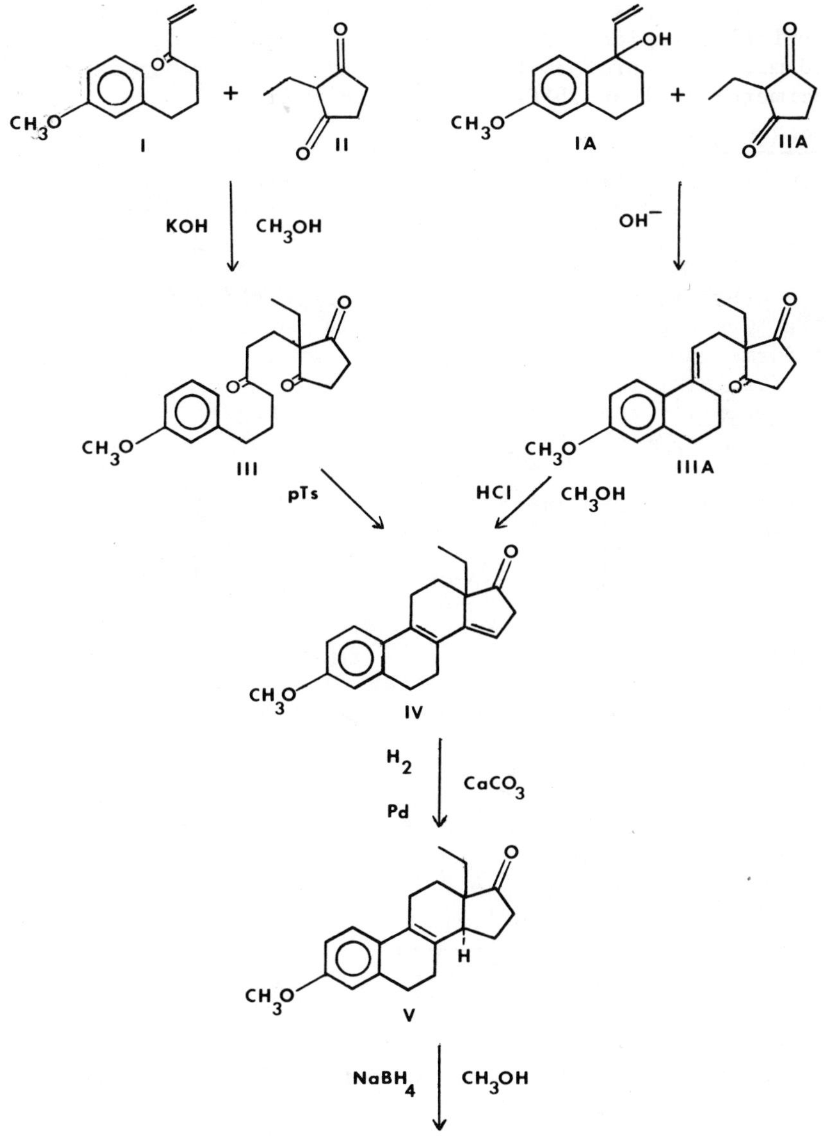

Figure 7 - Synthetic Routes for Norgestrel

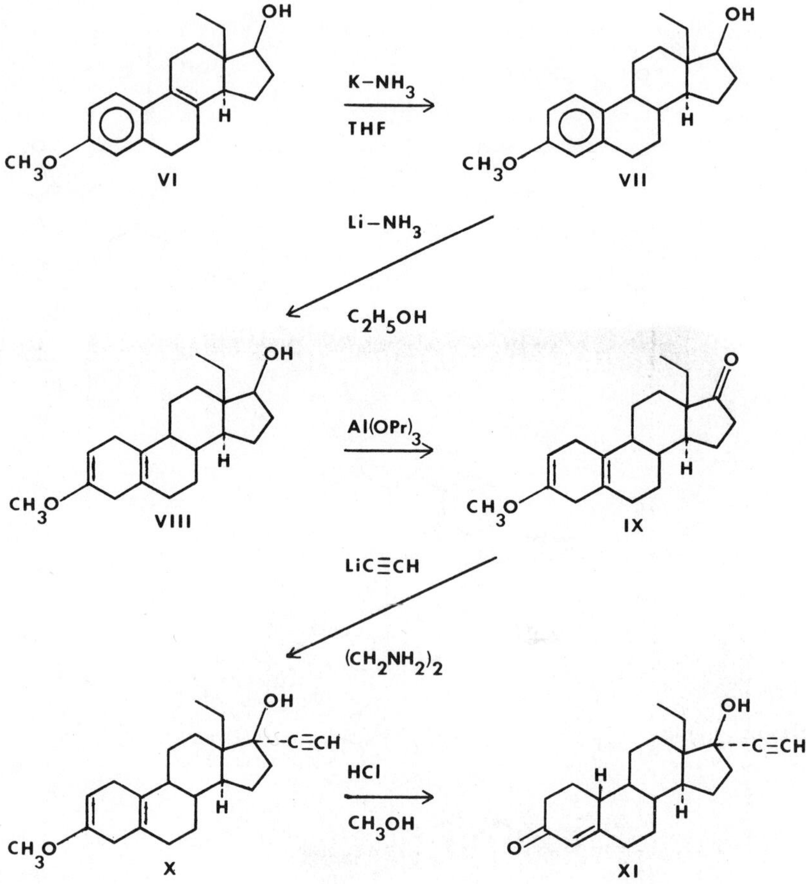

VI $\xrightarrow[\text{THF}]{\text{K-NH}_3}$ VII

VII $\xrightarrow[C_2H_5OH]{\text{Li-NH}_3}$ VIII

VIII $\xrightarrow{\text{Al(OPr)}_3}$ IX

IX $\xrightarrow[(CH_2NH_2)_2]{\text{LiC}\equiv\text{CH}}$ X

X $\xrightarrow[CH_3OH]{\text{HCl}}$ XI

Figure 7 - Synthetic Routes for Norgestrel (concluded)

position in conjugation with the 3-ketone. Subsequent work with Δ^4-3-keto progestational steroids in pharmaceutical systems under photolytic conditions[20,21] have further supported degradation of steroidal A-ring as described by Savard et.al. However, the specific decomposition products for the progestin, norgestrel, have not been reported in the literature.

5. Drug Metabolic Products

Administration of racemic norgestrel to humans has been shown by mass spectroscopy and other analytical techniques to yield a range of metabolites. The major metabolic product has been identified as tetrahydronorgestrel, 13β-ethyl-17α-ethynyl-5β-gonan-3α, 17β-diol; i.e., norgestrel reduced at the Δ^4-3-keto group in the A-ring.[6] By optical rotatory dispersion studies, this metabolite was shown to be entirely the optically pure d-isomer[22]. Other metabolites have been identified as 13β-ethyl-17α-ethynyl-5β-gonan-3β,17β-diol and mono-hydroxylated derivatives of norgestrel.[23,24]

6. Methods of Analysis

6.1 Elemental Analysis

The following data were obtained on norgestrel (Wyeth Reference Standard material, Lot #C-10484).

Element	% Theory	% Determined
C	80.73%	80.66%
H	9.03%	8.92%

6.2 Ultraviolet Spectrophotometric Analysis

Δ^4-3-ketosteroids produce a characteristic chromophore system with maximum absorption between 230 nm. and 270 nm.[25] The ultraviolet absorption maximum of norgestrel at about 240 nm. in 85% ethanol in water (v/v) has been utilized for the analysis of the steroid in tablet formulations containing estrogens.[26] In this procedure a sample equivalent to about 0.5 mg. norgestrel is stirred in 85% ethanol in water (v/v). This solution is diluted to obtain a final concentration of about 10 µg./ml. in the 85% ethanol. Following centrifugation to remove insoluble excipient material, the absorbance is spectrophotometrically measured and compared with a standard solution of norgestrel. Norgestrel has been isolated from interfering formulation components by an extraction of the steroid with organic solvents (viz., chloroform and methylene chloride) from an acidic aqueous solution.

A modified version of an automatic analyzer system procedure for a related Δ^4-3-ketosteroid[27] has been applied to the quantitative analysis of norgestrel.[17] This completely automated technique is capable of disintegrating a whole tablet, dissolving the active constituent, filtering it, diluting a portion of the clear filtrate to a desired volume and obtaining a complete ultraviolet absorption spectrum. The accuracy of the automated method is comparable to that of the manual spectrophotometric method for norgestrel.

The absorption characteristics exhibited by norgestrel in ethanol solution at about 240 nm. is due to the conjugated character of the α,β-unsaturated carbonyl group in the A-ring. Any alteration in the conjugated character of this system by prolonged heating at extreme temperatures or by prolonged irradiation with a strong ultraviolet light source will result in a corresponding loss in ultraviolet absorptivity at 240 nm. Consequently, this method will measure intact norgestrel in the presence of its photochemical and thermal degradation products and is, thus, stability-indicating.[26]

Reactions of the 3-keto group of progestational steroids with the common carbonyl reagents have provided the basis for several other spectrophotometric analytical methods. These techniques can be applied to the analysis of norgestrel. Evans and Gilliam observed that the absorption maxima of the thiosemicarbazones of saturated and α,β-unsaturated ketones could be determined in the presence of excess thiosemicarbazide.[28] Bush[29] and Talbot <u>et.al.</u>[30] have applied this technique to the determination of a series of Δ^4-3-ketosteroids which yield thiosemicarbazones with absorption maxima at 299 to 301 nm.

Görög[31] developed a procedure for the determination of Δ^4-3-ketosteroids which is based on the Claisen condensation of the active methylene groups of ketosteroids with diethyl oxalate leading to spectrophotometrically active glyoxalyl derivatives. The development of the chromophore was carried out at room temperature in a mixture of tertiary butanol and cyclohexane in the presence of sodium tertiary butoxide while the absorbance was measured in moderately acidic ethanol. The method is suitable for the characterization and quantitative determination of norgestrel which produces absorption maxima at about 244 nm. and 318 nm.[17]

Salicyloyl hydrazide reacts with Δ^4-3-ketosteroids to form characteristic hydrazones which absorb strongly in the ultraviolet[32] and may be used for the analytical determination of norgestrel.[17]

Görög developed a simple and rapid ultraviolet spectrophotometric method for the determination of Δ^4-3-ketosteroids in pharmaceutical preparations.[33] The method is based on the reduction of the C-3 carbonyl group with sodium borohydride, followed by the determination of the decrease of absorbance due to the reduction as measured by differential spectrophotometry. This procedure is applicable to the analysis of norgestrel in tablet formulations.[17] Ethinyl estradiol and common inactive tablet components do not interfere with the determination.

6.3 Colorimetric Analysis

6.31 Isonicotinic Acid Hydrazide

Norgestrel can be assayed utilizing the well-known isonicotinic acid hydrazide (INH) colorimetric method described by Umberger.[34] The chemistry of this technique is based on the formation of the isonicotinyl hydrazone of norgestrel from the reaction of the steroid with INH reagent in absolute alcohol acidified with hydrochloric acid. The Δ^4-3-keto group reacts quantitatively at room temperature in less than 1 hour. A stable colored species results which allows measurement of its absorption at about 380 nm. In the analysis of tablet formulations, norgestrel is extracted with chloroform from an acidic aqueous suspension of the tablets, INH reagent is added to an aliquot of the extract and the resultant yellow solution is spectrophotometrically measured. Estrogens [viz., 17α ethinyl estradiol and 17α-ethinyl estradiol-3-methyl ether (mestranol)] and inactive components typically found in oral estrogen-progestin combination dosage forms do not contribute to the color development; thus, do not interfere in the assay.[20,35]

The absorption characteristics exhibited by norgestrel in this colorimetric method are due to the conjugated character of the steroid A-ring with the formed hydrazone. Thus, this method will measure intact norgestrel in the presence of its photochemical and thermal degradation products for reasons analogous to those described for the ultraviolet spectrophotometric procedure.[17,20] Consequently, the procedure is also stability-indicating.

Russo-Alesi describes a completely automated INH reagent procedure for the content uniformity testing of corticosteroid tablet formulations.[36] This sensitive, accurate, and rapid automatic analyzer procedure has been adapted to the analysis of norgestrel in oral contraceptive tablet formulations.[17]

6.32 Dinitrophenylhydrazine

Steroids with an α, β-unsaturated carbonyl group were observed by Djerassi and Ryan to produce 2,4-dinitrophenylhydrazones with absorption maxima at 390 nm. in chloroform solution.[37] Gornall and MacDonald[38] have described a colorimetric procedure, applicable to norgestrel, in which the pale yellow color of the hydrazone derivative is converted to a stable, deep red resonating quinoidal ion species in alkaline solution.[39] The reaction with norgestrel is complete in less than 10 minutes and the resulting colored species is spectrophotometrically measured at about 450 nm.

In a related study, Nishina et.al.[40] report that Δ^4-3-ketosteroids will condense with p-nitrophenylhydrazine to yield the corresponding hydrazone, which produces a reddish-purple species in an alkaline solution of dimethylformamide. Norgestrel produces a λmax at 540 nm.

6.33 Blue Tetrazolium

The utility of the blue tetrazolium reagent in the analysis of α-ketol and non-ketol steroids is well documented.[41,42] Meyer and Lindberg[43] noted that Δ^4-3-ketosteroids, devoid of an α-ketol function, react with blue tetrazolium in the presence of a strong organic base to form the reddish-purple diformazan derivative with a λmax at 530 nm. Application of the tetrazolium technique to the selective analysis of norgestrel in anovulatory tablet formulations containing estrogens has been described by Smith et.al.[44] This colorimetric reaction has been adapted to an automatic analyzer procedure for the analysis of norgestrel in biopharmaceutical systems.[45]

6.34 2,6-Di-tert-butyl-p-cresol

A color reaction is obtained by the action of 2,6-di-tert-butyl-p-cresol on norgestrel in an alkaline medium.[46,47] The concentration of norgestrel in the resulting colored solution can be determined spectrophotometrically in the visible region.

6.4 Fluorometric Analysis

Cullen *et.al.*[18] describe a sensitive procedure, based on a sulfuric acid-induced fluorescence, for the analysis of norgestrel in tablets of low dosage, i.e., 15-75 μg. per tablet. Fluorogen formation, which is effected with an 85% sulfuric acid reagent, is measured at an emission λmax of 550 nm. with an excitation λmax of 418 nm. Specificity of the method with respect to the analysis of intact norgestrel in the presence of its photochemical and thermal degradation products was demonstrated by comparison to quantitative thin-layer chromatography assay values. This procedure has been completely automated to permit rapid content uniformity testing of the dosage form. The automatic analyzer procedure is capable of analyzing 15 samples per hour with a relative standard deviation of ± 1.4% at the 50 μg. norgestrel per tablet level.

Short and coworkers [1,48] applied fluorometry to the determination of norgestrel in biological fluids. Norgestrel is extracted from serum with methylene chloride and subsequently re-extracted into an 80% sulfuric acid in ethanol reagent which produces an intense fluorescence. The resulting fluorogen is measured at an emission λmax of 520 nm. and excitation λmax of 460 nm. This method can be applied to the analysis of norgestrel solutions containing 0.02 μg. per ml.

Bush and Sandberg[49] noted that paper chromatograms sprayed with sodium hydroxide developed an orange-yellow fluorescence specific for Δ^4-3-ketosteroids under UV irradiation. Subsequently, Abelson and Bondy[50] found that potassium *tert*-butoxide could produce the alkaline fluorogenic reaction, and is suitable for quantitative measurement of norgestrel in the 0.1-10.0 μg. per ml. level in ethanolic solution.[17]

6.5 Titrimetric Analysis

The ethinyl group of norgestrel reacts stoichiometrically with silver nitrate in tetrahydrofuran. The nitric acid produced can be titrated with 0.1 N sodium hydroxide to a potentiometric endpoint using a glass-calomel electrode system.[51,52]

6.6 Polarographic Analysis

Norgestrel exhibits a well-defined cathodic wave at the dropping mercury electrode in alkaline isopropanol. The halfwave potential of norgestrel in this system is about -1.5v and the diffusion current is linear with concentration over the range of 10^{-4} to 10^{-3} M.[9]

6.7 Chromatographic Analysis

Chromatography is used qualitatively for the identification and quantitatively for the determination of purity and stability of norgestrel.

6.71 Thin-Layer Chromatography

The various eluant and adsorbent systems used for thin-layer chromatography of norgestrel are given in Table V. The visualization techniques used for detection of norgestrel are also included in Table V.

6.72 Gas Chromatography

Norgestrel has been directly chromatographed on a 4 ft. x ¼ inch glass column packed with 0.5% QF-1 on Chromosorb G (80/100 mesh) utilizing a flame ionization detector.[55] A column temperature of 215°C. is employed with helium at 40 ml./min. as the carrier gas. This technique has been applied to the selective analysis of norgestrel at the 0.25 to 1.0 mg. dosage level in anovulatory tablet formulations. In the analysis norgestrel is extracted with chloroform from an acidic aqueous suspension of the tablets, the chloroform extract is concentrated by evaporation and subsequently chromatographed using cholesterol as the internal standard.

6.73 Column Chromatography

The quantitative analysis of norgestrel and structurally related progestational steroids in orally administered anovulatory formulations has been described utilizing a gel filtration column technique and ultraviolet spectrometry.[4] A synthetic polysaccharide (Sephadex LH-20) column with methanol-water (17:3) as the eluant permits the quantitative separation of norgestrel from formulation components. Estrogens (viz., ethinyl estradiol, mestranol, estradiol, and estradiol benzoate) and common tablet excipient materials were shown not to interfere in this assay procedure.

Table V

Thin-Layer Chromatographic System for Norgestrel

Solvent System	Adsorbent	Visualization Technique	R_f	Reference
A	Silica Gel GF	1, 2	0.37	53
B	Silica Gel GF	1, 2	0.45	53
C	Silica Gel GF	1, 2	0.59	53
D	Silica Gel GF	1, 2	0.53	53
E	Silica Gel G impregnated with DuPont Luminescent Chemical 609	1, 2	0.47	54
F	Silica Gel G impregnated with DuPont Luminescent Chemical 609	1, 2	0.46	54
G	Silica Gel G	3	0.54	26
H	Silica Gel G	4	0.73	1
I	MN Silica Gel G-HR/UV	1, 2	--	18

Solvent Systems

A	Benzene:methanol (95:5)
B	Benzene:acetone (80:20)
C	Chloroform:methanol (90:10)
D	Methylene chloride:methanol:water (150:9:0.5)
E	Petroleum ether:ethyl ether (10:90)
F	Petroleum ether:ethyl ether:diethylamine (40:60:10)
G	Chloroform:alcohol, U.S.P. (96:4)
H	N-hexane:ethyl acetate (1:3)
I	Benzene:chloroform (8:2)

Visualization Techniques

1	Ultraviolet light
2	Sulfuric Acid Spray Reagent
3	10% Ethanolic Phosphomolybdic Acid Spray Reagent
4	Antimony Trichloride Spray Reagent

7. References

1. P. M. Short, E. T. Abbs and C. T. Rhodes, Can. J. Pharm. Sci., 4, 8(1969).
2. L. Bellamy, "The Infrared Spectra of Complex Molecules," 2nd Ed., J. Wiley and Sons, Inc., New York, N. Y., 1964, Chapter 6.
3. R. M. Silverstein and G. C. Bassler, "Spectrometric Identification of Organic Compounds," 2nd Ed., J. Wiley and Sons, Inc., New York, N. Y., 1967, Chapter 3.
4. A. A. Fernandez and V. T. Noceda, J. Pharm. Sci., 58, 740(1969).
5. T. Chang and C. Kuhlman, Wyeth Laboratories, Inc., Personal Communication.
6. D. C. DeJongh, J. D. Hribar, P. Littleton, K. Fotherby, R. W. A. Rees, S. Shrader, T. J. Foell and H. Smith, Steroids, 11, 649(1968).
7. "United States Pharmacopeia," 18th Revision, Mack Publishing Co., Easton, Pa., 1970, pp. 935-936.
8. N. DeAngelis, Wyeth Laboratories, Inc., Personal Communication.
9. M. B. Freeman, Wyeth Laboratories, Inc., Personal Communication.
10. H. Smith, G. A. Hughes, G. H. Douglas, G. R. Wendt, G. C. Buzby, R. A. Edgren, J. Fisher, T. Foell, B. Gadsby, D. Hartley, D. Herbst, A. B. A. Jansen, K. Ledig, B. J. McLoughlin, J. McMenamin, T. W. Pattison, P. C. Phillips, R. Rees, J. Siddall, J. Siuda, L. L. Smith, J. Tokolics and D. H. P. Watson, J. Chem. Soc., 1964, 4472(1964).
11. V. Petrow, Chem. Rev., 70, 713(1970).
12. G. H. Hughes and H. Smith, U. S. Patent 3,478,106 (1969); U.S. Patent 3,442,920(1969).
13. A. L. Wilds and N. A. Nelson, J. Amer. Chem. Soc., 75, 5360(1953).
14. A. J. Birch and S. M. Mukherji, Nature, 163, 766 (1949).
15. R. V. Oppenauer, Org. Syn., 21, 18(1941).
16. P. D. Klimstra, J. Amer. Pharm. Educ., 10, 630(1970).
17. L. F. Cullen, Wyeth Laboratories, Inc., Personal Communication.
18. L. F. Cullen, J. G. Rutgers, P. A. Lucchesi and G. J. Papariello, J. Pharm. Sci., 57, 1857(1968).
19. K. Savard, H. W. Wotiz, P. Marcus and H. M. Lemon, J. Amer. Chem. Soc., 75, 6327(1953).

20. R. E. Graham, P. A. Williams and C. T. Kenner, J. Pharm. Sci., 59, 1152(1970).
21. W. E. Hamlin, T. Chulski, R. H. Johnson and J. G. Wagner, J. Amer. Pharm. Ass., Sci. Ed., 49, 253(1960).
22. K. Fotherby and C. A. Keenan, Acta Endocrinol. Suppl. 138, 83(1969).
23. P. Littleton and K. Fotherby, Acta Endocrinol. Suppl. 119, 162(1967).
24. S. Kamyab, K. Fotherby and G. Wilson, J. Biochem., 103, 14(1967).
25. T. Higuchi and E. Brochmann-Hanssen, "Pharmaceutical Analysis," Interscience Publishers, Inc., New York, N. Y., 1961, Chapter IV.
26. G. J. Papariello, Wyeth Laboratories, Inc., Personal Communication.
27. A. J. Khoury and L. J. Cali, Ann. N. Y. Acad. Sci., 153, 456(1968).
28. L. K. Evans and A. E. Gilliam, J. Chem. Soc., 1943, 565(1943).
29. I. E. Bush, Fed. Proc., 12, 186(1953).
30. N. B. Talbot, S. Ulick, A. Koupreianow and A. Zygmuntowicz, J. Clin. Endocrinol. Metab., 15, 301 (1955).
31. S. Görög, Anal. Chem., 42, 560(1970).
32. P. S. Chen Jr., Anal. Chem., 31, 292(1959).
33. S. Görög, J. Pharm. Sci., 57, 1737(1968).
34. E. J. Umberger, Anal. Chem., 27, 768(1955).
35. Y. P. John, J. Ass. Offic. Anal. Chem., 53, 831(1970).
36. F. M. Russo-Alesi, Ann. N. Y. Acad. Sci., 153, 511 (1968).
37. C. Djerassi and E. Ryan, J. Amer. Chem. Soc., 71, 1000(1949).
38. A. G. Gornall and M. P. MacDonald, J. Biol. Chem., 201, 279(1953).
39. F. D. Chattaway and G. R. Clemo, J. Chem. Soc., 1923, 3041(1923).
40. T. Nishina, Y. Sakai and M. Kimura, Steroids, 4, 255(1964).
41. W. J. Mader and R. R. Buck, Anal. Chem., 24, 666 (1952).
42. J. E. Sinsheimer and E. F. Salim, Anal. Chem., 37, 566(1965).
43. A. S. Meyer and M. C. Lindberg, Anal. Chem., 27, 813(1955).
44. R. V. Smith, T. H. Hassall and S. C. Liv, J. Ass. Offic. Anal. Chem., 53, 1089(1970).
45. P. M. Short and C. T. Rhodes, Steroids, 16, 217(1970).

46. E. P. Schultz and J. D. Neuss, Anal. Chem., 29, 1662(1957).
47. E. P. Schultz, M. A. Diaz and L. M. Guerrero, J. Pharm. Sci., 53, 1119(1964).
48. M. P. Short and C. T. Rhodes, Can. J. Pharm. Sci., 8, 26(1973).
49. I. E. Bush and A. A. Sandberg, J. Biol. Chem., 205, 783(1953).
50. D. Abelson and P. K. Bondy, Arch. Biochem. Biophys., 57, 208(1955).
51. "United States Pharmacopeia," 18th Revision, Mack Publishing Co., Easton, Pa., 1970, pp. 454-455.
52. J. G. Rutgers, Wyeth Laboratories, Inc., Personal Communication.
53. M. B. Simard and B. A. Lodge, J. Chromatogr., 51, 517(1970).
54. G. Moretti, G. Cavina, P. Paciotti and P. Siniscalchi, Il Farmaco, 27, 537(1972).
55. K. Dilloway, Wyeth Laboratories, Inc., Personal Communication.

The Literature Search was conducted up to July, 1974.

PHENFORMIN HYDROCHLORIDE

Joseph E. Moody

CONTENTS

Analytical Profile - Phenformin Hydrochloride

1. Description

1.1 Name, Formula, Molecular Weight

Phenformin hydrochloride is named in Chemical Abstracts as 1-phenethylbiguanide hydrochloride. Other names occasionally used are N'-β-phenethylformamidinyl-iminourea hydrochloride and DBI.

$$C_6H_5-CH_2-CH_2-NH-\underset{\underset{N-H}{\|}}{C}-NH-\underset{\underset{N-H}{\|}}{C}-NH_2 \cdot HCl$$

$C_{10}H_{16}N_5Cl$ Mol. Wt. - 241.73

1.2 Appearance, Color, Odor

Phenformin hydrochloride is a white or practically white, odorless, crystalline powder.

2. Physical Properties

2.1 Infrared Spectrum

The infrared spectrum of phenformin hydrochloride (USP Standard) is shown in figure 1. The spectrum was obtained on a Perkin-Elmer 621 Spectrophotometer from a KBr pellet. The structural assignments have been correlated with the following band frequencies.

Frequency (cm^{-1})	Assignment
3400 - 3250	N-H stretching
3200 - 3100	Amine salt N-H stretching
1660 - 1610	>C=N (α-β-unsaturated)
1590 - 1510	Amine salt bending
756 & 695	Phenyl (monosubstituted)

2.2 Nuclear Magnetic Resonance Spectrum

Phenformin hydrochloride displayed the NMR spectrum shown in figure 2. The sample was dissolved in dimethyl sulfoxide-d_6. The spectrum was measured on a Varian A-60 NMR Spectrometer with TMS as the internal reference. The following structural assignments have been made for figure 2.

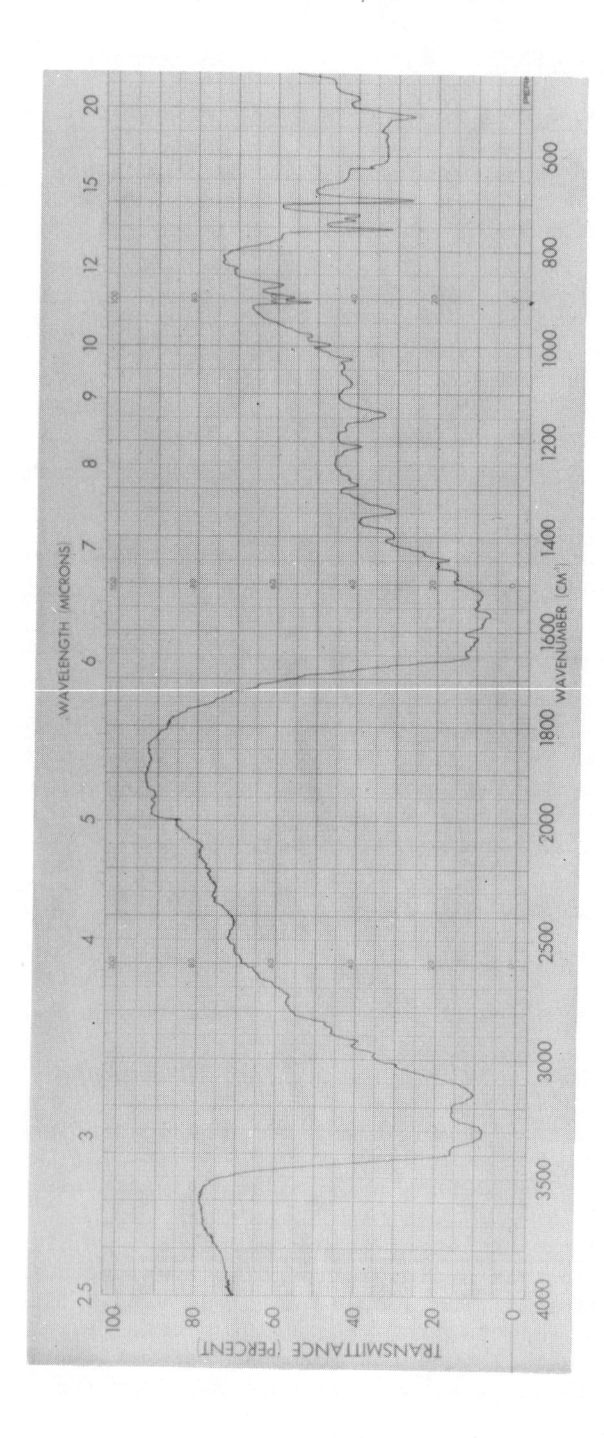

Figure 1. Infrared Spectrum of Phenformin Hydrochloride (USV House Standard batch 22574) taken in KBr Disc. Instrument: Perkin Elmer 621.

Chemical Shift (δ)	Assignment
2.90 and 3.35	ϕ -C$\underline{H}_2$-C$\underline{H}_2$-NH-
7.05	Exchangeable
7.30	Aromatic

2.3 Ultraviolet Spectrum

Shapiro (1) reported the absorption spectra for phenformin hydrochloride. The spectra are shown in Table 1. Samples were measured in acidic, neutral and alkaline medium as well as in methanol.

TABLE 1

Ultraviolet Spectra of Phenformin Hydrochloride

Solvent	λ max (nm)	E x 10^{-3}
H_2O	233	14.5
$1x10^{-3}$NHCl	233	11.2
$1x10^{-4}$NHCl	233	14.5
$1x10^{-1}$NNaOH	232	12.7
1 N NaOH	225 - 228	12.0
methanol	234	17.7

2.4 Mass Spectrum

Phenformin hydrochloride was found to be unstable under electron impact-mass spectrometry conditions. The major fragments peaks were observed at m/e 205, 146, 104, and 101 (2). The molecular ion did not appear.

2.5 Melting Range

Phenformin hydrochloride has a melting point of 176-178° (1).

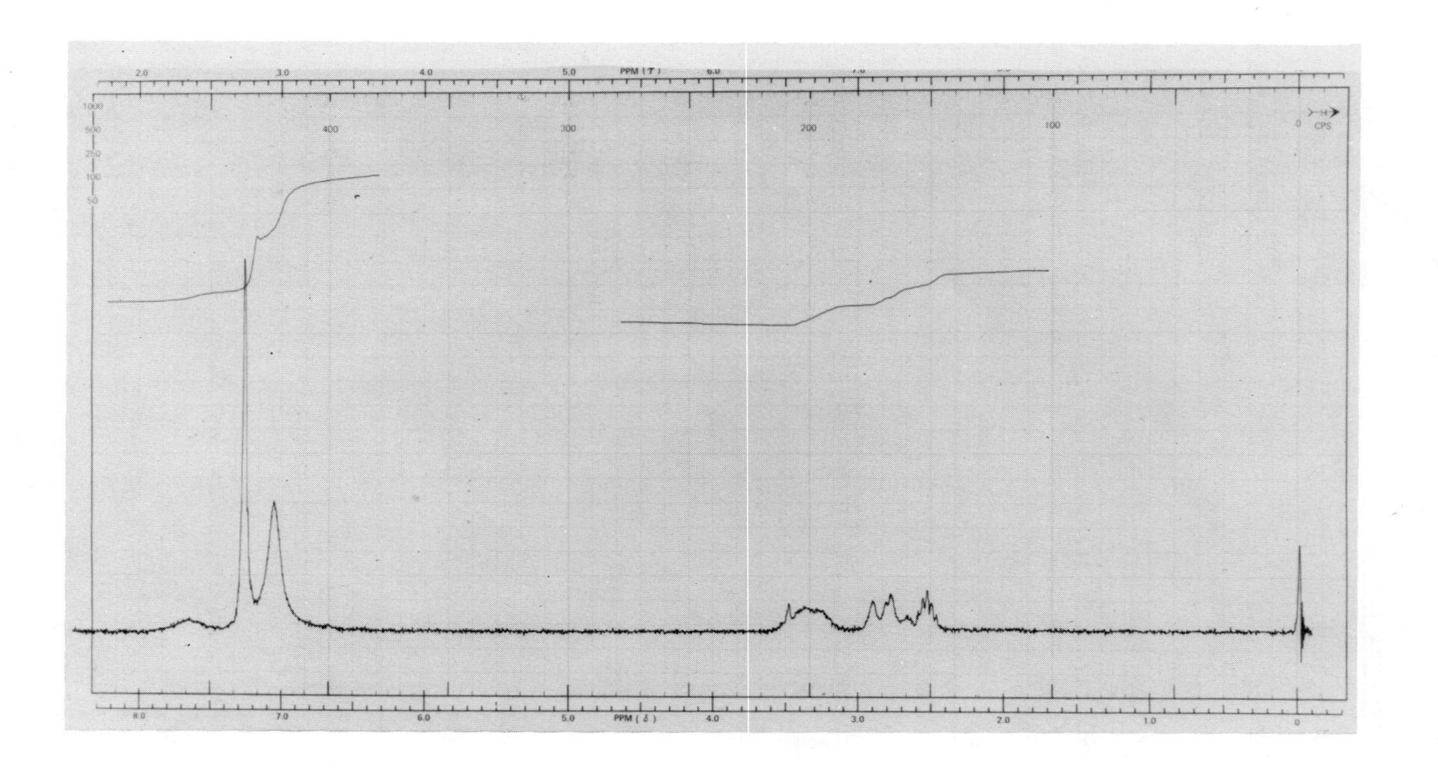

Figure 2. NMR Spectrum of Phenformin Hydrochloride (USV House Standard batch 22574) in deuterated DMSO. Instrument: Varian A-60.

2.6 Solubility

The solubility of phenformin hydrochloride was determined at room temperature in the following solvents.

Solvent	mg/ml	Solvent < 0.02 mg/ml
Water	83	Chloroform
Ethanol Abs.	34	Acetone
Ethanol 95%	64	Ethyl Acetate
Methanol	225	Ether
Isopropanol	3.8	Hexane
Dimethylformamide	415	

2.7 Distribution Coefficients

The distribution coefficients, P, of phenformin hydrochloride were determined in the oil/water systems, A and B. The values are shown below as a function of sodium hydroxide concentration (3). The partitioning of the drug into the organic phase increases with higher pH levels.

P_A	(NaOH)	P_B	(NaOH)
0.01	0.00	0.18	0.004
0.17	0.05	1.11	0.056
0.32	0.20	1.92	0.24
0.57	0.80	2.37	0.64

System A = methylene chloride/water

System B = (1:1) chloroform t-amylalcohol/water

2.8 Ionization Constant, pKa

Phenformin is a strongly basic substance and consequently exists in di and mono ionic forms. Ray (4) has reported the ionization constants, pKa' - 11.8 and pKa" - 2.7 at 32°. In our laboratory , the apparent pKa" - 3.1 was measured at 25° by potentiometric titration. The second ionization constant, pKa', could not be reliably measured in aqueous medium by our method. However, Garrett (3) has calculated an approximate pKa' of 13.0 from plots of the reciprocals of the apparent partition coefficients against the hydrogen ion concentration.

2.9 Differential Scanning Calorimetry

The thermal stability of phenformin hydrochloride was determined on a Perkin-Elmer Model DSC-1B Thermal Analyzer (2). The sample is stable when heated to 10° below its melting point for 30 minutes. A melt endotherm was observed at 175 - 179°. However, phenformin hydrochloride is unstable on heating above its melting point. The following data shows the heat of fusion and melting point of phenformin hydrochloride as determined by DSC Analysis on a sample of 99.8% purity.

Heat of Fusion - 39.8 cal/gram

9600 cal/mole

Melting Range - 175 - 179°C

Melting Point - 176.5°C

2.10 Thermogravimetric Analysis

Phenformin hydrochloride is weight stable through its melting point. At temperatures above 230°, the sample loses approximately 30% of its initial weight. The thermal analysis data was determined on a Perkin-Elmer TGS Thermalanalyzer (2).

3. Synthesis

Phenformin hydrochloride can be prepared by the addition of 2-phenylethylamine hydrochloride to dicyanodiamine at 150° (1). Phenformin hydrochloride is then recrystallized from the reaction mixture with isopropanol.

$$C_6H_5-CH_2CH_2NH_3Cl + H_2N-\overset{\overset{NH}{\|}}{C}-NHCN \xrightarrow[150^\circ]{} \text{Phenformin hydrochloride}$$

4. Stability

Phenformin can be recovered as the diacid salt after prolonged heating with 3 and 6N hydrochloric acid, 50% sulfuric acid or polyphosphoric acid. Phenformin hydrochloride is stable in dilute sodium hydroxide at room temperature. However, the compound hydrolyzes in strong sodium hydroxide to β-phenethylguanidine and small amounts of β-phenethylurea and β-phenethylamine (1).

5. Metabolism

Wick (5) identified the major metabolites of phenformin hydrochloride in the excreted urine of male rats as p-hydroxy - β - phenethylbiguanide and its glucuronide conjugate.

Beckmann (6) discovered that man also metabolizes phenformin hydrochloride to p-hydroxyphenformin. Nearly 33% of the dose was excreted as this metabolite.

Hall (7) studied the metabolism of phenformin with an *in vitro* liver system. The biotransformation to p-hydrox-phenformin was inhibited by a known metabolic inhibitor. This evidence and the tissue distribution of C^{14} labeled phenformin suggest that the liver is the major site of metabolism.

6. Methods of Analysis

6.1 Elemental Analysis

Element	% Theory	% Reported
C	49.7	49.6
H	6.7	6.7
N	29.0	29.3
Cl	14.6	14.3

6.2 Ultraviolet

Phenformin hydrochloride can be analyzed by measuring its absorbance in water at 233 nm (8).

6.3 Fluorometric

Bailey (9) treated solutions of phenformin hydrochloride with alkaline ninhydrin and measured the fluorescence of the phenformin-ninhydrin complex at 518 nm on a Turner fluorometer. The fluorescent complex showed excitation peaks at 304 and 392 nm. The activation monochrometer was set at 390 nm. This method can detect 0.025 μg/ml of phenformin.

6.4 Colorimetric Methods

Phenformin hydrochloric forms an extractable (1:1) ion pair with bromthymol blue. Garrett (10) found that the ion pair is extracted quantitatively in methylene chloride from aqueous solution. The absorbance of the ion pair was then measured in methylene chloride at 413 nm. Garrett discovered a more sensitive procedure by treating the phenformin-bromthymol blue ion pair with excess tetrabutylammonium hydroxide to form a (2:1) tetrabutylammonium-bromthymol blue ion pair. The absorbance of this (2:1) ion pair was determined at 630 nm. This latter method was applied to the analysis of phenformin in human urine. Minor interference from nicotine in some urine samples was observed.

Shepherd and McDonald (11) developed a colorimetric assay based on the color reaction of phenformin hydrochloride with the 1-naphthol-diacetyl reagent. The quantity of phenformin was determined from the absorbance of the colored solution at 565 nm. Beer's Law was obeyed up to a concentration of 20 µg of phenformin per ml of solution. This method was applied to the analysis of phenformin in biological fluids (12). The U. S. Pharmacopeia (13) describes this colorimetric procedure.

To achieve the specificity required of modern analytical methods, these colorimetric procedures should be coupled with a suitable chromatographic separation technique.

6.5 Chromatographic Methods

Phenformin hydrochloride can be determined by the following chromatographic methods.

6.51 Paper Chromatography

The following data show the solvent system and response factors, (Rf) for phenformin hydrochloride. For the qualitative detection of phenformin, the chromatograms can be sprayed with the location reagents, pentacyanoaquoferriate (PCF), Sakaguchi, and 1-naphthol-diacetyl reagents (14, 15). 1-Naphthol-diacetyl reagent is the most sensitive, with a detection limitation of 0.05 µg phenformin.

Solvent system	Rf	Reference
Acetone:water 1:1	0.76	9,16
Isopropanol:water:ammonium hydroxide 40:9:6	0.73	16
Pyridine:butanol:water 1:1:1	0.79	16
Ethyl acetate:ethanol:water 6:3:1	0.76	13,16
Butanol:water:ammonium hydroxide 4:5:1	0.49	16
Butanol saturated with water	0.43	16

6.52 Thin Layer Chromatography

A TLC system was developed by Bailey (9) for phenformin hydrochloride on silica gel GF glass plates. The TLC plates were eluted with the upper layer of equilibrated n-butanol:acetic acid:H_2O (40:5:55). The migration rate was approximately 14 cm in 5 hours with an Rf of 0.45.

Phenformin hydrochloride is quantitatively determined in our laboratory on cellulose plates which are pre-washed in the developing solvent, isopropanol:H_2O:acetic acid, 80:20:5. After development in the above solvent system, phenformin hydrochloride is extracted in methanol and the absorbance of the solution determined at 233 nm.

6.53 Paper Electrophoresis

Bailey (9) has developed an electrophoresis method using Whatman No. 3 paper at 400 volts. With 1N acetic acid, phenformin showed a migration rate of 8.25 cm in 1 hour. Phenformin was extracted from the paper chromatogram and determined by u.v. spectroscopy in the usual manner.

6.54 Gas Liquid Chromatography

Wickramasinghe and Shaw (17) have reported the thermal instability of phenformin hydrochloride to gas chromatographic conditions. A single peak of retention time about 22 minutes was observed on a glass spiral column (122 cm long) packed with 0.5% Carbowax 20M on glass beads (80-100 mesh) at a column oven temperature of 225°. The analysis of the single GC peak on a LKB-9000 gas chromatograph-mass spectrometer showed prominent GC-MS signals at m/e 230, 139, 110, 91 and 68. This GC-MS data is consistent with the s-triazine structure shown below.

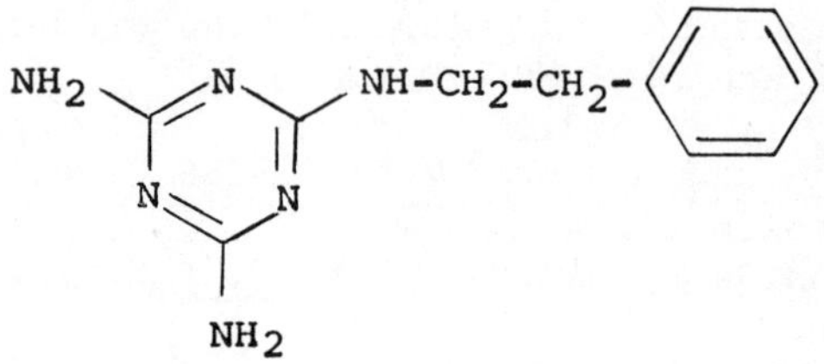

NH_2 N NH-CH_2-CH_2-

N N

NH_2

7. References

1. S.L. Shapiro, V.A. Parrino and L. Freedman, J. Am. Chem. Soc., 81, 2220 (1959).

2. F. Tischler, Ciba-Geigy Pharm. Corp., personal communication.

3. E.R. Garrett, J. Tsau, and P.H. Hinderling, J. Pharm. Sci., 61, 1411 (1972).

4. P. Ray, Chem. Rev., 61, 313-359 (1961)

5. P.J. Murphy and A.N. Wick, J. Pharm. Sci., 57, 1125 (1968).

6. R. Beckmann, Ann. N.Y. Acad. Sci., 148, 820-832 (1968); Chem. Abstr., 69, 50656p (1968).

7. H. Hall, G. Ramachander and J.M. Glassman, Ann. N.Y. Acad. Sci., 1968, 148, 601-611; Chem. Abstr., 69, 65872e (1968).

8. U. S. Pharmacopeia XVIII, p. 488.

9. R.E. Bailey, Clin. Biochem., 3, 23-31 (1970).

10. E.R. Garrett and J. Tsau, J. Pharm. Sci., 61, 1404 (1972).

11. H.G. Shepherd, Jr., and H.J. McDonald, Clinical Chem., 4, 496 (1958).

12. L. Freedman, M. Blitz, E. Gunsberg, and S. Zak, J. Lab. Clin. Med., 58, 662-666 (1961).

13. U. S. Pharmacopeia XVIII, p. 487.

14. R. E. Bailey and D. A. Durfee, J. Chromatogr., 16, 546 (1964).

15. I. Smith, "Chromatographic and Electrophoretic Techniques", Interscience, New York (1960) p. 226.

16. J. L. Faymon, C. J. Stewart and A. N. Wick, Appl. Therap., 4, 378-381 (1962).

17. J. Wickramasinghe and S. R. Shaw, J. Chromatogr., 71, 265-273 (1972).

PROCAINAMIDE HYDROCHLORIDE

Raymond B. Poet and Harold Kadin

TABLE OF CONTENTS

1. Description

1.1 Name, Formula, Molecular Weight

Procainamide hydrochloride is benzamide, p-amino-N-[(2-diethylamino)ethyl]monohydrochloride with Chem. Abstr. registry number 614-39-1.

Among the generic and trivial names for this compound are procaine amide hydrochloride, amidoprocain and procainamide. Common trade names are Pronestyl, Procamide, Procardyl, Novocamid, Novocainamid and Supicaine Amide.

$$H_2N-C_6H_4-C(=O)-NHCH_2CH_2N(CH_2CH_3)_2 \cdot HCl$$

$C_{13}H_{21}N_3O \cdot HCl$ Molecular Weight 271.8

1.2 Appearance, Color, Odor

Procainamide hydrochloride is a white-to-tan, odorless crystalline powder.

2. Physical Properties

2.1 Spectral Properties

2.11 Infrared Spectra

Samples of procainamide hydrochloride were dispersed in a potassium bromide pellet or in mineral oil to obtain the infrared spectra given in Figures 1a and 1b from a Perkin-Elmer Model 21 prism instrument[12]. The potassium bromide and mineral oil spectra are essentially identical.

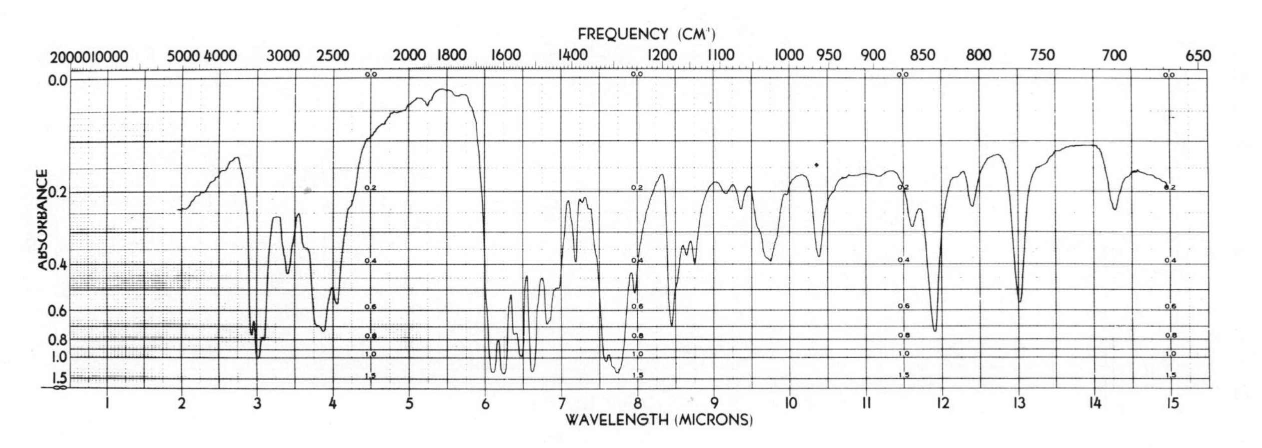

Figure 1a. Infrared Spectrum of Procainamide Hydrochloride from KBr pellet. Instrument: P.E. Model 21 Infrared Spectrophotometer.

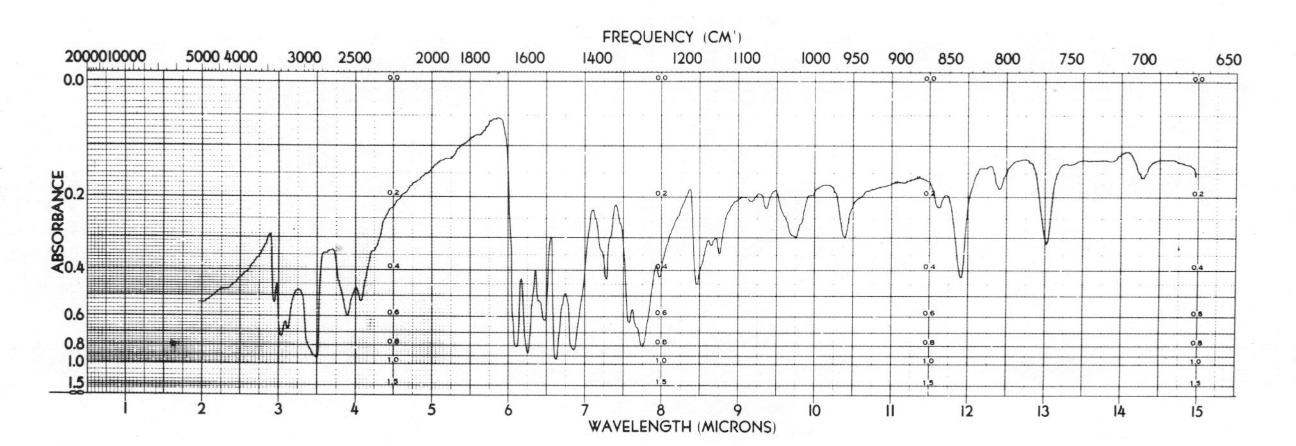

Figure 1b. Infrared Spectrum of Procainamide Hydrochloride from mineral oil mull. Instrument: P.E. Model 21 Infrared Spectrophotometer.

Structurally significant bands were interpreted by Toeplitz[12] as follows (see Figures 1a and 1b).

Wavelength (cm^{-1})	Interpretation
3389,3279,3205	NH_2 and NH
2564,2457	HCl
1639	Amide C=O
1538	Secondary Amide
1600	NH_2 and Aromatic C=C
1511	Aromatic C=C
840	Para-Substituted Aromatic

2.12 Nuclear (Proton) Magnetic Resonance Spectra

Cohen[59] obtained both 60- and 100-MHz spectra of procainamide and its hydrochloride. The 60-MHz NMR spectra were obtained on a Varian Associates A-60 spectrometer of CD_3OD solutions containing tetramethylsilane as an internal reference. The 100-MHz NMR spectra were obtained on a Varian Associates XL-100 spectrometer deuterium-locked to the solvent ($CDCl_3$, CD_3OD, or d_5-pyridine) containing tetramethylsilane as internal reference.

Because procainamide hydrochloride is not soluble in chlorinated hydrocarbons, the 60-MHz NMR spectrum was obtained in tetradeuteromethanol (Figure 2); this spectrum is compared with that of the free base in the same solvent (Figure 3). Although the spectra are qualitatively the same, the spectrum of the hydrochloride salt differs because of protonation at the alkylamine site, resulting in a great downfield shift of the protons on the methylene groups directly attached to the aliphatic amine nitrogen and a

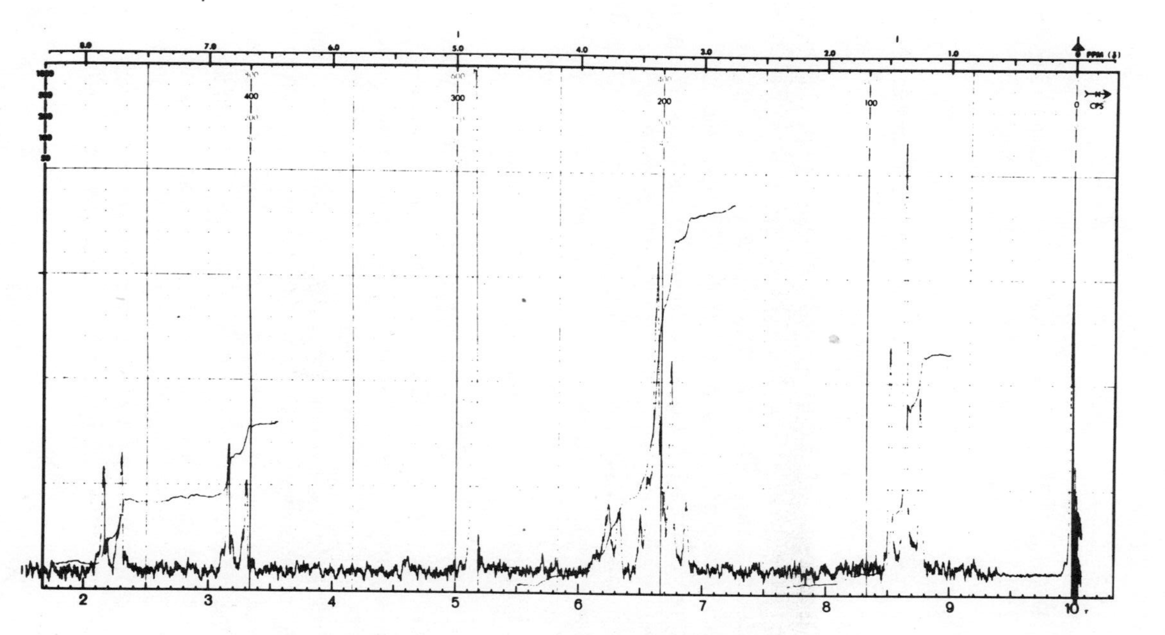

Figure 2. 60-MHz NMR Spectrum of Procainamide Hydrochloride in Tetradeuteromethanol. Instrument: Varian Associates A-60 Spectrometer. Tetramethylsilane as Internal Standard.

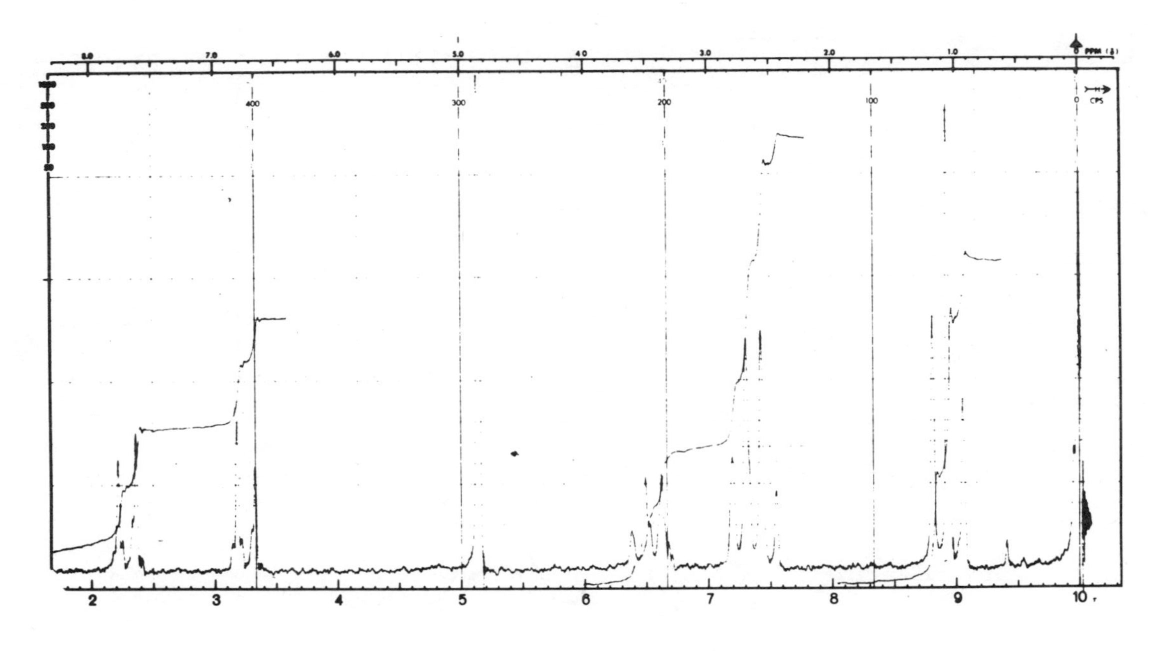

Figure 3. 60-MHz NMR Spectrum of Procainamide Base in Tetra-deuteromethanol. Instrument: Varian Associates A-60 Spectrometer. Tetramethylsilane as Internal Standard.

Table 1. NMR Spectral Assignments

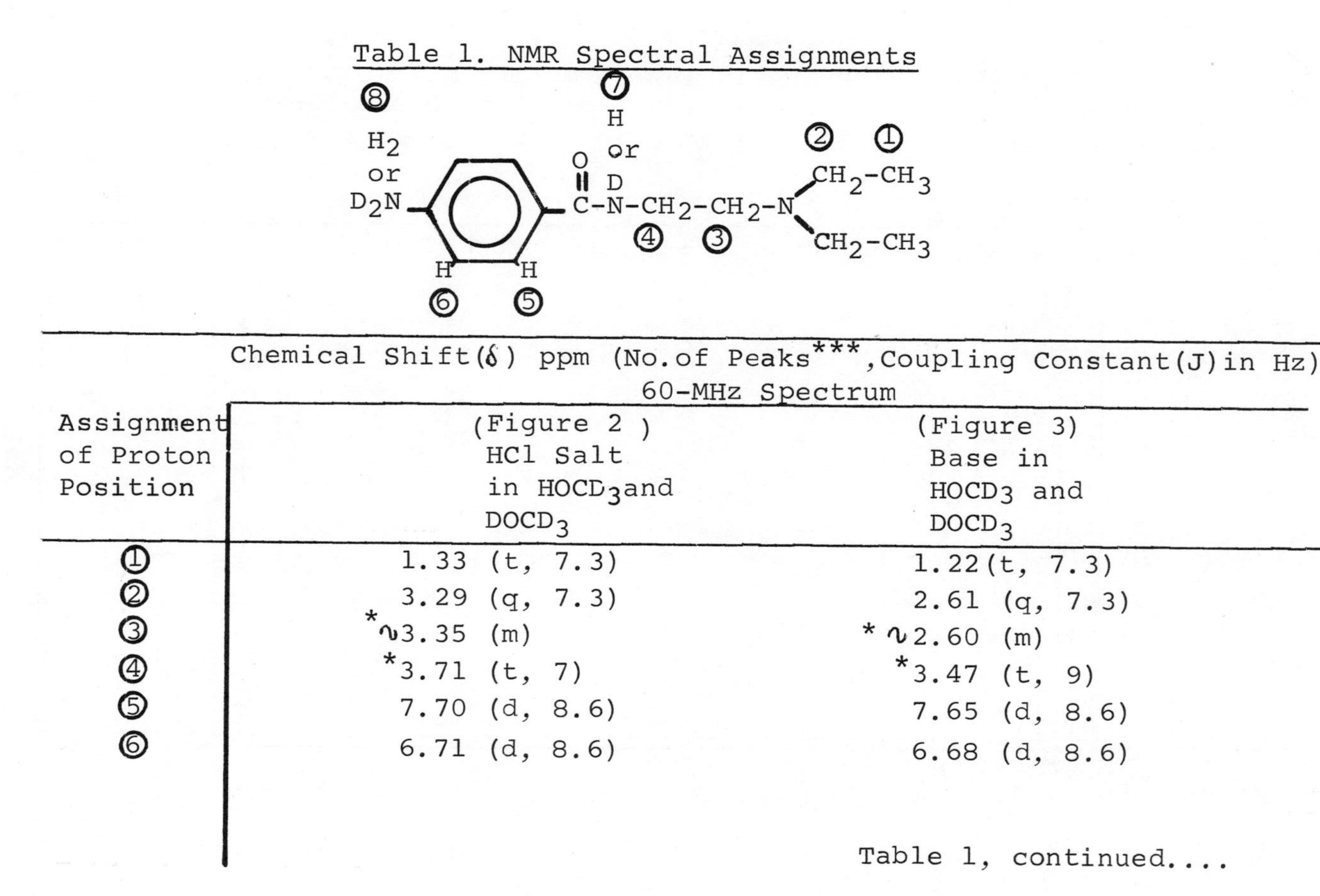

Assignment of Proton Position	Chemical Shift (δ) ppm (No. of Peaks***, Coupling Constant (J) in Hz) 60-MHz Spectrum: (Figure 2) HCl Salt in $HOCD_3$ and $DOCD_3$	(Figure 3) Base in $HOCD_3$ and $DOCD_3$
①	1.33 (t, 7.3)	1.22 (t, 7.3)
②	3.29 (q, 7.3)	2.61 (q, 7.3)
③	*∿3.35 (m)	*∿2.60 (m)
④	*3.71 (t, 7)	*3.47 (t, 9)
⑤	7.70 (d, 8.6)	7.65 (d, 8.6)
⑥	6.71 (d, 8.6)	6.68 (d, 8.6)

Table 1, continued....

Table 1, continued...

Assignment of Proton Position	Chemical Shift(δ)ppm(No. of Peaks***, Coupling Constant(J)in Hz) 100-MHz Spectrum		
	(Figure 4) Base in $HOCD_3$ and $DOCD_3$	(Figure 5) Base in $CDCl_3$	(Figure 6) Base in Perdeuteropyridine
①	1.08(t, 7)	1.04 (t, 7)	0.95 (t, 7)
②	2.61 (q, 7)	2.57 (q, 7)	2.48 (q, 7)
③	**2.66 (m)	2.62 (m)	2.67 (t, 7)
④	**3.48 (m, 8)	3.46 (m)	3.71 (q, 7)
⑤	7.58 (d, 8.5)	7.59 (d, 8)	7.13 (d, 8)
⑥	6.65 (d, 8.5)	6.59 (d, 8)	5.93 (d, 8)
⑦	-	6.79 (Broad)	6.79 (Broad)
⑧	-	4.04 (Broad)	4.04 (Broad)

*Hydrogens not distinguishable.

**Hydrogens are distinguishable.

***No. of peaks is designated t = triplet, d = doublet, q = quartet, and m = multiplet.

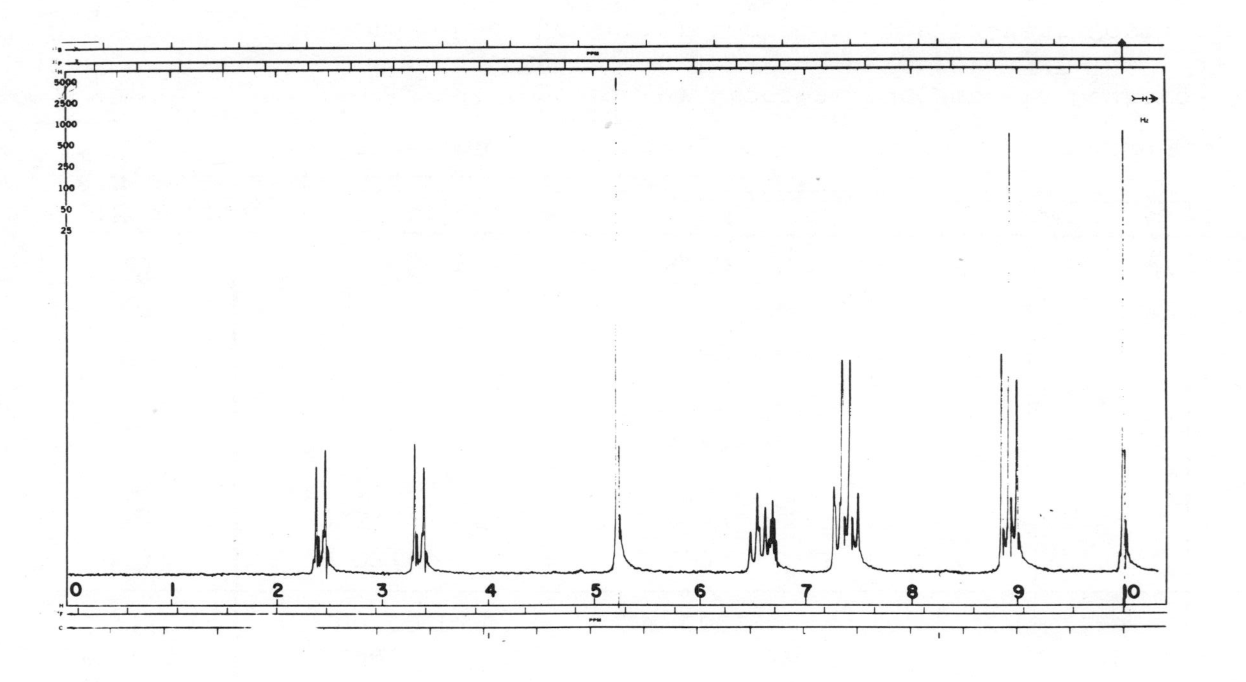

Figure 4. 100-MHz NMR Spectrum of Procainamide Base in Tetradeuteromethanol. Instrument: Varian Associates XL-100 Spectrometer, Deuterium-Locked to the Solvent Containing Tetramethylsilane as Internal Reference.

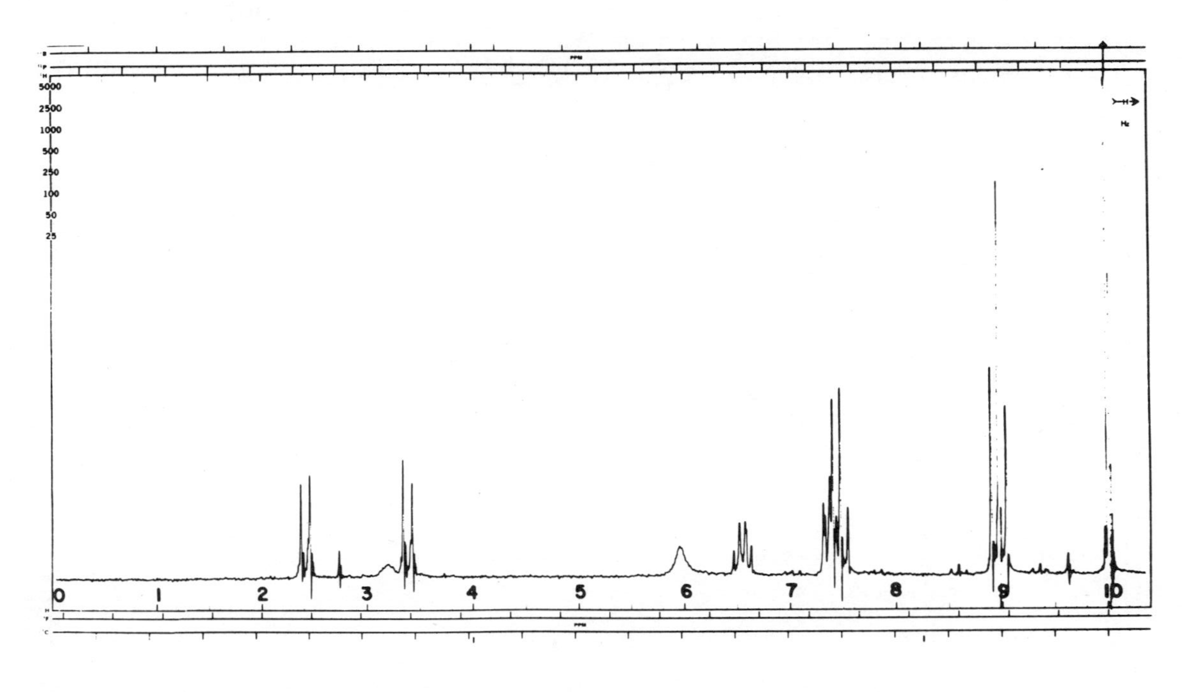

Figure 5. 100-MHz NMR Spectrum of Procainamide Base in Deuterochloroform. Instrument: Varian Associates XL-100 Spectrometer, Deuterium-Locked to the Solvent Containing Tetramethylsilane as Internal Reference.

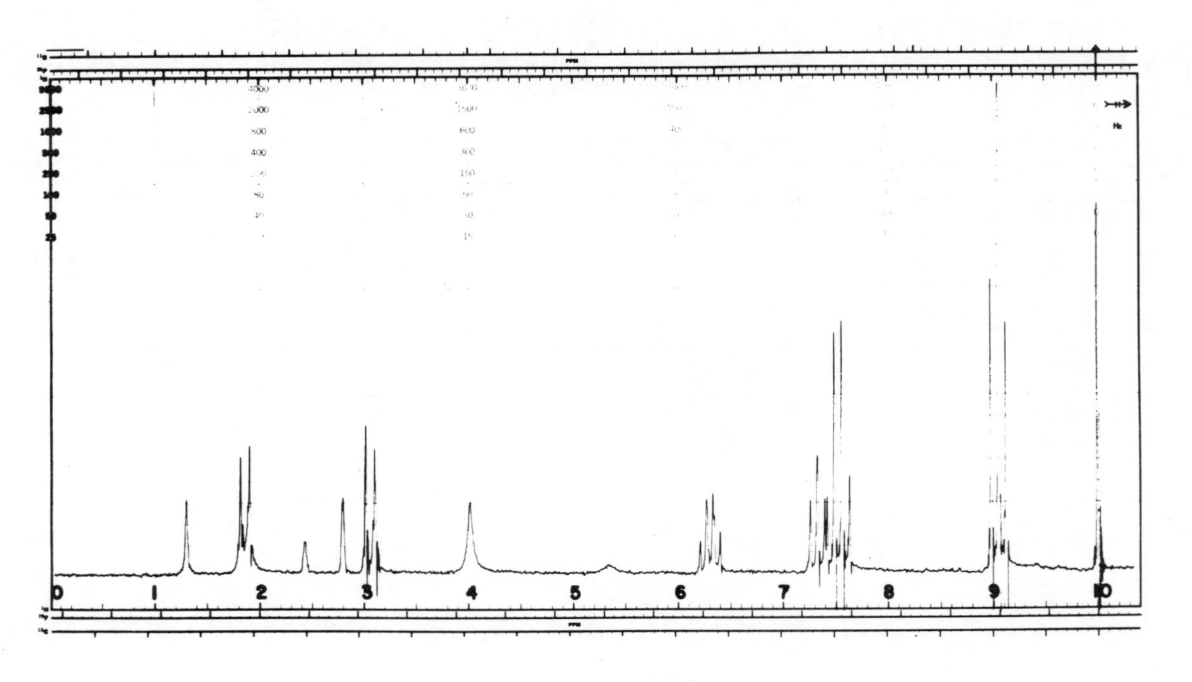

Figure 6. 100-MHz NMR Spectrum of Procainamide Base in Perdeuteropyridine. Instrument: Varian Associates XL-100 Spectrometer, Deuterium-Locked to the Solvent Containing Tetramethylsilane as Internal Reference.

slight downfield shift of the protons on the β-carbons. That the site of protonation is not on the aromatic amine is demonstrated by the similarity of the aromatic proton resonances of the two compounds. Protonation of the aromatic amine would have resulted in a large downfield shift of the *ortho* protons and a small downfield shift of the *meta* protons, because the $-NH_3^+$ group is not as strong an electron-donating group as is the $-NH_2$ group. The assignments of the chemical shifts shown in Table 1 are given in Delta values (δ) and the coupling constants, J, are given in Hz with t-triplet, d-doublet, q-quartet and m-multiplet.

It is evident that the 100-MHz spectrum of the base in CD_3OD yields additional resolution (Figure 4 and Table 1). The 100-MHz NMR spectrum of procainamide base (Figure 5) shows the amine and amide protons. There are additional lines for the methylene protons attached to the amide nitrogen due to coupling of the amide proton, which is $J \sim 7$Hz. The assignments are shown in Table 1. When the 100-MHz NMR spectrum of procainamide base is obtained in perdeuteropyridine, considerable changes in chemical shifts occur that result in separation of all the proton resonances (Figure 6). The assignment of the spectrum is depicted in Table 1.

2.13 Ultraviolet Spectra

Dunham[57] had indicated that the ultraviolet spectra of procainamide hydrochloride should be measured in 0.1N ammonium hydroxide in either methanol or ethanol. Spectra for the compound dissolved in either solvent alone can have variable wavelengths of maximum absorbance, as indicated by the following spectral data obtained with a Cary 15 Spectrophotometer.

Solvent	λ max (nm)	E (1%, 1 cm)
MeOH containing	282	667
0.1N NH_4OH	282	670
	282	676 (Note)
MeOH	291	694
MeOH	291	690
MeOH containing		
10^{-4} N HCl	293	682
10^{-3} N HCl	293	399
EtOH, 95%	286	656
EtOH, 95%	286	654
EtOH, 95%	282	658
containing		
0.1N NH_4OH	282	654

Note: The spectrum of this solution, determined again 5 hr. later, showed λ max = 282 nm, E (1%,1 cm) = 670.

The data indicate that the wavelength of maximum absorbance is pH dependent. Both methanolic and ethanolic solutions of procainamide hydrochloride had a λ max of 282 nm in 0.1N ammonium hydroxide, although the E(1%,1cm) varied between the two solvents. Addition of acid to the methanolic solution shifts the maximum to 293 nm. Spectra of the compound dissolved in either solvent alone can show λ max values that are quite variable, depending on the possible contribution of slight residual acid or base on the glassware or from impurities in the batch of solvent used to make the solutions and measurements. When 10^{-4} N hydrochloric acid in methanol is used as solvent, there is a slight decline (about 1.5%) in absorbance, with a 2-nm upward shift in λ max, relative to the absorbance in methanol alone. However, when the acid concentration is increased to 10^{-3} N, the absorbance declines appreciably (about 42%), relative to absorbance in methanol alone. This decline has been ascribed to protonation of the aromatic-NH_2 group[55,65].

The effects of the substitution of amino and carbonyl groups on the shift of the absorption of the benzene ultraviolet chromophore have been discussed[51]. The use of ultraviolet spectrophotometry for determination of procainamide hydrochloride has been described[51,55] (see Section 6.3).

2.14 Mass Spectra

Cohen[59] has reported the following concerning the mass spectra of procainamide hydrochloride.

The low-resolution mass spectrum (see Figure 7) is dominated by the m/e 86 ion (base peak) resulting from the cleavage of the bond beta to the tertiary amine nitrogen. This cleavage is anticipated in the fragmentation of amines. The M+ of the compound is observed at m/e 235. High-resolution peak matching, Table 2, and high-resolution data acquisition, Table 3, were employed to determine the composition of each fragment ion. Assignment of the fragmentation is shown below[3,5].

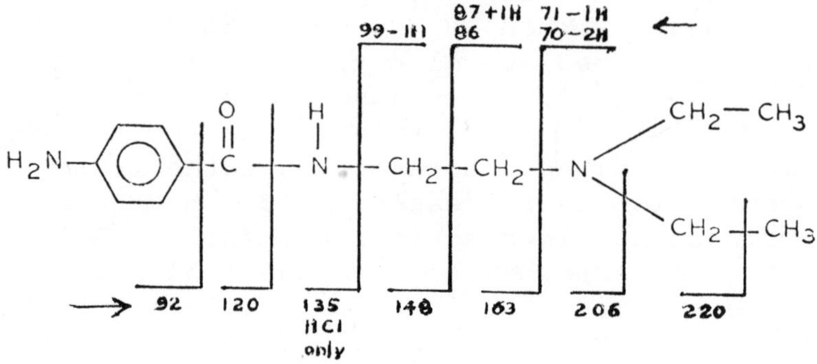

The low-and high-resolution mass spectra were obtained on an AEI MS-902 Spectrometer equipped with a frequency-modulated analog tape recorder. The low-resolution mass spectrum was processed on a Digital Equipment Corporation PDP-11 computer using Squibb programs[62], while the high-resolution mass spectrum was processed by Womack[63] on an EAI computer. Peak-matched data were obtained from the AEI MS-902 spectrometer equipped with a wide-ratio accessory.

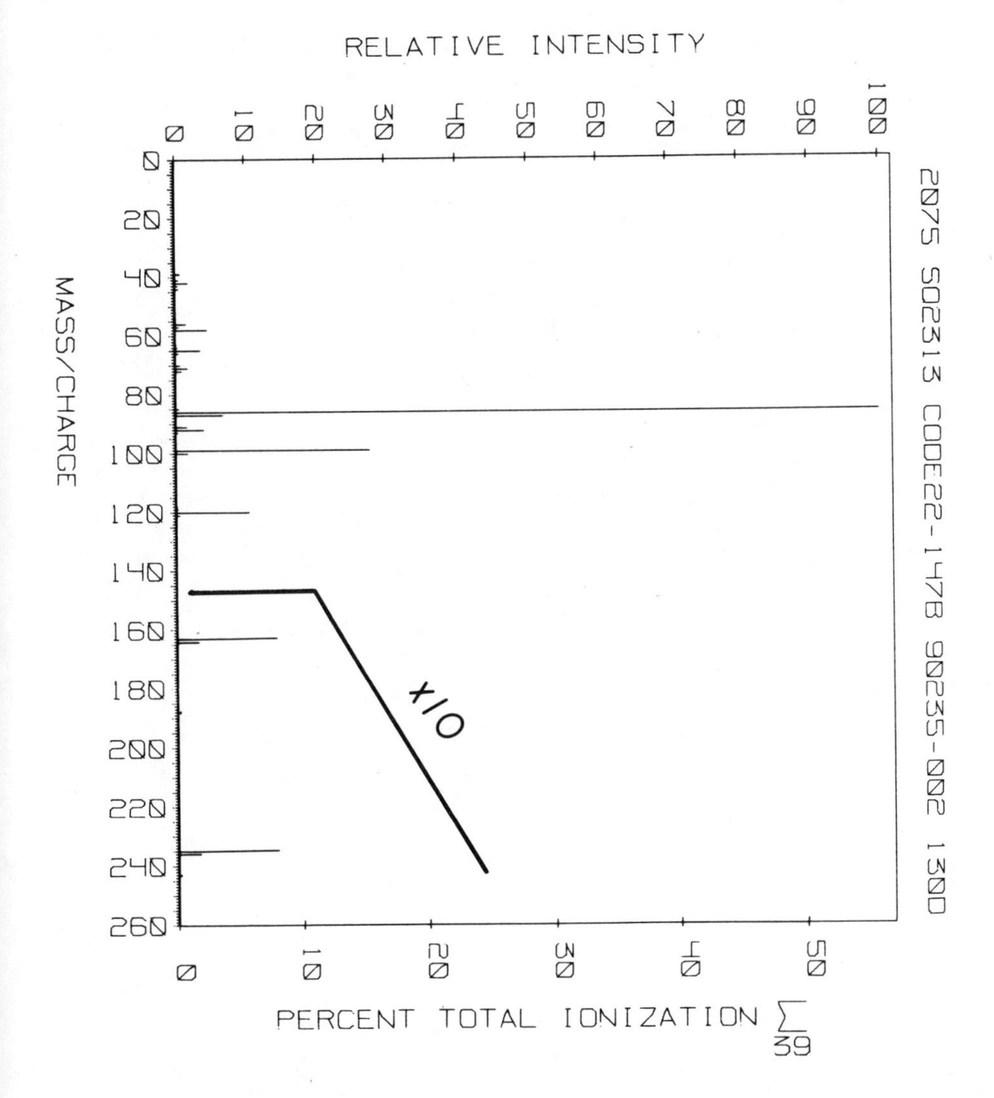

Fig. 7 Low Resolution Mass Spectrum of Procainamide Hydrochloride. Instrument: AEI MS-902 Spectrometer Equipped with Frequency Modulated Tape Recorder; Spectrum Processed on Digital Equipment Corporation PDP-11 Computer

Table 2
High-Resolution Peak-Match Data of Procainamide

	Found	Calcd.	C	H	N	O
M+	235.1685	235.1685	13	21	3	1
	206.1293	206.1293	11	16	3	1
	205.1215	205.1215	11	15	3	1
	189.1266	189.1266	11	15	3	0
	188.1188	188.1188	11	14	3	0
	174.1031	174.1031	10	12	3	0
	163.0871	163.0872	9	11	2	1
	148.0678	148.0637	8	8	2	1
	136.0627	136.0637	7	8	2	1
	120.0451	120.0449	7	6	1	1
	99.1041	99.1048	6	13	1	0
	92.0507	92.0500	6	6	1	0
	86.0963	86.0970	5	12	1	0
	71.0735	71.0735	4	9	1	0
	70.0653	70.0657	4	8	1	0

Table 3
High-Resolution Mass Spectrum of Procainamide[a]

	Found	Calcd.	C/C^{13}	H	N	O	Rel. Int.
M+	235.1667	235.1685	13/0	21	3	1	4
	233.1530	233.1529	13/0	19	3	1	4
	189.1241	189.1266	11	15	3	0	4
	188.1137	188.1188	11	14	3	0	1
	187.1151						1
	164						5
	149.0705	149.0715	8	9	2	1	4
	148.0700	148.0637	8	8	2	1	5
	137.0646	137.0715	7	9	2	1	3
	136.0597	136.0637	7	8	2	1	7
	121.0479	121.0482	6/1	6	1	1	12
	120.0458	120.0449	7	6	1	1	141
	119.0368	119.0371	7	5	1	1	2
	100.1078	100.1064	5/1	13	1	0	20
	99.1031	99.1048	6	13	1	0	242
	93.0554	93.0533	5/1	6	1	0	6
	92.0488	92.0500	6	6	1	0	57
	91.0421	91.0422	6	5	1	0	8
	87.0974	87.1003	4/1	12	1	0	78
	86.0944	86.0970	5	12	1	0	999
	72.0799	72.0813	4	10	1	0	22
	71.0732	71.0735	4	9	1	0	25
	70.0662	70.0657	4	8	1	0	13
	65.0395	65.0391	5	5	0	0	55
	58.0687	58.0657	3	8	1	0	90
	57.0625	57.0578	3	7	1	0	11
	56.0501	56.0500	3	6	1	0	36

[a] High-Resolution FM analog tape processed by Dr. J. Womack, EAI, Princeton, N.J.

A similar mass-spectral fragmentation pattern was reported by Atkinson et al[39]. These authors commented that the most prominent ion (m/e 86 ion) results from the ability of the free electron pair on the tertiary amine nitrogen atom to stabilize a positive charge on the adjacent carbon atom. Atkinson et al.[39] ascribed the m/e 99 fragment to a McLafferty rearrangement[66].

2.15 Fluorescence

Procainamide has been reported[19] to exhibit maximum fluorescence at 385 nm, with maximum activation at 295 nm, at pH 11. The spectrophotofluorometer used employs a xenon arc source emitting a continuum from 200 to 800 nm as the activating light source. This instrument yields spectra uncorrected for the non-linear intensity of emission of the xenon arc and the uneven photomultiplier response over this continuum. This may explain the same maximum activation at 295 nm, but a different fluorescence maximum at 360 nm, at pH 11, that was reported by Koch-Weser[8].

Refer to sections 6.5 and 7.0 for applications of this physical property.

2.2 Crystal Properties

2.21 Differential Thermal Analysis

Differential thermal analysis (DTA) of procainamide hydrochloride by Valenti[30] yielded a sharp endotherm at 169°C. A DuPont 900 Thermoanalyzer programmed for a temperature rise of 15° per min produced the thermogram of Figure 8.

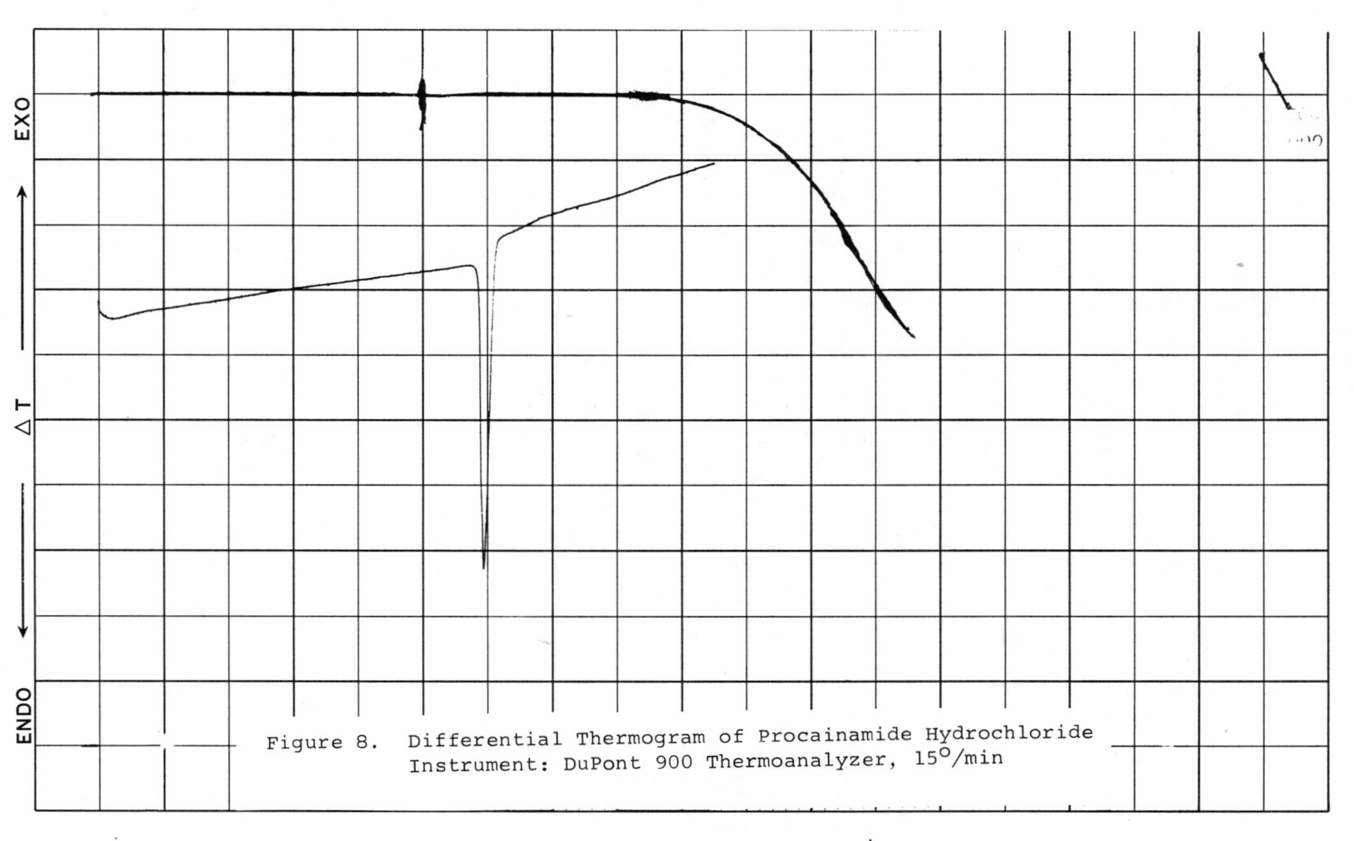

Figure 8. Differential Thermogram of Procainamide Hydrochloride
Instrument: DuPont 900 Thermoanalyzer, 15°/min

2.22 Differential Scanning Calorimetry (D.S.C.)

Valenti[30] also reported the D.S.C. purity of the same lot of procainamide hydrochloride to be 99.94 mole percent. A Perkin Elmer D.S.C. Model 1B was used.

2.23 Melting Range

The melting range for U.S.P. procainamide hydrochloride is specified as 165.0 to 169.0°C[1]. The sample of procainamide hydrochloride used for most of the studies reported herein has a U.S.P. melting range between 168.4 and 169.2°C. This narrow, high-melting range signifies excellent purity, which is confirmed by the sharp 169°C D.T.A. endotherm and the 99.94% D.S.C. purity (Sections 2.21 and 2.22). Various preparations of procainamide hydrochloride are reported to have the following melting ranges:

Melting Range (°C)	Author(s)	Ref.No.
165-9	K. Iwaya and K. Yoshida	98
165-7	M. Yamazaki, Y. Kitagawa, S. Hiraki, and Y. Tsukamoto	85
164	G. Ghielmetti	99
167	H. Najer and M. Joannic	86
166-7	V. Hack and E. Koppová	87
166-6.5	Z. Ledóchowski, M. Bogucka, A. Ledóchowski, and A. Chimiak	89
167-8	E. Pavlovska, J. Lukac and M. Borovicka	100
165-8	K. Radushkevich	72
165-8	P. Egli	93
165-8	Y. Tashika and M. Kuranari	84

2.24 X-Ray Powder Diffraction

The X-Ray powder diffraction of procainamide hydrochloride was obtained by Ochs[35] at a voltage of 25 kv and a current of 10 ma, utilizing a Phillips X-Ray Powder Diffractometer. The sample was irradiated by a copper source at 1.54A°. Data derived from the spectrum are listed in Table 4; the intensity of each peak in the spectrum is shown relative to the incident radiation (I_o), as I/Io x 100, at twice the angle of incidence or reflection of the radiation (2θ) for the distance, d, between the atomic planes in the crystal.

Table 4

X-Ray Powder Diffraction Pattern of Procainamide Hydrochloride

Instrument:

Phillips X-Ray Powder Diffractometer

'd' (A^o)*	I/Io x 100**	I(2θ)***
8.30	25.26	24
7.40	12.63	12
6.43	7.37	07
5.76	10.53	10
5.40	14.74	14
5.26	100.00	95
4.82	14.74	14
4.52	40.00	38
4.30	27.37	26
4.10	30.53	29
4.01	90.53	86
3.92	66.32	63
3.74	49.47	47
3.48	52.63	50
3.34	18.95	18
3.23	8.42	08
3.14	18.95	18
2.98	28.42	27
2.95	14.74	14
2.86	8.42	08
2.77	20.00	19
2.50	10.53	10
2.36	7.37	07

*d (interplanar distance) = $\frac{n\lambda}{2 \sin \theta}$

$\lambda = 1.539 A^o$

** Based on highest intensity of 100.

*** Twice the angle of incidence or reflection.

2.25 Hygroscopicity

Walton[67] reported that solid procainamide hydrochloride is readily converted to a solution in an atmosphere of 65% relative humidity.

2.3 Solution Data

2.31 Solubility

Valenti[30] reported the following solubilities (U.S.P. definition) of procainamide hydrochloride in various solvents, at room temperature.

Solvent	Solubility	Parts of Solvent/One Part of Solute
Water	very soluble	less than 1
0.1N Hydrochloric acid	very soluble	less than 1
0.1N Sodium hydroxide	very soluble	less than 1
95% Ethanol	soluble	10 to 30
Propylene Glycol	sparingly soluble	30 to 100
Chloroform	slightly soluble	100 to 1000
Acetone	slightly soluble	100 to 1000
Ether	insoluble	more than 10000
Benzene	insoluble	more than 10000
Hexane	insoluble	more than 10000

The solubility of procainamide hydrochloride in acetone at 32°C was reported to be 3.43 mg per g of solvent[52].

Radushkevich[72] reported the solubility of procainamide hydrochloride in absolute ethanol to be 6% at 20° and 70% at the boiling point of ethanol. Procainamide hydrochloride was reported to be insoluble in 1,2-dichloroethane or in diethyl ether[72].

2.32 pKa

The pKa of procainamide hydrochloride, determined by a titrimetric procedure, was 9.24 (± 0.10)[77]. The pKa refers to the following type of dissociation:

$$R\text{-}NH_3^+ \rightleftharpoons R\text{-}NH_2 + H^+$$

2.33 Surface Chemistry

The dielectric potentials and surface tensions of solutions of procaine and of procainamide hydrochloride were compared over a wide pH range[97]. At pH 10.3 and 10.5 the following data on dielectric potential were obtained:

Concentration and Drug	pH	Dielectric Potential (mv)
0.0066M Procaine	10.5	477
0.02M Procainamide	10.3	268

Although the solution of procainamide was three times as concentrated as that of procaine, the dielectric potential of the former was less. At lower pH's, the dielectric potential of the procainamide solution was also less than that of procaine, although the difference was not as pronounced as at pH 10.3 and 10.5.

Similarly, the surface tension for the procainamide solution was less than that for procaine, over the pH range tested.

The authors[97] concluded that the toxicity of procainamide is less than that of procaine because procainamide is less surface active.

2.34 Complex Formation

Complex formation by procaine and procainamide with adenosine triphosphate (ATP) in aqueous solution was studied by optical rotation and NMR methods[96]. These studies indicated that hydrophobic stacking of the aromaticity, rather than horizontal hydrogen bonding, is implicated in the formation of these complexes. At pH 7.0, ATP binds one drug molecule tightly or two molecules weakly. The formation constants for tightly bound complexes as determined by both optical rotation and NMR measurements, were in good agreement. The weak-association constants were obtained by trial-and-error procedure.

The complex-formation constants were:

ATP Complex Strength	Procaine	Procainamide
Tight	$1430 \pm 380 \underline{M}^{-1}$	$3.60 \pm 90 \underline{M}^{-1}$
Weak	$1.1 \pm 0.7 \times 10^{5} \underline{M}^{-2}$	$4.7 \pm 2.8 \times 10^{3} \underline{M}^{-2}$

According to Thymm et al.[96] the activity of these drugs on nerve membranes may be ascribed to complexing with membrane-bound ATP.

3. Synthesis

p-Nitrobenzoic acid is converted to the acid chloride by reaction with sulfonyl chloride. p-Nitrobenzoyl chloride is then reacted with diethylaminoethylamine. The resultant nitro condensation product is then treated with hydrochloric acid and reduced to yield the final product[85].

$$O_2N-C_6H_4-COOH + SOCl_2 \longrightarrow O_2N-C_6H_4-COCl$$

$$+ NH_2CH_2CH_2N(C_2H_5)_2 \longrightarrow O_2N-C_6H_4-CONHCH_2CH_2N(C_2H_5)_2$$

$$\xrightarrow[\text{red.}]{HCl} H_2N-C_6H_4-CONHCH_2CH_2N(C_2H_5)_2 \cdot HCl$$

Alternative syntheses of procainamide have been described by several workers[72,86-90,95,98]. Preparation of procainamide by electrolytic reduction of N-(2-diethylaminoethyl)-p-nitrobenzamide has also been reported[84].

The synthesis of ^{14}C-procainamide hydrochloride, reported by Egli[93], utilized p-aminobenzoic acid labelled in the carboxyl carbon. The synthesis outlined above was followed. A purity analysis was performed by radioscanning a thin-layer chromatogram of the final product, using the support and solvent system II of Fairbrother and Shand[68] as described in section 6.42. The final product contained 4.83 μCi/mg, with less than 0.5% of radiochemical impurities. Egli[93] reported that it was possible to maintain the product at only 0.5% impurity by keeping the specific activity as low as possible and growing the crystals as large as possible; this procedure apparently minimizes the surface oxidation that is common with labelled compounds.

4. Stability

Bulk samples of procainamide hydrochloride were stored at 25 and 50°C for as long as 1 year. Samples stored at these temperatures showed no significant differences in intravenous or oral toxicity in mice. The purity, measured by titration with perchloric acid, ranged from 98.9-100.1% over the period of one year. Procainamide hydrochloride bulk appeared to be stable under the conditions employed[91,92].

5. Analysis of Impurities, Degradation Products and Residual Intermediates

N-(2-diethylaminoethyl)-*p*-nitrobenzamide, an intermediate in the synthesis of procainamide, has been determined polarographically[73]. The pH 4.0 buffer contained 250 ml of 1*M* sodium acetate, 750 ml of 1*M* acetic acid, 7.46 g of potassium chloride and 530 mg of dodecyltrimethyl ammonium chloride. A dropping-mercury electrolysis cell was used in which the average half-wave reduction potential (vs. Hg) was found to be -0.405 volts. Use of a gravimetric method[69] has also been reported, in which the nitrobenzamide compound was extracted into ether from alkaline solution; the solvent was removed by evaporation and the residue was dried and weighed.

The determination of *p*-nitrobenzoic acid, a possible process contaminant, was accomplished gravimetrically[69] after extraction from acid solution into ether and removal of the solvent by evaporation. The stability of *p*-nitrobenzoic acid can be monitored by thin-layer chromatography[83] on silica gel plates (with indicator) with a solvent system prepared by mixing 10 mls water with 20 mls of U.S.P. ethanol then diluting to 1000 mls with methyl isobutyl ketone.

Concentrations of N,N-diethylethylenediamine (a precursor diamine) in the range of 0.01-0.3% have been determined in procainamide hydrochloride by a colorimetric method[70] by Whigan and Kadin. In this test, based upon a report by J. Bartos[94], a 50-mg sample of procainamide hydrochloride is dissolved in dimethylformamide and reacted with a 0.1% solution of ascorbic acid, dissolved in the same solvent, by heating the mixture for 10 min in a boiling-water bath. The absorbance of the colored reaction product is measured at 530 nm. The color reaction between the diamine and ascorbic acid is enchanced by the presence of procainamide. A similar enhancement by p-aminobenzoic acid (PABA) had been observed. Therefore, 30 mg of PABA were added to both the reagent blank and the diamine standard to correct for this enhancement.

As little as 0.1% N,N-diethylethylenediamine in procainamide could also be determined quantitatively by paper electrophoresis and densitometry[71]. Samples and standards were spotted 12 cm from the anodic edge of 12 x 36 cm Whatman 3MM paper. Electrophoresis in a pyridinium-formate buffer was carried out for 30 min at a voltage gradient of 10 V/cm. The sheet was air dried, sprayed with a ninhydrin reagent containing sym-collidine, and developed for 3 hr. The colored spots were evaluated by densitometry.

p-Aminobenzoic acid in solutions of procainamide has been detected by circular paper chromatography[60]. After development with ammonia-saturated butanol, the paper was dried and sprayed with sodium nitrite in 5% hydrochloric acid, followed by 1-naphthol and ethanolic potassium hydroxide. Procainamide and its cleavage product give red circular zones.

p-Aminobenzoic acid, a potential cleavage product, has been separated by electrophoresis on paper[78], eluted in distilled water, then determined coulometrically. p-Aminobenzoic acid may be separated from procainamide in an extraction system[79]. The alkalinized sample is extracted with chloroform and the *p*- aminobenzoic acid remaining in the aqueous phase as the sodium salt can be determined spectrophotometrically (Section 6.3).

The phase-solubility of procainamide in acetone at 23 and 32°C has been used to determine the purity of the compound[52,80]. Purity was also evaluated by Differential Scanning Calorimetry (Section 2.22).

6. Analytical Tests and Methods

6.1 Elemental Analysis

$C_{13}H_{22}N_3OCl$

Element	Theory	Reported[64]
C	57.45%	57.45
H	8.16%	8.16
N	15.46%	15.46
O	5.89%	5.88
Cl	13.04%	13.05

6.2 Identification Tests

A test based on the preparation of benzoyl procainamide and the determination of its melting point is described in the U.S.P.[1].

Clarke has reported results for microcrystals and Vitali's color tests[11]. Crystal-reaction identification tests have been described[58]. A number of spot-plate tests to identify procainamide have been reported[13,15,17,18]. Identification has been achieved by paper and test-tube chromatography,[16,14] as well as by gas chromatography[41]. Color reactions with barbituric acid and 2-thiobarbituric acid have been reported[33].

Infrared spectroscopy (Section 2.11), spectrophotometry (Section 6.3), paper and thin-layer chromatography(Section 6.42), spectrophotofluorometry (Section 6.5) and gas chromatography (Section 6.43) provide alternate methods for purposes of identification.

6.3 Spectrophotometric Methods

Procainamide and its decomposition product *p*-aminobenzoic acid have been determined by measuring their ultraviolet absorption[55] (see Section 2.13).

The color developed upon the reaction of procainamide with vanillin in acid solution has been used to measure its concentration[48,53,54]. A reaction with thyme camphor after diazotization has also been reported[49]. Reaction with sulphonphthalein dyes[50] has been used to give a colored product for measurement.

Refer to Section 7 for spectrophotometric methods applicable to the determination of procainamide in biologic fluids and tissues.

6.4 Isolation and Chromatographic Methods

6.41 Solvent Extraction

Procainamide has been isolated by solvent extraction (see Sections 5.0 and 7.0) for subsequent determination by a variety of methods (see Sections 6.43, 6.6 and 7).

6.42 Paper and Thin-Layer Chromatography

Procainamide has been assayed by paper chromatographic methods[22,47,60]. Roberts[47] using S&S 597 filter paper, developed chromatograms for 16 hr in the solvent-rich upper phase of a solvent system composed of equal volumes of isobutanol and 0.05N phosphoric acid. The zones on the developed chromatograms were located by their ultraviolet absorbance.

A number of thin-layer chromatographic systems have been used to examine procainamide qualitatively[20,21,22,38] and quantitatively[22,36,40,68,82,93]. The quantitative TLC systems[36,82] can separate procainamide from such known impurities as *p*-aminobenzoic acid and *p*-nitrobenzoic acid. The most useful thin-layer systems are given in the next table.

6.43 Gas Chromatography

Gas chromatography has been used to separate and quantitate procainamide (see Section 7.0), as well as for identification of the compound (see Section 6.2).

Thin-Layer Chromatographic Systems

Ref.	Support	Solvent Composition
38	Silica Gel GF250	Benzene:Ethyl Acetate:Ethanol: Ammonium Hydroxide (15:15:5:1)
36,38,40	Silica Gel GF250	Benzene:Ethyl Acetate:Methanol: Ammonium Hydroxide(160:80:160:1)
82	Kieselgel GF254(Merck)	Benzene:Ethyl Acetate:Methanol: Ammonium Hydroxide (20:10:20:1)
93,68	Silica Gel QlF250 (Quantum Industries)	I. Benzene:Ammonium Hydroxide: 1,4-Dioxane(10:5:80) II. Isopropanol:Chloroform:Ammonium Hydroxide(45:45:5)

6.44 Ion Exchange Chromatography

The use of ion-exchange chromatography for the separation of procainamide from mixtures has been reported[24,29,74,75].

6.5 Spectrophotofluorometric Methods

Fluorescence of procainamide in the ultraviolet has been reported[19]. It exhibited an uncorrected activation maximum at 295 nm, with an uncorrected fluorescence maximum to 385 nm at pH 11. The minimal detectable concentration was reported to be 0.01 μg/ml.

The concentration of procainamide in biologic fluids has been measured spectrophotofluorometrically[8,43] (see Section 7.0).

6.6 Titrimetric Methods

Procainamide has been determined by titration with ICl[26]. The most generally useful methods for the determination of procainamide have been those involving titration with nitrite[23,25,27,28,42] and with perchloric acid[37,42].

Kadin[61] recently reported that procainamide, as well as some other aromatic amines, readily undergo partial acetylation in glacial acetic acid containing acetic anhydride. The acetylated aromatic amines are not titratable with acetous perchloric acid. The addition of acetic anhydride to glacial acetic acid is commonly employed to remove water when the acetous perchloric acid is constituted. Indeed, quite satisfactory titrations may be obtained when the acetic anhydride is not added to the acetous perchloric acid. However, reagent grade glacial acetic acid may already contain offending acetic anhydride. Kadin[61] has shown that acetic anhydride interference can be eliminated by prior reaction of the titration solvent with an aromatic amine. He titrated an aliquot of solvent, which had been cleared of acetic anhydride through reaction with the aromatic amine benzocaine, to neutrality just prior to dissolving the procainamide sample for titration. However, he strongly recommended abandonment of this rather cumbersome nonaqueous titration in favor of a simpler, more selective nitrite titration of procainamide using the stable internal indicator ferrocyphen.

Procainamide, isolated by extraction into chloroform from ammoniacal aqueous solution, was dissolved in dilute hydrochloric acid after removal of the solvent by evaporation. Procainamide was titrated with 0.1<u>M</u> sodium nitrite by oxidation-reduction potentiometry, using platinum electrodes[31]. The direct conductometric titration of procainamide hydrochloride with sodium hydroxide has been described[34]. The usual change in the slope of the conductance response with volume of the sodium hydroxide titrant occurs after the neutralization of the amine hydrochloride. Intersection of the two linear plots, representing the changes in conductance before and after neutralization, then accurately delineates the endpoint.

6.7 Microbiological Methods

A microbiological assay of procainamide with *Acetobacter suboxydans* has been reported[32].

7. Analysis in Biological Fluids and Tissues

Concentrations of procainamide in biological media have been measured colorimetrically after diazotization and coupling to the Bratton-Marshall reagent (N(1-naphthyl)-ethylene-diamine)to form a colored product. These methods have involved protein precipitation[45] or solvent extraction[2,43,46,56] to isolate procainamide prior to the colorimetric measurement.

Concentrations of procainamide in blood and urine were estimated after reaction of the drug with 4-dimethylaminocinnamaldehyde[81] to form a colored product.

A rapid procedure for measuring plasma procainamide concentrations by gas chromatography has been reported by Atkinson et al[39]. They used a 2 m x 2 mm i.d. glass coil packed with 0.2% OV-17 on a 100/120 mesh glass bead (Corning GLC-110). The temperature of the injector port was 230°C, that of the column 225°C, and of the flame ionization detector, 240°C. A flow rate of 30 ml/min N_2 carrier gas and H_2 was used. The air-flow rate for the detector was 300 ml/min. Procainamide was extracted from alkaline solution into methylene chloride, with p-amino-N-(2-dipropyl-aminoethyl)benzamidine HCl as the internal standard. The residue resulting from the evaporation of this methylene chloride solution was dissolved in ethyl acetate and injected into the chromatograph.

The determination of procainamide in serum[8,43] by use of its fluorescent properties has provided an additional tool. Procainamide was extracted from alkalinized salt-saturated serum into benzene containing 1.5% isopentyl alcohol.

It was re-extracted into dilute hydrochloric acid. Its fluorescence was measured after the pH had been adjusted to 11 by the addition of sodium hydroxide.

A colorimetric method for the determination of N-acetylprocainamide, a metabolite of procainamide, has been reported by Poet[44]. In this method[2,46], N-acetylprocainamide is extracted from alkalized serum into benzene containing 1.5% isoamyl alcohol. It is then re-extracted from the solvent into 1N-hydrochloric acid. An aliquot is heated in a boiling-water bath to hydrolyze any N-acetylprocainamide present. N-acetylprocainamide is then determined as the difference in procainamide content of the sample extract before and after hydrolysis, as measured by the previously reported colorimetric method[2,46].

The metabolism and distribution of procainamide hydrochloride in man and dog was studied by Mark et al[2]. They reported that 50-60% of a single intravenous dose administered to man was excreted unchanged in the urine; that about 2-10% was accounted for as p-aminobenzoic acid; and that the drug did not accumulate when given in repeated oral doses. Procainamide concentrations in these studies were measured after isolation of the drug from alkalized biological material by extraction into benzene, augmented by saturation of the aqueous phase with sodium chloride. The drug was returned to dilute acid, diazotized, and coupled with N(1-naphthyl)ethylene diamine. The resulting colored dye was determined spectrophotometrically (see Section 7.0). The specificity of the method was demonstrated by the calculation of comparative distribution ratios[76].

Dreyfuss *et al.*[3,4] studied the biotransformation of procainamide hydrochloride in rhesus monkey, man, and dog, using ^{14}C-labelled drug. In the dog, 50-67% of the ^{14}C activity excreted in the urine was unchanged procainamide; four metabolites were recognized. In monkey, 22-49% of the ^{14}C activity was excreted in the urine as unchanged procainamide; two metabolites were recognized. N-acetyl procainamide, isolated and identified by a combination of TLC, NMR and mass spectrometry, was a major metabolite excreted by monkey; although present in man this metabolite did not appear to occur in dog. The combination of these analytical techniques helped to provide essential data in these metabolic studies. These studies also suggested that compounds containing readily acetylated aromatic amino groups should be studied in primates rather than in dogs, but preferably in both species. Weily and Genton[6] have reported on the routes and mechanism of procainamide excretion; the effect of urine pH; and the effects of flow rate and the presence of renal or hepatic impairment on the plasma half time and on the renal and hepatic clearance and excretion of procainamide.

The pharmacokinetics of procainamide hydrochloride in man has been studied extensively[7,8,9]. Koch-Weser demonstrated that plasma concentrations between 4 and 8 mg/L are found after the usually effective therapeutic doses. He showed that knowledge of plasma concentrations is helpful in establishing the optimal dosage in individual patients. Procainamide concentrations were determined spectrophotometrically[8] and spectrophotofluorometrically[8] by methods that are well within the capability of hospital laboratories (see Section 7).

A recent report[10] indicated that N-acetyl procainamide was detected in the plasma of two patients receiving procainamide. This compound was identified by a combination of TLC, gas chromatography and mass spectrometry. This metabolite was demonstrated to have anti-arrhythmic activity in mice. The report concluded that the observed anti-arrhythmic activity of procainamide in man may be due, in part, to its conversion to N-acetylprocainamide (see also Section 7).

8. References

1. United States Pharmacopeia, XVIII, pgs. 8,541(1970).

2. L. C. Mark, H. J. Kayden, J. M. Steele, J. R. Cooper, I. Berlin, E. A. Rovenstine, and B. B. Brodie, J. Pharmacol. Exp. Ther., 102, 5-15(1951).

3. J. Dreyfuss, J. T. Bigger, A. I. Cohen and E. C. Schreiber, Clin. Pharmacol. Ther. 13, 366-71(1972).

4. J. Dreyfuss, J. J. Ross, and E. C. Schreiber, Arzneim-Forsch.,21(7), 948-51(1971).

5. A. I. Cohen,Squibb Institute, Personal Communication.

6. H. S. Weily and E. Genton, Arch. Intern. Med., 130, 366-69(1972).

7. J. Koch-Weser, Ann. N.Y. Acad. Sci., 179, 370-82(1971).

8. J. Koch-Weser and S. W. Klein, J. Amer. Med. Ass., 215, 1454-60(1971).

9. E. G. Giardina, R. H. Heissenbuttel and J. T. Bigger, Circulation, 42, Supp. 3, 156(1970).

10. D. E. Drayer and M. M. Reidenberg, Pharmacologist, 15, 154(1973).

11. E. G. C. Clarke, J. Pharm. Pharmacol., 8, 202-6(1956).

12. N. H. Coy and B. Toeplitz, Squibb Institute, Personal Communication.

13. P. Cooper, Pharm. J., 177, 495-6(1956).

14. R. Fischer and N. Otterbeck, Sci.Pharm., 26, 76-8(1958),Chem.Abstr., 53,4657g (1959).

15. G. I. Luḱ-Yanchikova, Med. Prom.S.S.S.R., 15 No. 8, 43-5(1961), Chem.Abstr.,56, 2512c(1962).

16. L. O. Kirichenko, Farm. Zh.(kier), 17 (15), 68-72(1962),Chem.Abstr., 61, 4155e (1964).

17. L. M. Atherden, Pharm. J., 195, 115(1965).

18. N. P. Yavorskii, V. S. Koval and M. M. Starushchenko, Farm. Zh.(Kiev), 20 (5), 31-7(1965) (Ukrain), Chem.,Abstr., 64, 9512f(1966).

19. S. Udenfriend, D. E. Duggan, B.M. Vasta, J. Pharmacol. Exp. Ther., 120, 26-32(1957)

20. J. A. Fresen, Pharm. Weekbl., 102 (28), 659-78(1967).

21. H. Amal, E. Gursu and S. Demir, Int. Symp. Chromatogr. Electrophoresis, 5th 1968 (Pub. 1969) 441-6(FR) Ann.Arbor-Humphrey Sci. Pub. Inc., Ann Arbor, Mich., Chem. Abstr., 72, 28191(1970).

22. K. Macek and J. Vecerkova, Pharmazie, 20(10), 605-16(1965) (GER), Chem. Abstr., 64, 527c (1966).

23. L. N. Guseva, N. I. Vestfal and M. I. Kuleshova, Farmatsiya (Moscow), 16 (3), 46-51(1967) (Russ.), Chem. Abstr., 67, 67668(1967).

24. G. A. Vaisman and M. M. Yampol'Skaya, Aptechn. Delo., 11(5), 38-41(1962), Chem. Abstr., 60, 10476b(1964).

25. R. Vasiliev, A. Cosmin, M. Mangu and I. Burnea, Rev. Chim. (Buchar), 13,239 (1962)., Chem.Abstr.,57,16752a(1962).

26. Ya. I. Kadyrov and O. Alimkhanov, Uzbeksk. Khim. Zh., 9 (1), 31-3(1965) (Russ.), Chem. Abstr., 63, 2852a (1965).

27. W. Wisniewski and T. Kindlik, Acta. Pol. Pharm., 25 (1), 55-8(1968) (Pol.), Chem. Abstr., 68, 107930(1968).

28. W. Wisniewski and T. Kindlik, Acta. Pol. Pharm., 26 (1), 39-43(1969) (Pol.), Chem. Abstr., 70, 109201(1969).

29. D. S. Yaskina and I. Nguyen-Bah, Aptechn. Delo., 12(1), 76-7(1963), Chem. Abstr., 61, 13131g(1964).

30. V. Valenti, Squibb Institute, Personal Communication.

31. W. Wisniewski and T. Kindlik, Acta. Pol. Pharm., 23 (6), 517-22(1966) (Pol.), Anal. Abstr., 15, 1039(1968).

32. A. Rappe, S. Baur and G. Mauquoy, Ann. Pharm. Fr., 27(11), 655-62(1969) (Fr.), Chem. Abstr., 73, 28977(1970).

33. G. I. Kudymov, A. A. Kiseleva and M. V. Mokrouz, Tr. Permsk. Farm. Inst., No. 3, 107-9(1969) (Russ.), Chem. Abstr., 75, 40532(1971).

34. B. P. ArtAmonov and S. L. Maiofis, Farmatsiya (Moscow), 20 (1), 36-41(1971) (Russ.), Chem. Abstr., 75, 40535(1971).

35. Q. Ochs, Squibb Institute, Personal Communication.

36. P. Taylor and H. Roberts, Squibb Institute, Personal Communication.

37. G. Hart, Squibb Quality Control, Personal Communication.

38. P. Taylor and H. Roberts, Squibb Institute, Personal Communication.

39. A. J. Atkinson, M. Parker and J. Strong, Clin. Chem., 18(7), 643-46(1972).

40. H. R. Roberts and P. Taylor, Squibb Institute, Personal Communication.

41. B. S. Finkle, E. J. Cherry and D. M. Taylor, J. Chromatogr. Sci., 9,393 (1971).

42. H. Kadin and E. J. Jenkins, Squibb Institute, Personal Communication.

43. J. Koch-Weser, S. W. Klein, L. L. Foo-Canto, J. A. Kastor and R. W. Desanctis, N. Eng. J. Med., 281, 1253-60 (1969).

44. R. B. Poet, Squibb Institute, Personal Communication.

45. S. Bellet, S. E. Zeeman and S. A. Hirsh, Am. J. Med., 13, 145-157(1952).

46. R. B. Poet and J. Kowald, Squibb Institute, Personal Communication.

47. H. R. Roberts, Squibb Institute, Personal Communication.

48. G. I. Luḱ-Yanchikova and V. N. Bernshtein, U.S.S.R. Patent No. 141, 486 (16,10,61), Anal. Abstr., 9, 2492(1962).

49. W. Wisniewski and T. Kindlik, Diss.Pharm. Pharmacol., 18 (5), 529-32(1966) (Pol.), Chem. Abstr., 67, 14908(1967).

50. M. Horioka, Yakugaku Zasshi, 77, 200-6 (1957), Chem. Abstr., 51, 8367b(1957).

51. J. Kracmar and J. Kracmarova, Cesk. Farm. 15 (3), 121-9(1966) (Czech.), Chem. Abstr., 65, 3668b(1966).

52. D. Dicksius, Squibb Institute, Personal Communication.

53. G. I. Luk-Yanchikova and V. N. Bernshtein, Peredovye. Metody. Khim. Tekhnol, i Kontrolya. Proiz. Sb., 257-60 (1964) (Russ.), Chem. Abstr., 62, 12977b(1965).

54. V. N. Bernshtein and G. I. Luk-Yanchikova Uch. Zap. Pyatigor. Gos. Farmatsevt. Inst., 5, 165-9(1961), Chem. Abstr., 59, 2592g(1963).

55. M. D. Denisov, Farm. Zh. (Kiev), 19 (3), 60-5 (1964) (Ukrain), Chem. Abstr. 64, 3286a(1966).

56. G. Pitel and T. Luce, Ann. Pharm. Franc., 23 (11), 673-81(1965) (Fr.), Chem. Abstr. 64, 14775a (1966).

57. J. M. Dunham, Squibb Institute, Personal Communication.

58. T. Van der Wegen, Pharm. Weekbl., 103, 173-197 (1968).

59. A. Cohen, Squibb Institute, Personal Communication.

60. J. Zimmer, Krankenhaus-Apotheker, 7, 26(1975), Chem. Abstr., 52, 8463d(1958).

61. H. Kadin, J. Pharm. Sci., 63, 919(1974).

62. P. J. Black and A. I. Cohen, paper presented at Twentieth Annual Congress on Mass Spectrometry and Allied Topic, Dallas, Texas (6/4-9/72).

63. J. Womack, E.A.I., Princeton, N.J., Personal Communication.

64. J. Alicino and J. Hydro, Squibb Institute, Personal Communication.

65. E. Feldman, J. Am. Pharm. Assoc. Sci.Ed. 47, 676(1958).

66. K. Biemann, "Mass Spectrometry: Organic Chemical Applications", McGraw-Hill Book Co., Inc., New York(1962).

67. R. Walton, Squibb Institute, Personal Communication.

68. J. Fairbrother and S. Shand, Squibb International Development Laboratories, Personal Communication.

69. G. Brewer, Squibb Institute, Personal Communication.

70. D. Whigan and H. Kadin, Squibb Institute, Personal Communication.

71. O. Kocy, Squibb Institute, Personal Communication.

72. K. Radushkevich, Mater. Obmenu. Peredovym. Opytom.Nauch.Dostizhen. Khim-Farm.Prom.,No.1,22-6(1958),Chem.

Abstr., 54, 17316i (1960).

73. O. Kocy, Squibb Institute, Personal Communication.

74. G. L. Starobinets and I. P. Koka, Vestsi. Akad. Navuk. Belarus. SSR Ser. Khim. Navuk., (1967) (4) 16-21 (Russ.), Chem. Abstr., 68, 89839(1968).

75. G. L. Strobinets and S. D. Kul'kina, Vestsi. Akad. Navuk. Belarus. SSR. Khim. Navuk., (1971) (2) 46-50 (Russ.), Chem. Abstr., 75, 25276 (1971).

76. B. Brodie and S. Udenfriend, J. Biol. Chem., 158, 705 (1945).

77. H. Jacobson and C. Schaefer, Squibb Institute, Personal Communication.

78. K. Kalinowski and Z. Zwierzchowski, Acta. Polon. Pharm., 15, 175-8 (1958), Chem. Abstr., 52, 17617b (1958).

79. M. Dolliver, Squibb Institute, Personal Communication.

80. D. Dicksius, Squibb Institute, Personal Communication.

81. C. K. Parekh, U.S. Patent 3, 716,336 (Cl,23/230B; G 0ln), 13 Feb. 1973, Chem. Abstr., 79, 15563(1973).

82. J. Fairbrother, Squibb International Development Laboratory, Personal Communication.

83. T. Soh, Squibb Quality Control, Personal Communication.

84. Y. Tashika and M. Kuranari, J. Pharm. Soc. Japan, 73, 1069-71(1953), Chem. Abstr., 48, 12027g (1954).

85. M. Yamazaki, Y. Kitagawa, S. Hiraki and Y. Tsukamoto, J. Pharm. Soc. Japan, 73, 294-7(1953), Chem.Abstr., 48,2003i(1954).

86. H. Najer and M. Joannic, Ann. Pharm. Franc, 13, 556-9(1955), Chem. Abstr., 50, 9331d (1956).

87. V. Hach and E. Koppová, Česk. Farm., 5, 582-3(1956), Chem. Abstr., 51, 8693c (1957).

88. K. Iwaya and K. Yoshida, Ann. Rep. Shionogi. Res. Lab., #2, 40-1(1952), Chem. Abstr., 51, 10424h (1957).

89. Z. Ledóchowski, M. Bogucka, A. Ledóchowski and A. Chimiak, Przem. Chem., 38, 91-2 (1959), Chem. Abstr., 53, 19953d (1959).

90. T. Shen, L. Yu, C. Chin, I. Wang and W. Luh, Yao Hsueh T'ung. Pao., 8,319-21 (1960), Chem. Abstr.,59, 2699h (1963).

91. M. Blaich and A. Restivo, Squibb Institute, Personal Communication.

92. M. Blaich and A. Restivo, Squibb Institute, Personal Communication.

93. P. Egli, Squibb Institute, Personal

Communication.

94. J. Bartos, Ann. Pharm., Franc, 22, 383 (1964).

95. J. Orloski, Squibb Institute, Personal Communication.

96. P. T. Thymm, R. Luchi, H. L. Conn., J. Pharmacol. Exp. Ther., 164, 239-251 (1968).

97. J. Kruk, Zesz. Nauk. Uniw. Jagiellon. Pr. Chem., No. 12, 111-14(1967) (Eng.), Chem. Abstr., 69, 89883 (1968).

98. K. Iwaya and K. Yoshida, Japan, 4325 ('52), Oct. 22, Chem. Abstr., 48, 5218i (1954).

99. G. Ghielmetti, Farmaco (Pavia) Ed. Sci. 9, 384-6 (1954), Chem. Abstr., 49, 7521a (1955).

100. E. Pavlovska, J. Lukac and M. Borovicka, Czech., 114, 447(Cl. C 07c) April 15, (1965), Chem. Abstr., 64, 6575c (1966).

RESERPINE

Roger E. Schirmer

CONTENTS

1. Description

Reserpine is 1,2-didehydro-2,7-dihydro-11,17α-dimethoxy-3β,20α-Yohimban-16β-carboxylic acid methyl ester, 18β-trimethoxybenzoate ester.

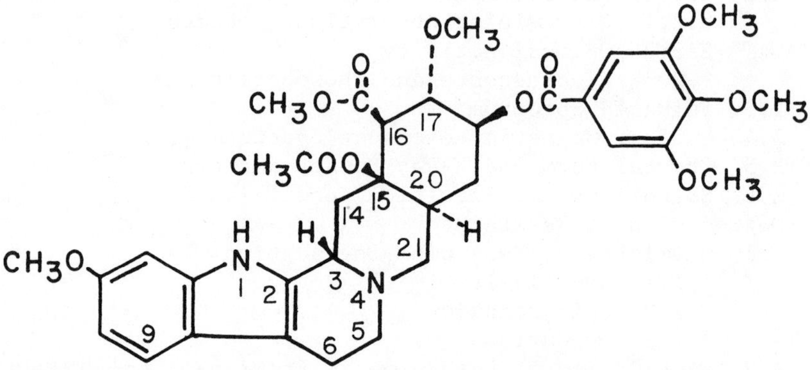

Reserpine is generally obtained by extraction of the roots of certain species of Rauwolfia (Apocynaceae), principally R. Serpentina and R. Vomitoria,[1] or by synthesis.[2-5] It occurs as a white or pale buff to slightly yellow, odorless, crystalline substance. It is weakly basic, with a pK of 6.6.

2. Physical Properties

2.1 Melting Point

Reserpine melts at 264-265°C with decomposition.[1] Hochstein et. al.[6] have reported that reserpine melts at 284-286°C in an evacuated tube, and that the apparent melting point is less dependent on heating rate when measured under vacuum.

The melting points of several salts of reserpine are listed in Table 1,[7] and melting points for several stereoisomers have been included in Table 2.

Table 1

Melting Points of Several Reserpine Salts[7]

Salt	Empirical Formula	Melting Point*
hydrochloride	$C_{33}H_{40}N_2O_4 \cdot HCl \cdot H_2O$	224°
sulfate	$C_{33}H_{40}N_2O_9 \cdot H_2SO_4$	242-244°
perchlorate	$C_{33}H_{40}N_2O_9 \cdot HClO_4$	238-239°
nitrate	$C_{33}H_{40}N_2O_9 \cdot HNO_3$	235°
oxalate	$C_{33}H_{40}N_2O_9 \cdot C_2H_2O_4 \cdot 1/2H_2O$	206°
methylbromide	$C_{33}H_{40}N_2O_9 \cdot CH_3Br$	271-272°
maleate	$C_{33}H_{40}N_2O_9 \cdot C_4H_4O_4$	226-227°
picrate	$C_{33}H_{40}N_2O_9 \cdot H_2O$	183-186°

Table 2

Sodium D-line Rotations for Reserpine and Several of Its Configurational Isomers

Compound	Isomer*	Melting Point °C	$[\alpha]_D$	Conditions	Ref
reserpine	-	264-265	-118°	$CHCl_3$, 23°C	7,8
			-164°	pyridine, 26°C	7
			-168°	dimethyl-foramide, 26°C	7
			-100°	dilute acetic acid, 25°C	16
			-125°	Dioxane, 25°C	16
16-epireserpine	16α	180	+ 44°	$CHCl_3$	1
18-epireserpine	18α	141-145	+ 38°	$CHCl_3$	10
16-epi-17-epi-reserpine (neoreserpine)	16α,17β	163-170	+ 29°	$CHCl_3$	11
3-isoreserpine	3α	152-165	-164°	$CHCl_3$	8,9
18-epiisoreserpine	18α	245-248		$CHCl_3$	10

*The entries in this column give the configuration of the carbons which differ from the configuration in reserpine.

2.2 Electronic Spectra

2.2.1 Ultraviolet Absorption Spectrum

The ultraviolet spectra of indole alkaloids generally consist of three bands, one band in the region of 225 nm and two bands, often unresolved, in the region of 280 nm. These bands have been assumed to have the same origin as the low energy transitions of benzene, and so the benzene nomenclature[12] is used for the indole chromophore.[13] As shown in Figure 1, all three bands are distinct in the spectrum of reserpine. The $^1Ba \leftarrow {^1A}$ band maximum occurs at 216 nm (ϵ=55,700), the $^1La \leftarrow {^1A}$ band at 267 nm (ϵ=15,700) and the $^1L_b \leftarrow {^1A}$ band at 296 nm (ϵ=9,660).[14] The absorption maxima and extinctions are essentially identical in chloroform and methanol.

2.2.2 Optical Activity

Carbons 3, 15, 16, 17, 18, and 20 of reserpine are asymmetric, resulting in 64 possible configurational isomers. Reserpine itself has a rotation of -118° (sodium D line) at 23°C in chloroform.[7,8] The D-line rotations for reserpine and several of its isomers are listed in Table 2. The circular dichroism (CD) and magnetic circular dichroism (MCD) spectra of reserpine have been reported[15] and are reproduced in Figure 2. The magnetic circular dichroism is of special analytical interest because the sign pattern can be used to distinguish indole alkaloids from alkaloids bearing the indoline or oxindole chromophores: the latter chromophores show a positive effect at low wavelength (^1L_a transition) and negative effects at higher wavelength (^1L_b transition), whereas indole alkaloids have negative effects at low wavelength and positive effects at higher wavelength.

2.2.3 Fluorescence and Phosphorescence

The absorption band at 296 nm in reserpine gives rise to strong fluorescent emission at 375 nm.[17,18,19] In water, the fluorescence intensity is maximum at low pH:[17] a spectrum in water at pH 1.0 is reproduced in Figure 3. Fluorescence is typical of compounds closely related to reserpine, including alkaloids which occur naturally along with reserpine and degradation products of reserpine (*vide infra*). The fluorescence characteristics of several of these are summarized in Table 3. In reserpine, deserpidine, and rescinnamine, it has been shown that light energy absorbed by the indole chromophore is emitted from the trimethoxybenzoate chromophore.[20,21] The intramolecular singlet-singlet exci-

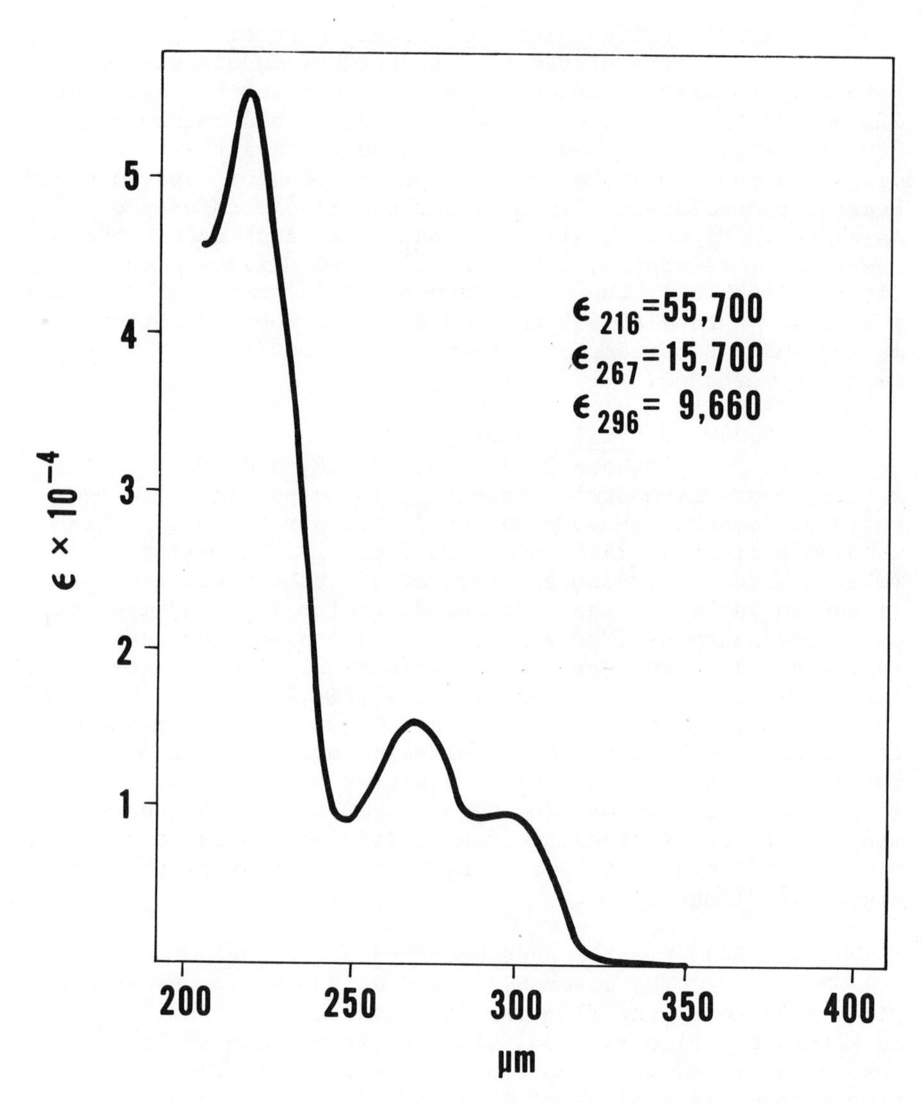

Figure 1. The Ultraviolet Absorption Spectrum of Reserpine

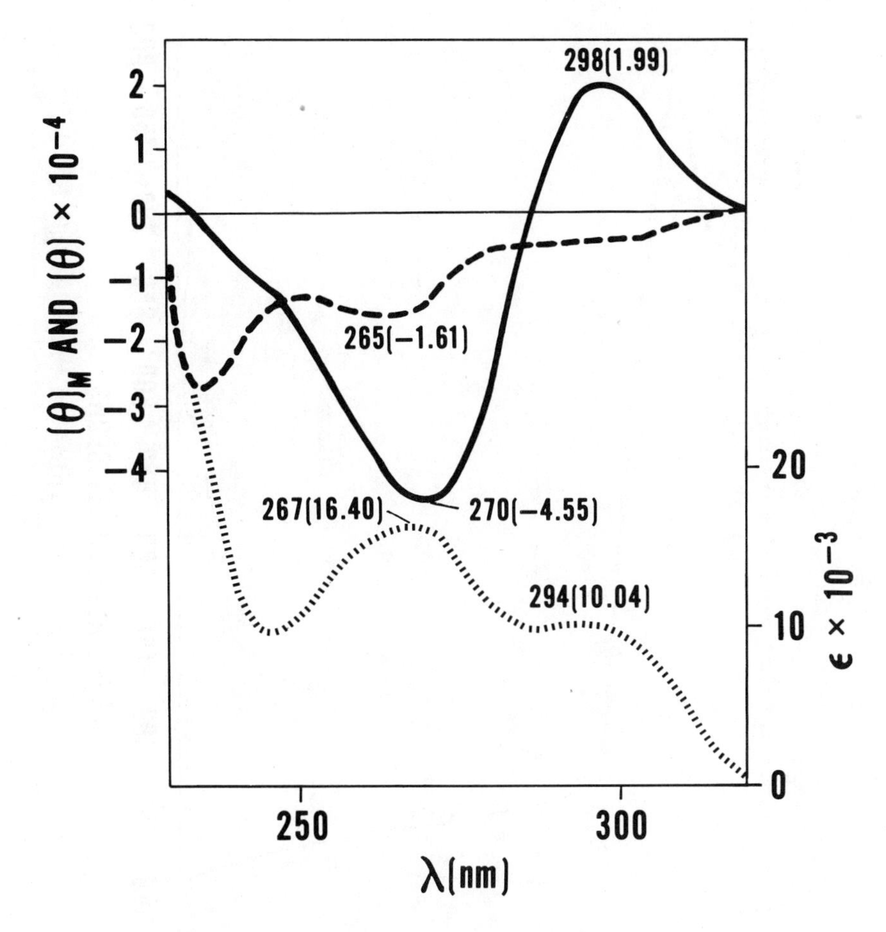

Figure 2. Magnetic Circular Dichroism (-), Circular Dichroism (----), and Absorption Spectra (.....) of Reserpine in Methanol

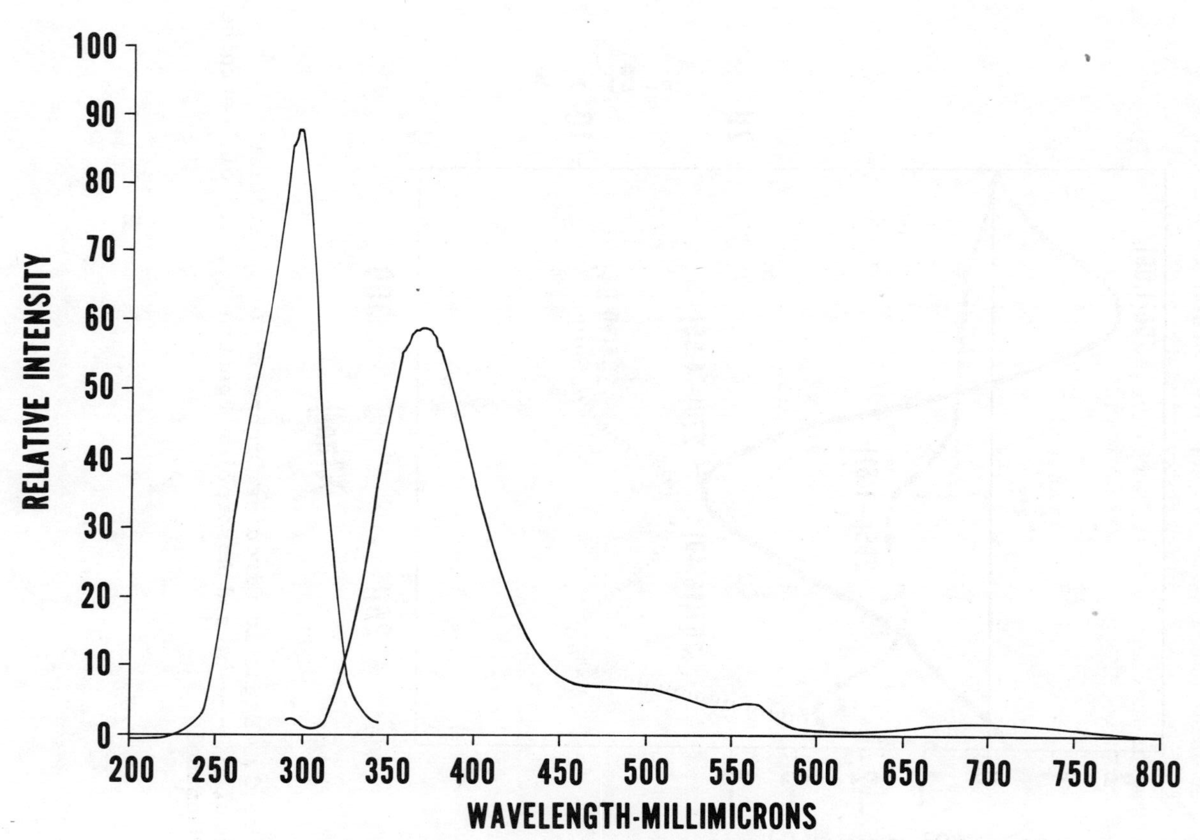

Figure 3. Fluorescence Spectrum of Reserpine in Water at pH 1.

tation transfer occurs with 100% efficiency, and involves transfer from the indole S_1 (π,π^*) state to the ^1L_b (π,π^*) state of the trimethoxybenzoate acceptor. The energy transfer appears to occur by an exchange mechanism[22] rather than a resonance mechanism.[23]

Table 3

Fluorescence Characteristics of Compounds Related to Reserpine[19]

Compound	Excitation Maximum	Emission Maximum	Relative Sensitivity
Reserpine	280	360	6
3-iso Reserpine	390	510	18
Dehydroreserpine perchlorate	390	510	75
Tetradehydroreserpine chloride	340	440	50
Rescinnamine	310	440	2
Rescinnamine N-oxide	310	340	2-3
Deserpidine	280	360	15
Reserpine N-oxide	280	360	8
Methyl Reserpate	300	360	19
Reserpinine	300	360	2
Methyl-0-(35-dimethoxy-4-hydroxybenzoyl reserpate)	300	360	8
Syrosingopine	300	360	4
Syrosingopine N-oxide	290	350	6
Trimethoxybenzoic Acid	280	360	2
Trimethoxycinnamid Acid	300	400	<1
Indole	280	340	36
Harmaline HCl	390	490	150

Phosphorescent emission from reserpine occurs at about 460 nm.[20]

2.3 Infrared Spectrum

The infrared spectrum of reserpine has been reported by several workers.[24-29] The spectrum in chloroform is reproduced in Figure 4, and the assignments of several bands in the spectrum are as follows:

3480 cm^{-1}	> N-H stretch
2840-3030 cm^{-1}	> C-H stretch
1732 cm^{-1}	> C=O stretch, acetyl group
1713 cm^{-1}	> C=O stretch, trimethoxybenzoate group

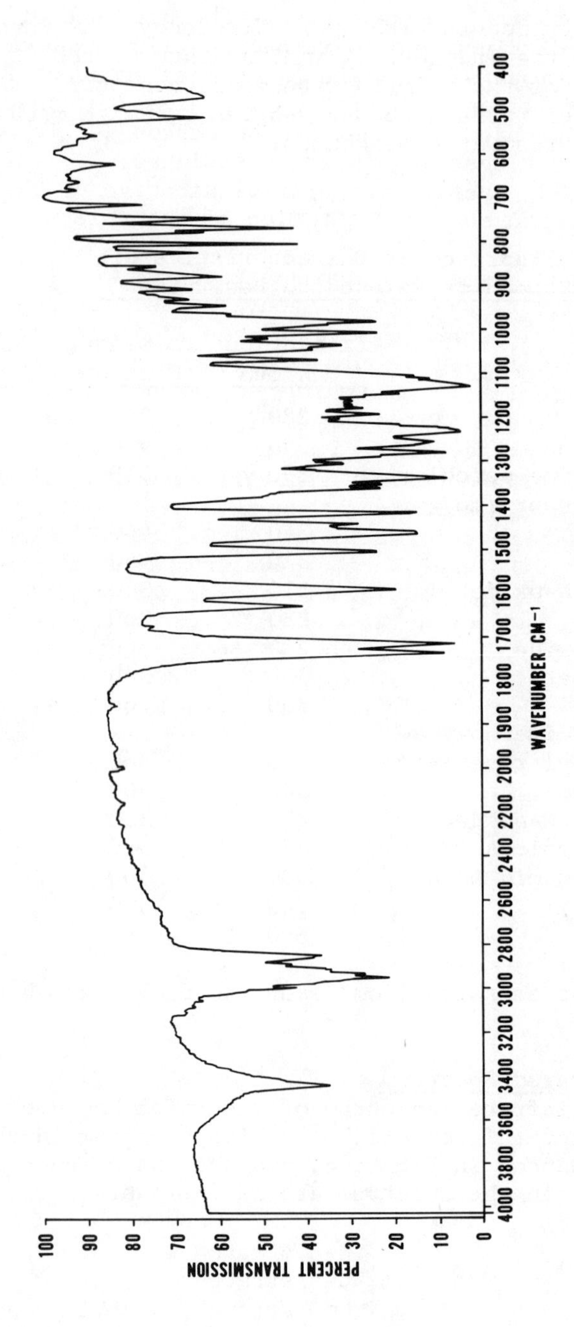

Figure 4. The Infrared Spectrum of Reserpine in Chloroform

At one time Wenkert[30] suggested that the presence of two or more bands or distinct shoulders between 2700-2900 cm^{-1} on the low wave number side of the major band at about 2900 cm^{-1}, indicated an α hydrogen on carbon 3, whereas the absence of these bands indicated a β hydrogen on carbon 3. It has since been shown[28] that these bands are associated with axial hydrogens at C-3, and are therefore indicative of the conformation rather than the configuration of the molecule.

2.4 Nuclear Magnetic Resonance Spectrum

The n.m.r. spectra of methyl 3-isoreserpate, methyl reserpate, methyl neoreserpate, and the 3',4',5'-trimethoxybenzoate esters of each of these compounds have been reported by Rosen and Shoolery.[8] The 60 MH_2 spectrum of reserpine is reproduced in Figure 5, and the assignments of several of the resonances in the spectrum are given in Table 4.

As axial protons are generally found at higher fields in the n.m.r. spectrum than the corresponding equatorial protons, the resonances of the C-3 and C-18 protons may be used to distinguish reserpine from isoreserpine or neoreserpine. In the most stable conformation of reserpine, the C-3 proton is equatorial and the C-18 proton is axial. In 3-isoreserpine, however, the C-3 proton is axial and is shifted to high enough field that it is obscured by the $-OCH_3$ resonances: the axial C-18 resonance is essentially unchanged. In neoreserpine the C-3 proton is again axial (therefore not discernable in the spectrum) and the C-18 proton is equatorial and therefore shifted to lower field than in reserpine (5.83 p.p.m. compared with 5.05 p.p.m., respectively).

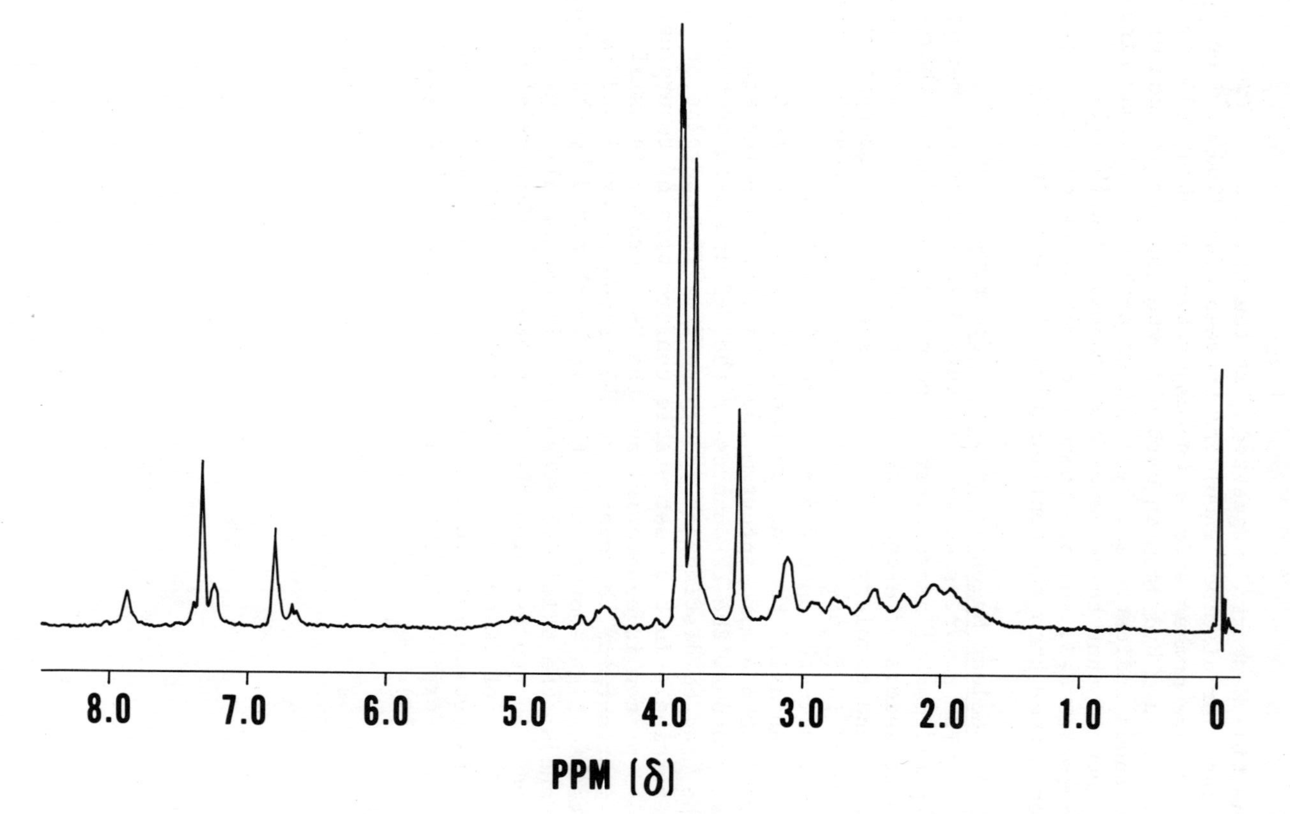

Figure 5. The 60 MH_z NMR Spectrum of Reserpine in $CDCl_3$. See Table 5 for Assignments.

Table 4

Assignment of the N.M.R. Spectrum of Reserpine in $CDCl_3$

Functional Group	Chemical Shift (p.p.m. from tetramethylsilane)
NH	7.85 p.p.m. (broad singlet)
Indole aromatic H (C-9,C-10,C-12)	Two groups of lines at about 6.7 and 7.3 p.p.m.
Trimethoxybenzoic acid aromatic H	7.34 p.p.m. (sharp singlet)
C-18 H	5.05 p.p.m. multiplet
C-3 H	4.43 p.p.m. multiplet
Trimethoxybenzoic acid, $-OCH_3$	3.92 p.p.m. singlet
C-11, $-OCH_3$	3.79 p.p.m. singlet
C-16, $-\overset{O}{C}-OCH_3$	3.79 p.p.m. singlet
C-17, $-OCH_3$	3.46 p.p.m. singlet

2.5 Crystal Form and X-Ray Powder Pattern

Crystallographic data on reserpine has been reported by Rose.[31] The crystals are monoclinic with axial ratios, a:b:c, of 1.654:1:1.537 and B = 115°8'. They occur as blades or needles elongated parallel to b and lying on the orthopinacoid 100 or basal pinacoid 001 . The crystals sometimes show the clinodome 011 and the hemipyramid: blades are generally terminated by the clinopinacoid 010 .

The density of the crystals was determined to be 1.298 g/cm^3 by flotation and 1.293 g/cm^3 by x-ray. The refractive indexes of the crystal at 25°C for 5893Å radiation are as follows: α=1.51, β=1.568 ± 0.002 and γ=1.687 ± 0.004.

The unit cell contains two molecules and has dimensions A=14.57Å, b=8.81Å, and c=13.5Å. The powder pattern is summarized in Table 5.

Hamilton[32] has reported that reserpine is one of the largest acentric molecules whose structure has been determined by direct x-ray techniques, but he did not report the details of this study.

Table 5

X-Ray Powder Pattern of Reserpine[31]

d	I/I_1	hkl	d(calculated)
13.14	0.40	100	13.11
12.22	0.40	001	12.18
7.44	0.20	101	7.48
7.19	1.00	011	7.14
5.73	0.40	111	5.70
5.34	0.40	210	5.26
5.02	0.40	012	5.01
4.78	0.60	102	4.80
4.49	0.80	103	4.45
4.21	0.40	112	4.21
4.18	0.20	021	4.15
3.71	0.40	202	3.74
3.56	0.40	401	3.57
3.47	0.20	103	3.48

2.6 Solubility and Partition Coefficients

Reserpine is only soluble to the extent of 6 mcg/ml in 0.1 N HCl at 37°C, and its solubility in aqueous solvents decreases at high pH. Reserpine is much more soluble in less polar solvents, with a solubility of about 0.56 mcg/ml in ethanol and 167 mg/ml in chloroform at room temperature.[7]

Distribution coefficients for reserpine and several related alkaloids have been reported by Hochstein et. al.[6] and are reproduced in Table 6.

Table 6

Distribution Coefficients for Reserpine and Related Alkaloids[6]

Compound	Solvent System A	Solvent System B	Solvent System C
Reserpine	9.0	0.01	8.0
Yohimbine	9.0	0.01	1.5
Rauwolscine	9.0	0.01	0.7
Ajmalicine	11.0	0.01	8.0
Heterophyllin	20.0	0.01	8.0
Serpentine	0.01	11-13	7
Ajmaline	0.01	11-13	1

System	Solvent 1	Solvent 2
A	benzene	.05M pH 7 phosphate buffer
B	50% CH_3OH in H_2O	$CHCl_3$
C	n-butanol	15% CH_3COOH in H_2O

Distribution Coefficient = conc. solvent 1/conc. solvent 2

3. Degradation of Reserpine

3.1 Chemistry of Reserpine Degradation

3.1.1 Hydrolysis

The reserpine molecule contains two ester groups, both of which are susceptible to hydrolysis, as shown in Figure 6.

Hydrolysis of the trimethoxybenzoate ester group occurs much more rapidly than hydrolysis of the methyl ester group.[26,33,34] As with all esters, the hydrolysis of reserpine is catalyzed by both acid and base with basic catalysis generally being most efficient.[35] Although the full pH profile of reserpine hydrolysis has apparently not been studied, one would expect the rate of hydrolysis to be minimum in the pH range of 3-4.

3.1.2 Epimerization

Although reserpine has six asymmetric carbons, only C-3 is inverted easily enough for isomerization to be a significant route of degradation. Epimerization of reserpine to 3-isoreserpine occurs principally in strong acid solution.[36-38] Gaskell and Joule[39] have shown that epimerization under these conditions is initiated by protonation of C-2 with subsequent ring opening to give an intermediate in which C-3 is planar and properly oriented for efficient reclosure of the system to 3-isoreserpine.[39] Bayer[40] has reported that epimerization of reserpine in solution in chloroform is also promoted by heat and light.

Hakkesteegt[38] has reported that the ratio of epimerization to hydrolysis was 2.2, 5.6 and 8.1 for solutions of reserpine at pH 1.3, 2.2 and 3.0, respectively, after storage of the solutions for 48 hours at 100°C.

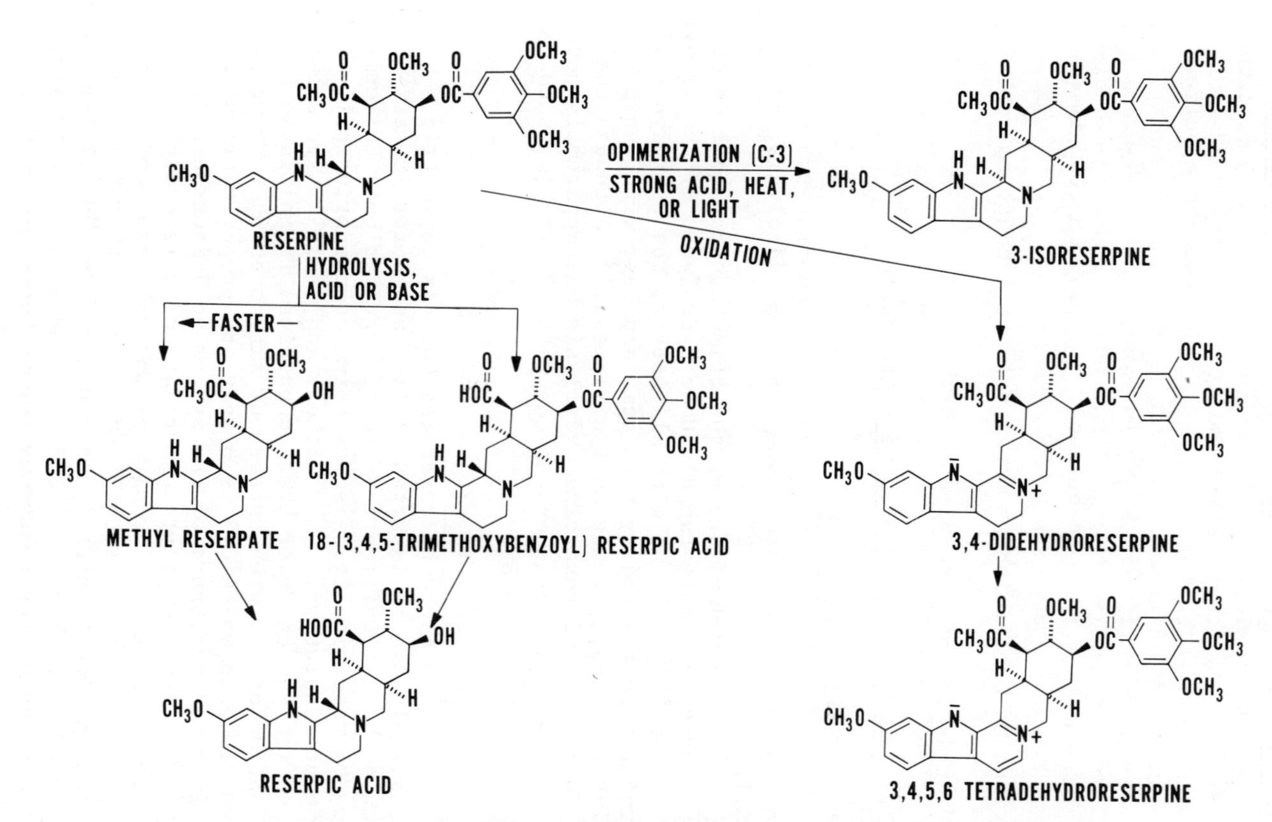

Figure 6. Principle Routes for Degradation of Reserpine

3.1.3 Oxidation

Oxidation of reserpine may be induced either photolytically or chemically, and it is usually the major route of degradation for reserpine. Ljungberg[41] found that photo-oxidation of reserpine first produces a yellow substance with yellow-green fluorescence, but that further photolysis gives rise to a reddish brown solution with blue fluorescence (due to lumireserpine) which eventually returnes through red and orange to yellow. Banes <u>et</u>. <u>al</u>.[42] first suggested that the initial oxidation product was 3,4 di-dehydroreserpine, and this was later confirmed by Krebs and Futscher.[43] Lumireserpine was shown by Hakkesteegt[44] to be 3,4,5,6-tetradehydroreserpine, a result which has been re-confirmed recently by Wright and Tang.[45] Hakkesteegt[34] has suggested that some of the non-fluorescent brown and yellow products formed during oxidative degradation of reserpine are polymers. The oxidations of reserpine are also summarized in Figure 6.

The oxidation of reserpine is rapid in the presence of light and oxygen. The oxidation is also catalyzed by acid[34], and by certain metal ions. The efficiency of several metal ions in catalyzing the oxidation of reserpine decreases in the following order: Cu Mn Fe Al.[46]

3.2 Degradation of Reserpine in Pharmaceutical Dosage Forms

Hydrolysis of reserpine appears to be very unlikely in solid dosage forms[34], and there apparently have been no reports in the literature of significant levels of hydrolysis in reserpine injectables or elixirs.

Bayer[40] has reported that reserpine does undergo epimerization in both solution and crystalline pure drug under the influence of heat and light, but did not provide any information on the extent of conversion observed. Weis-Fogh[47] studied the degradation of a 0.25% reserpine solution at pH 2.5 under the influence of sunlight. After three months at room temperature, he found that the reserpine content had decreased to 70% of initial, and that there was 5% 3,4-didehydroreserpine present. Weis-Fogh felt that the remaining 25% of the reserpine was present largely as 3-iso-reserpine, but this conclusion was probably incorrect because his assay method would have measured most of the iso-reserpine along with the reserpine. Hakkesteegt[38] was unable to find 3-isoreserpine in crystalline reserpine or in reserpine tablets from one to seven years old and estimated from his kinetic studies that it would take 26 years for 10% of the reserpine in a pH 3.0 solution to be lost through combined hydrolysis and epimerization at 20°C. Wright and Tang[45] found evidence of 3-isoreserpine in several commercial brands of reserpine tablets by TLC, but it may be concluded that epimerization is not likely to be a significant route of degradation for reserpine in most pharmaceutical preparations.

Oxidation appears to be the most important route of degradation for reserpine in formulations. Banes[42] examined several reserpine injections, elixirs and tablets. All injections and elixirs examined contained significant levels of oxidation products, whereas only one of the eight lots of tablets examined showed any evidence of these products. Haddesteegt[34] found levels of oxidation products as high as 35% in one lot of reserpine tablets, and more recently, Wright and Tang[45] found that 3,4-didehydroreserpine was present in all commercially available reserpine tablets that they examined (3,4,5,6-tetradehydroreserpine was not detected in any of them).

4. Metabolism of Reserpine

The distribution, excretion, and metabolism of reserpine have been studied in the mouse[48-51,69,70], rat[49,52-59,71-74], dog[49,52,61], monkey[52], rabbit[49,62], guinea pig[54,63], man[64-67], and Busycon canaliculatum.[75]

The metabolism of reserpine is qualitatively similar in all species[49,53], with oxidative demethylation of the 4 position of the 3,4,5-trimethoxybenzoyl group[50,53,68] and hydrolysis of the trimethoxybenzoate ester linkage[49,53,60,66] being the major identified routes of metabolism. The metabolic products therefore include methyl reserpate[50,52], 3,4,5-trimethoxybenzoic acid[48-50,60,66,67] and its glucuronide or sulfate conjugates[67], syringoyl methyl reserpate[54], and syringic acid.[54] Sheppard[63] has made special note of the absence of 3-isoreserpine, reserpine N-oxide, didehydroreserpine, and tetradehydroreserpine in guinea pig brain at doses which produced measurable quantities of unmetabolized reserpine in these tissues.

5. Identification of Reserpine

Reserpine can be identified by its NMR, IR, and UV spectra and by melting point (See section 2). Reserpine forms a rose-pink color upon addition of about 1 mg. to a few tenths of a 1-2% solution of vanillin in acid: the color will deepen upon standing or gentle heating.[76-77] A green color is produced when 0.5 mg. of reserpine is treated with 5 mg dimethylaminobenzaldehyde, 0.2 ml glacial acetic acid, and 0.2 ml of sulfuric acid: the color will change from green to red upon addition of 1 ml of glacial acetic acid.[77] 1 mg of reserpine treated with 0.1% solution of sodium molybdate in sulfuric acid produces a yellow color which turns to blue within about two minutes.[77] If 1 mg of reserpine is dissolved in 0.5 ml of dilute acetic acid and 5 drops of sodium chloride test solution are added, a white precipitate forms.[78] A yellow color accompanied by greenish fluorescence develops in a solution of 1 mg reserpine in 5 ml of chloroform upon addition of 5 ml of a 10% solution of trichloroacetic acid in chloroform.[78] Identification of reserpine by formation of crystalline derivatives[79], formation of eutectic mixtures[80], and by use of a variety of other color tests[81-86] have been reported.

6. Elemental Analysis

Carbon	65.12%
Hydrogen	6.62%
Nitrogen	4.60%
Oxygen	23.66%

7. Chromatographic Methods of Analysis

7.1 Thin Layer Chromatography

A variety of thin layer systems have been developed for reserpine and a number of these are summarized in Table 7. Reagents used for detection and identification of reserpine on the plate are summarized in Table 8. Thin layer systems have also been described in references 102-111.

Quantitation of reserpine following thin layer chromatography is described in references 50, 92, 96-101, 104, 107 and 110-111.

7.2 Paper Chromatography

Several paper chromatography systems for reserpine are summarized in Table 10 and methods for visualizing reserpine on the paper are summarized in Table 11. Paper chromatography systems for reserpine are also discussed in references 119-127. Becker[127] has applied elatography to reserpine, elatography being the technique in which a sample is spotted on the paper, treated with an appropriate reagent, and the reaction products are then separated chromatographically.

Quantitation of reserpine by ultraviolet and infrared spectrophotometry, colorimetry, titration, and polarography after isolation by paper chromatography are discussed in references 116 and 125.

Table 7

Thin Layer Chromatography Systems for Reserpine

	Solvent System	Sorbent	Rf	Application and Comments	Reference
1.	$CHCl_3$:Me_2CO:NH_4OH, 80:20:1	Silica gel-gypsum plates pretreated at 120°C for 30 minutes		Separation from other tranquilizers	87
2.	$CHCl_3$:Me_2CO, 85:15	Alumina	0.60	Separation from other alkaloids of R. serpentina	88, 90, 94
3.	$CHCl_3$:EtOH:Me_2CO, 90:5:5	Alumina	0.89	Separation from other alkaloids of R. serpentina	88, 90
4.	$CHCl_3$:Me_2CO:Et_2NH, 5:4:1	Silica Gel G	0.72	Separation of Rauwolfia and opium alkaloids	89-91
5.	$CHCl_3$:Et_2NH, 9:1	Silica Gel G	0.80	Separation of Rauwolfia alkaloids	90, 91
6.	Cyclohexane:$CHCl_3$:ET_2NH, 5:4:1	Silica Gel G	0.25	Separation of Rauwolfia alkaloids	50, 90-92
7.	Cyclohexane:$CHCl_3$, 3:7 and 0.05% Et_2NH	Alumina G	0.35	Separation of Rauwolfia alkaloids	90, 91
8.	Methanol	Silica Gel G, 0.1N NaOH impregnated	0.69	Separation of Rauwolfia alkaloids	90, 91

	Solvent System	Sorbent	Rf	Application and Comments	Reference
9.	Heptane:Me-CO-Et, 1:1 in atmosphere of ammonia	Cellulose, formamide impregnated	0.59	Separation of Rauwolfia alkaloids	90, 93
10.	Heptane:Me-CO-Et: MeOH, 60:30:10	Whatman SG 41 Silica Gel	0.25	Separation of 10 Rauwolfia alkaloids	95
11.	Heptane:Me-CO-Et: MeOH, 60:30:10	Kieselgel G (Silica Gel)	0.32	Separation of 10 Rauwolfia alkaloids	95
12.	Heptane:Me-CO-Et: MeOH, 60:30:10	Aluminum oxide G	0.63	Separation of 10 Rauwolfia alkaloids	95
13.	Heptane:Me-CO-Et:n-BuOH, 60:30:10	Whatman SG 41 (Silica Gel)	0.47	Separation of 10 Rauwolfia alkaloids	95
14.	Heptane:Me-CO-Et:n-BuOH, 60:30:10	Kieselgel G (Silica Gel)	0.52	Separation of 10 Rauwolfia alkaloids	95
15.	Heptane:Me-CO-Et:n-BuOH, 60:30:10	Aluminum oxide G	0.78	Separation of 10 Rauwolfia alkaloids	95
16.	Heptane:Me-CO-Et: Pyridine, 70:15:15	Whatman SG 41 (Silica Gel)	0.07	Separation of 10 Rauwolfia alkaloids	95
17.	Heptane:Me-CO-Et: Pyridine, 70:15:15	Kieselgel G (Silica Gel)	0.05	Separation of 10 Rauwolfia	95

	Solvent System	Sorbent	Rf	Application and Comments	Reference
18.	Heptane:Me-CO-Et: Pyridine, 70:15:15	Aluminum oxide G	0.90	Separation of 10 Rauwolfia alkaloids	95
19.	n-BuOH:glacial HOAc:H_2O, 4:1:1	Whatman SG 41 (Silica Gel)	0.64	Separation of 10 Rauwolfia alkaloids	95
20.	n-BuOH:glacial HOAc:H_2O, 4:1:1	Kieselfel G (Silica Gel)	0.75	Separation of 10 Rauwolfia alkaloids	95
21.	n-BuOH:glacial HOAc:H_2O, 4:1:1	Aluminum oxide G	0.97	Separation of 10 Rauwolfia alkaloids	95
22.	Isooctane:Et_2O: xylene:EtOAc, 45:40:15:5	Silica Gel G	0.00	Separation of Rauwolfia alkaloids for quantitative analysis	96, 97
23.	Ethylene dichloride: EtOAc:n-BuOH, 60:30:10	Silica Gel G	0.58	Separation of Rauwolfia alkaloids for quantitative analysis	96, 97
24.	Me_2CO:Pet. Ether: CCl_4:isooctane, 35:30:20:15	Silica Gel G	0.32	Separation of Rauwolfia alkaloids for quantitative analysis	96, 97
25.	Isooctane:Me_2CO:n-BuOH, 58:33.6:8.4	Silica Gel G	0.40	Separation of Rauwolfia alkaloids for quantitative analysis	96

	Solvent System	Sorbent	Rf	Application and Comments	Reference
26.	Me_2CO:Pet. Ether: glacial HOAc, 45:45:10	Silica Gel G	0.22	Separation of Rauwolfia alkaloids for quantitative analysis	96, 97
27.	Me_2CO:MeOH:glacial HOAc, 70:25:5	Silica Gel G	0.80	Separation of Rauwolfia alkaloids for quantitative analysis	96, 97
28.	Pet. Ether:Me_2CO: Et_2NH, 70:20:10	Silica Gel G	0.60	Separation of Rauwolfia alkaloids for quantitative analysis	96
29.	Me_2CO:MeOH:Et_2NH, 70:20:10	Silica Gel G	0.96	Separation of Rauwolfia alkaloids for quantitative analysis	96
30.	Me_2CO:CCl_4:Pet. Ether, 45:45:30	Silica Gel G	0.43	Separation of Rauwolfia alkaloids for quantitative analysis	97
31.	H_2O:EtOH:$CHCl_3$, 56:42:2	Cellulose	1.00	Detection of cholinesterase inhibitors at low levels	98
32.	C_6H_6:EtOH, 9:1	Aluminum oxide		Quantitative analysis	100
33.	n-BuOH:Me-CO-Et: H_2O, 65:25:25	Silica Gel G	0.95	Separation of reserpine from its degradation products	34

	Solvent System	Sorbent	Rf	Application and Comments	Reference
34.	$CHCl_3$:Me_2CO, 70:30	Silica Gel G	0.35	Separation of reserpine from other drug substances	101

Table 8

Visualization of Reserpine on Thin Layer Plates

No.	Treatment	Result	Reference
1	Iodine Vapor		87
2	Dragendorff Reagent	brown spot	88
3	Acetylchloride, then heat		89
4	1% Ceric Sulfate in 10% H_2SO_4	greenish brown, turning brown after 5 minutes at 105°C	95
5	Frohde's Reagent (Sulphomolybdic acid)	yellow-brown	95
6	5% Ferric Chloride in 50% HNO_3	yellow-green turning green-brown after 5 minutes at 105°C	95
7	Iodoplatinate	pink	95
8	0.5% Phosphomolybdic acid in 50% HNO_3	yellow-green	96
9	1% Ammonium Vanadute in 50% HNO_3	yellow-green	96

No.	Treatment	Result	Reference
10	Spray with source of cholinesterase (human blood plasma suitable), followed by 1 part 0.6% bromthymol blue in 0.1N NaOH and 15 parts aqueous 1% acetylcholine chloride	blue spots on yellow background	98, 99
11	500 mg p-dimethylaminobenzaldehyde in 50 ml conc. H_2SO_4	greenish-black spot	101

Table 9

Paper Chromatography Systems for Reserpine

No.	Solvent Systems	Immobile Phase	Rf	Application and Comments	Reference
1.	C_6H_6	50% ethanolic formamide	.96	Distinguish reserpine from synthetic precursors and related compounds. Detection limit, 1 mcg.	112, 116
2.	C_6H_6	50% ethanolic formamide containing 5% ammonium formate	.75	Distinguish reserpine from synthetic precursors and related compounds. Detection limit, 1 mcg.	112
3.	C_6H_6:C_6H_{12}, 1:1	50% ethanolic formamide	.61	Distinguish reserpine from synthetic precursors and related compounds. Detection limit, 1 mcg.	112
4.	C_6H_6:C_6H_{12}, 1:1	50% ethanolic formamide containing 5% ammonium formate	.20	Distinguish reserpine from synthetic precursors and related compounds. Detection limit, 1 mcg.	112
5.	C_6H_{12}	50% ethanolic formamide containing 5% ammonium formate	.00	Distinguish reserpine from synthetic precursors and related compounds. Detection limit, 1 mcg.	112

No.	Solvent Systems	Immobile Phase	Rf	Application and Comments	Reference
6.	Shake isooctane: C_6H_6:Formamide, 100:50:5 together, discard lower layer, add 2 parts C_6H_{12} and filter. Saturate chamber with NH_3 vapor	Me_2CO: formamide, 100:30	.56	Identification of reserpine and related compounds in drug formulations	113, 114
7.	C_6H_6:C_6H_{12}, 1:1	Formamide:MeOH, 70:30		Estimation of reserpine in Rauwolfia root	115
8.	C_6H_6:C_6H_{12}, 1:1 saturated with propylene glycol	Propylene glycol: MeOH;HOAc, 50:50:1		Quantitative analysis of raw materials	116
9.	Et-CO-Me:Me_2CO: HCOOH:H_2O, 40:2:1:6	Whatman No. 1	.86		117
10.	Et-CO-Me:Me_2NH:H_2O, 921:2:77	Whatman No. 1	.95		117
11.	i-Bu-CO-Me:HCOOH: H_2O, 10 parts ketone saturated with 1 part 4% formic acid	Whatman No. 1	.44		117

No.	Solvent System	Immobile Phase	Rf	Application and Comments	Reference
12.	$CHCl_3$:MeOH:HCOOH:H_2O, 10 parts $CHCl_3$ saturated with a mixture of 1 part MeOH and 1 part 4% HCOOH	Whatman No. 1	.91		117
13.	C_6H_6:Et-CO-Me:HCOOH:H_2O, 9 parts C_6H_6 and 1 part Et-CO-Me saturated with 1 part 2% HCOOH	Whatman No. 1	.12		117
14.	C_6H_6:HCOOH:H_2O, 10 parts C_6H_6 saturated with 1 part 2% formic acid	Whatman No. 1	.05		117
15.	Acetic Acid: 5% aqueous sodium acetate, 10:90				
	a. Shake with n-BuOH, added in small portions, until saturation of the aqueous phase is just achieved	Whatman No. 542 paper	.34	Detect less than 1 mcg	118

No.	Solvent System	Immobile Phase	Rf	Application and Comments	Reference
	b. Same as a except replace n-BuOH with i-pantanol	Whatman No. 542 paper	.34	Detect less than 1 mcg	118
16.	n-BuOH:C_6H_6: equal parts 1.5N NH_4OH and 1.5N $(NH_4)_2CO_3$, 80:5:15	Whatman No. 1	.90	Radio-assay for reserpine in biological samples	48

Table 10

Visualization of Reserpine on Paper Chromatograms

No.	Treatment	Result	Reference
1.	Observe under low or high pressure Hg lamp	Strong green fluorescence	112
2.	Observe under low pressure Hg lamp after spraying with .0025% solution of fluorescein in 0.5M ammonia	Strong green	112
3.	Observe under UV lamp after spraying with 3% solution of sodium nitroprusside in 50% trichloroacetic acid	Strong green	112
4.	Spray with mixture containing 5 ml HCl, 95 ml ethanol, and 1 g p-dimethylaminobenzyldehyde, heat briefly at 100°C, respray with nitroprusside solution (No. 3 above), and heat again at 100°C for 3-5 minutes	Brown spot, reaction weak	112
5.	Dragendorff Reagent (2% potassium bismuth tetraiodide in 0.01N HCl)		117

7.3 Column Chromatography

System	Description	Reference
1.	Column: Solka-Floc and Celite 545, 1:1 Eluting Solvent: 5N acetic acid Application: Quantitative determination - pharmaceutical preparations.	128
2.	Column: A 200 x 22 mm id column packed in four layers is used. 1) bottom layer - 1 g celite +0.5 ml of fresh 2% $NaHCO_3$, 2) 1 g celite +0.5 ml of fresh 0.5% citric acid solution, 3) 0.5 g celite +0.5 ml H_2O, and 4) 1 mg equivalent of reserpine sample and 1 ml dimethylsulfoxide and 2 g celite. Eluting Solvent: Chloroform Application: Quantitative determination of reserpine in formulations.	129-131
3.	Column: 1) Bottom layer consisting of 1.5 g celite 545 and 0.4 ml ethanol and 1 ml 2% $NaHCO_3$, and 2) top layer of 10 g celite 545 and 10 ml of a solution prepared by dissolving 1.05 g citric acid in water, diluting to 50 ml with water, and adding 20 ml ethanol. Eluting Solvent: Shake together 100 ml $CHCl_3$, 200 ml isooctane, 100 ml water and 40 ml ethanol. Discard aqueous layer and filter. Application: Quantitative determination of reserpine in formulations.	113, 114
4.	Column: Celite 450 and formamide layer resulting from shaking together 15 ml heptane, 110 ml $CHCl_3$, 1 ml morpholine, and 25 ml formamide. Eluting Solvent: $CHCl_3$ layer obtained as described under "column."	132, 133

System	Description	Reference
	Application: Separation and determination of reserpine, deserpidine, and rescinnamine.	
5.	Column: Acid activated aluminum oxide	134
	Eluting Solvent: Methanol	
	Application: Separation and quantitation of reserpine in Rauwolfia extracts.	
6.	Column: Dowex 50-X-2 cation exchanger	135
	Eluting Solvent: Methanol/ammonia, 4/1	
	Application: Quantitation of reserpine in formulations.	
7.	Column: Aluminum oxide equilibrated with $CHCl_3$.	136
	Eluting Solvent: Ethanol/$CHCl_3$, 1/99	
	Application: Separation of reserpine from weakly basic alkaloids in R. Serpentina extract.	
8.	Column: Alumina	137
	Eluting Solvent: Gradient elution with benzene containing 0-40% ethanol.	

7.4 Electrophoresis

Reserpine has been separated from rescinnamine, reserpic acid and serpentine by electrophoresis on paper with 5N acetic acid as the electrolyte.[138,155] Under the conditions employed (8 volts/cm, 1.2 mA, 5 hours) rescinnamine was not completely separated from reserpine. Electrophoretic separations of reserpine from other Rauwolfia alkaloids were also reported in references 6, 121 and 136.

7.5 Countercurrent Distribution

The three solvent systems described in Table 7 have been used to separate reserpine from other Rauwolfia alkaloids by countercurrent distribution.[6] Kidd and Scott[139] also reported several solvent systems, but found the use of diethyl ether: chloroform (3:1) as the mobile phase and a pH 3.1 buffer (16.3 g citric acid and 16.1 g $Na_2HPO_4 \cdot 12H_2O$ in 1 liter water) as the stationary phase gave the best separation of the alkaloids of R. Serpentina and R. Vomitoria. Other countercurrent systems may be found in references 140-142.

8. Titrimetric Determination

Reserpine has been determined by titration in $CHCl_3$ solution using 0.1N $HClO_4$ in dioxane as titrant and detection of the endpoint with methyl red, methyl yellow, or potentiometrically.[143,144] Titrations with anionic surfactants, such as sodium lauryl sulfate, have also been reported.[145,146]

Reserpine has been determined titrimetrically in formulations after an appropriate extraction. Sun[147] extracted with 0.3N citric acid, alkalyzing with NH_4OH, back extracting into $CHCl_3$, drying, taking up into .01N H_2SO_4, and finally back-titrating with 0.01N NaOH using methyl red indicator. Sakurai[148] extracted reserpine with $CHCl_3$ from injection solutions alkalinized with NH_4OH and then titrated with 0.002N p-toluenesulfonic acid dissolved in ethylene glycol: 2-propanol (1:1).

Reserpine has also been determined by precipitation as its tetraphenylborate salt, dissolution of the precipitate in acetone, and titration with $AgNO_3$ using silver electrodes;[149] and by micro-Zeisel determination of its six methoxy groups.[150]

9. Electrochemical Analysis

In acidic media, alkaloids containing the 6-methoxyindole nucleus undergo a one electron oxidation which probably involves insertion of a hydroxyl group into the aromatic portion of the molecule. If the molecule contains available nitrogen with an unshared pair of electrons, a two electron oxidation occurs with formation of an N-oxide.[151] Reserpine does not undergo polarographic reduction.[152]

Reserpine has been analyzed coulometrically by reaction with chlorine generated electrochemically from HCl. The electrochemical coefficient was 0.0007885 mg/ma-sec.[153]

10. Spectrophotometric Analysis

10.1 Infrared

The infrared absorption of reserpine in the 5.0-6.5 micron region has been used for quantitation of this compound in formulations.[154] The band intensity was measured following extraction of the reserpine into chloroform.

10.2 Ultraviolet

The ultraviolet absorption of reserpine can be used for quantitation,[116,155-158,182-183] but the possibility of interference from related alkaloids, excipients, and degradation products requires that the reserpine be isolated from these other substances prior to measurement. The separation from roots and crude preparations has been accomplished by extensive extraction[156-157]; from formulations by paper chromatography (System No. 8, Table 9)[116]; and from crude extracts, raw materials, and formulations by electrophoresis on paper using 5N acetic acid as the electrolyte.[155,158] The absorbance of the final sample is generally measured at 268 nm. Page[183] has reported a semi-automated procedure for reserpine in tablets using a Technicon Auto-Analyzer® system with both ultraviolet and colorimetric detection.

Ultraviolet spectrophotometry has also been evaluated for the determination of foreign alkaloids in reserpine preparations.[159]

10.3 Colorimetric

The most commonly used colorimetric procedures for reserpine involve oxidation of the compound to 3,4-didehydroreserpine with nitrite and measurement of the absorbance of the oxidation product at about 390 nm. The reaction is carried out in solutions containing 5-15 μg/ml reserpine in either methanol or ethanol acidified with sulfuric acid.[160-166] This procedure has been applied to formulations[76,160,165], crude Rauwolfia preparations[161,162,166], and animal feeds.[169] An automated version of this assay for determining reserpine in tablets has been reported by Page.[183]

An alternate procedure employs oxidation of reserpine with nitrite in acetic acid followed by extraction into chloroform.[167-169] The absorbance of the chloroform layer is measured at 465 nm. This procedure is suitable for 50-300 μg of reserpine, and has the advantage that hydrolysis products of reserpine do not interfere.[167] A variation of this procedure in which the sample is treated with amyl nitrite rather than with acetic acid - sodium nitrite reagent has also been reported.[170]

Reserpine has also been analyzed colorimetrically by reaction with vanillin (absorbance of 0.1 at 532 nm with 17 μg/ml),[171,172] aminopyrimidine,[173] xanthydrol (measure 50-500 μg at 500 nm),[174,175] phenylisocyanate,[176] iodine,[177] and sodium glyoxalate - $FeCl_3$.[178]

10.4 Ion-Pair Extraction with Spectrophotometric Detection

Booth[179] reported a procedure for the analysis of reserpine in formulations by extracting the reserpine into chloroform from pH 4.0 phosphate buffer as an ion-pair with bromcresol purple. A final solution concentration of 2.8 μg/ml gave an absorbance of 0.100 at 402 nm. Procedures employing bromcresol green[180] and methyl orange[181] in place of bromcresol purple have also been reported.

10.5 Fluorescence

Since 3,4-didehydroreserpine is strongly fluorescent (Section 2.2.3), the sensitivity of the colorimetric methods employing oxidation of reserpine to this product (Section 10.3) can be increased by using fluorimetric detection. Nitrite oxidation and fluorimetric determination have been used for analysis of tablets[184] and for feeds containing reserpine at the ppm level.[185-188] An automated single-tablet assay based on this procedure has also been reported.[189,190]

Reagents other than nitrite have been used to develop fluorescence, including hydrogen peroxide,[158,190-192] selenious acid,[193,194] p-toluenesulfonic acid in acetic acid,[195] sulfovanadic acid,[196-199] and vanadium pentoxide.[200-203] The vanadium pentoxide procedure has been automated for use in single tablet assays.[200,202] The selenious acid procedure has been used to determine reserpine in biological samples.[193,194]

Kollistratos[204] has reported that $ThCl_4$ greatly enhances the fluorescence of reserpine in ethanol, methanol, acetone, and dioxane, but did not determine the mechanism responsible for the enhancement.

11. Analysis of Reserpine in Biological Systems

Reserpine levels have been determined fluorimetrically in tissue samples. Poet[193] and Hess[194] adjusted the pH of samples to 8.5 with borate buffer, extracted with heptane or petroleum ether, back-extracted into dilute sulfuric acid and finally developed fluorescence using the selenious acid oxidation procedure. Glaszko[52] and Maronde[64] adjusted the pH of plasma and urine samples to 3.7-4.0 with citrate buffer, extracted with ethylene dichloride, and developed fluorescence with sodium nitroprusside followed by hydrogen peroxide. Zsoter[67] analyzed plasma and urine samples using a modification of Jakovljevic's[195] fluorimetric procedure.

The most widely used methods for determining reserpine in tissues are those employing ^{14}C or ^{3}H labeled reserpine. Specificity is obtained in the radioactive determinations by prior extraction,[53,63] thin layer chromatography,[50,56,58,59,60,66,67,69,92] or paper chromatography.[48,49,54,65,66] Other procedures may be found in the references cited in Section 4.

The bioavailability of reserpine administered orally as a coprecipitate with polyvinylpyrrolidone or with bile acids has been determined using the blepharoptotic activity of the preparation as a measure of availability.[205-211] Ptosis was rated using the scale devised by Rubin *et. al*.[212] These studies included examination of the dissolution rates of the coprecipitates and the relationship between dissolution rate and blepharoptotic activity.

References

1. E. Schlittler, Rauwolfia Alkaloids with Special Reference to the Chemistry of Reserpine, In "The Alkaloids" (R.H.F. Manske, ed.), Vol. VIII, Academic Press, New York, 1965.
2. R.B. Woodward, F.E. Bader, H. Bickel, A.J. Frey, and R.W. Kierstead, J. Am. Chem. Soc. 78, 2023, 2657 (1956).
3. R.B. Woodward, F.E. Bader, H. Bickel, A.J. Frey, and R.W. Kierstead, Tetrahedron 2, 1 (1958).
4. L. Velluz, B. Muller, R. Joly, G. Nomine, A. Allais, J. Warnant, R. Bucourt, and J. Jolly, Bull. Soc. Chim. France p. 145 (1958)
5. L. Velluz, G. Muller, R. Joly, G. Nomine, J. Mathieu, A. Allais, J. Warnant, J. Valls, R. Bucourt, and J. Jolly, Bull. Soc. Chim. France p. 673 (1958).
6. F.A. Hochstein, K. Murai, and W. H. Boegemann, J. Am. Chem. Soc. 77, 3551 (1955).
7. "The Merck Index," 8th Edition, Merck & Co., Inc., Rahway, New Jersey (1968).
8. W.E. Rosen and J.N. Shoolery, J. Am. Chem. Soc. 83, 4816 (1961).
9. H.B. MacPhillamy, C.F. Huebner, E. Schlittler, A.F. St. Andre, and P.R. Ulshafer, J. Am. Chem. Soc. 77, 4335 (1955).
10. M.M. Robison, R.A. Lucas, H.B. MacPhillamy, R.L. Dziemian, I. Hsu, R.J. Kiesel, and M.J. Morris, Abstr. Papers Am. Chem. Soc. 139th Meeting, St. Louis, Missouri, 1961, p. 3N.
11. W.E. Rosen and J. M. O'Connor, J. Org. Chem. 26, 3051 (1961).
12. J.R. Platt, J. Chem. Phys. 17, 484 (1949).
13. J.R. Platt, J. Chem. Phys. 19, 101 (1951).
14. "Physical Data of Indole and Dihydroindole Alkaloids," Eli Lilly and Co., Indianapolis, Indiana (1964).
15. G. Barth, R.E. Linder, E. Bunnenberg, and C. Djerassi, Helv. Chim. Acta 55, 2168 (1972).
16. W.H. McMullen, H.J. Pazdera, S.R. Missan, L.L. Ciaccio, and T.C. Grenfell, J. Am. Pharm. Assoc. 44, 446 (1955).
17. S. Udenfriend, D. Duggan, B. Vasta, and B. Brodie, J. Pharmacol. Exptl. Therap. 120, 26 (1957).
18. G.G. Guilbault, "Practical Fluorescence," p. 325, Marcel Dekker, Inc., New York (1973).
19. R.P. Haycock, P.B. Sheth, and W.J. Mader, J. Am. Pharm. Assoc. 48, 479 (1959).
20. R.D. Rauh, T.R. Evans, and P.A. Leermakers, J. Am.

Chem. Soc. 90, 6897 (1968).
21. R.D. Rauh, T.R. Evans, and P.A. Leermakers, J. Am. Chem. Soc. 91, 1868 (1969).
22. D.L. Dexter, J. Chem. Phys. 21, 836 (1953).
23. T. Forster, Disc. Faraday Soc. 271, 1 (1959).
24. C. Djerassi, M. Gorman, A.L. Nussbaum, and J. Reynoso, J. Am. Chem. Soc. 75, 5446 (1953).
25. A. Furlenmeier, R. Lucas, H.B. MacPhillamy, J.M. Mueller, and E. Schlittler, Experentia 9, 331 (1953).
26. L. Dorfman, A. Furlenmeier, C.F. Huebner, R. Lucas, H.B. MacPhillamy, J.M. Mueller, E. Schlittler, R. Schwyer, and A.F. St. Andre, Helv. Chim. Acta 37, 59 (1954).
27. N. Neuss, H.E. Boaz, and J.W. Forbes, J. Am. Chem. Soc. 76, 2463 (1954).
28. W.E. Rosen, Tet. Letters p. 481 (1961).
29. P. Baudet, Cl. Otten, and E. Cherbuliez, Helv. Chim. Acta 47, 2430 (1964).
30. E. Wenkert and D.K. Roychaudhuri, J. Am. Chem. Soc. 78, 6417 (1956).
31. H.A. Rose, Anal. Chem. 26, 1245 (1954).
32. W.C. Hamilton, Sc. 169, 135 (1970).
33. M. Langejan and H.F.L. Liefferink, Pharm. Weekblad 91, 847 (1956).
34. Th.J. Hakkesteegt, Pharm. Weekblad 105, 829 (1970).
35. M. Kaern and M. Tønnesen, Farm. Revy. 57, 553 (1958).
36. E. Wenkert and L.H. Liu, Experentia 11, 302 (1955).
37. H.B. MacPhillamy, L. Dorfman, C.F. Huebner, E. Schlittler, and A.F. St. Andre, J. Am. Chem. Soc. 77, 1071 (1955).
38. Th.J. Hakkesteegt, Pharm. Weekblad 105, 801 (1970.
39. A.J. Gaskell and J.A. Joule, Tetrahedron 23, 4053 (1967).
40. J. Bayer, Pharmazie 13, 468 (1958).
41. F. Ljungberg, Farm. Tidskr., 62, 693 (1958).
42. D. Banes, J. Wolff, H.O. Fallscheer, and J. Carol, J. Am. Pharm. Assoc. 45, 710 (1956).
43. K.G. Krebs and N. Futscher, Arzneimittel-Forsch. 10, 75 (1960).
44. Th.J. Hakkesteegt, Pharm. Weekblad 103, 297 (1968).
45. E. Wright and T.Y. Yang, J. Pharm. Sc. 61, 299 (1972).
46. H. Potter and R. Voigt, Pharmazie 22, 436 (1967).
47. O. Weis-Fogh, Pharm. Acta Helv. 35, 442 (1960).
48. P. Numerof, M. Gordon, and J. M. Kelley, J. Pharmacol. Exptl. Therap. 115, 427 (1955).
49. H. Sheppard and W.H. Tsien, Proc. Soc. Exptl. Biol. Med. 90, 437 (1955).

50. R.E. Stitzel, L.A. Wagner, and R.J. Stawarz, J. Pharmacol. Exptl. Therap. 182, 500 (1972).
51. L.A. Wagner, Subcellular distribution of [^{3}H]-reserpine. Diss. Abstr. Int. B 31, 2870 (1970).
52. A.J. Glazko, W.A. Dill, and L.M. Wolf, J. Pharmacol. Exptl. Therap. 118, 377 (1956).
53. H. Sheppard, R.C. Lucas, and W.H. Tsien, Arch. internat. pharmacodyn. 103, 256 (1955).
54. H. Sheppard, W.H. Tsien, E.B. Sigg, R.A. Lucas, and A.J. Plummer, Arch. internat. pharmacodyn. 113, 160 (1957).
55. R.J. Stawarz, Factors influencing the microsomal metabolism of tritium-labelled reserpine in vitro. Diss. Abstr. Int. B 33, 5990 (1973).
56. L. Manara, P. Carminati, and T. Mennini, Eru. J. Pharmacol. 20, 109 (1972).
57. S.J. Enna and P.A. Shore, Biochem. Pharmacol. 20, 2910 (1971).
58. L. Manara and S. Garattini, Eru. J. Pharmacol. 2, 139 (1967).
59. L. Manara, T. Mennini, and P. Carminati, Eur. J. Pharmacol. 17, 183 (1972).
60. L.A. Wagner and R.E. Stitzel, J. Pharm. Pharmacol. 24, 396 (1972).
61. E.A. DeFelice, Experentia 13, 373 (1957).
62. S.M. Hess, P.A. Shore, and B.A. Brodie, J. Pharmacol. Exptl. Therap. 118, 84 (1956).
63. H. Sheppard, W.H. Tsien, A.J. Plummer, E.A. Peets, B.J. Giletti, and R.A. Schulert, Proc. Soc. Exptl. Biol. Med. 97, 717 (1958).
64. R.F. Maronde, J. Haywood, D. Feinstein, and C. Sobel, J. Am. Med. Assoc. 184, 7 (1963).
65. P. Numerof, A.J. Virgona, E.H. Cranswick, T. Cunningham, and N.S. Kline, Psychiat. Res. Rep., Wash. 9, 139 (1958).
66. A.R. Maass, B. Jenkins, Y. Shen, and P. Tannenbaum, Clin. Pharmacol. Ther. 10, 366 (1969).
67. T.T. Zsoter, G.E. Johnson, G.A. Deveber, and H. Paul, Clin. Pharmacol. Ther. 14, 325 (1973).
68. T.K. Adler and M.E. Latham, Proc. Soc. Exptl. Biol. Med. 73, 401 (1950).
69. L.A. Wagner and R.E. Stitzel, J. Pharm. Pharmacol. 21, 875 (1969).
70. G.F. Placidi, G. Ciccone, A. Moschino, E. Gliozzi, and G.B. Cassano, J. Nucl. Biol. Med. 16, 32 (1972).
71. M.M. Dhar, J.D. Kohli, and S.K. Srivastava, J. Sc. Ind. Research (India) 14C, 179 (1955).

72. M.M. Dhar, J.D. Kohli, and S.K. Srivastava, Indian J. Pharm. 18, 293 (1956).
73. J.D. Kohli, M.M. Dhar, and S.K. Srivastava, J. Sci. Ind. Research (India) 15C, 167 (1956).
74. S.K. Srivistava, M.M. Dhar, and J.D. Kohli, J. Sci. Ind. Research (India) 16C, 73 (1957).
75. M. Mirolli, Sc. 149, 1503 (1965).
76. The United States Pharmacopoeia XVIII.
77. British Pharmacopoeia 1973.
78. The Pharmacopoeia of Japan, 7th Edition, Part 1 (1961).
79. C. Korczak-Fabierkiewicz and G.H.W. Lucas, Proc. Can. Soc. Forensic Sci. 4, 233 (1965).
80. M. Brandstaetter-Kuhnert, A. Kofler, R. Hoffman, and H. C. Rhi, Sci. Pharm. 33, 205 (1965).
81. H. Kaneko, Chem. Pharm. Bull. (Tokyo) 6, 318 (1958).
82. A. Hofmann, Helv. Chim. Acta 37, 314, 849 (1954).
83. R. Cortesi and H. Laubie, Bull. soc. pharm. Bordeaux 93, 116 (1955).
84. R. San-Martin-Casamada, Rev. Real Acad. Farm. Barcelona, No. 12, 1 (1968).
85. R. Bonino, Rev. assoc. bioquim. Argentina 20, 229 (1955).
86. S. Cheng, J. Chinese Chem. Soc. (Taiwan) Ser II 4, 112 (1957).
87. T. Fuwa, T. Kido, and H. Tanaka, Yakuzaigaku 25, 138 (1965).
88. M. Ikram, G.A. Miana, and M. Islam, J. Chromatog. 11, 260 (1963).
89. A. Kaess and C. Mathis, Ann. Pharm. Franc 23, 739 (1965).
90. E. Stahl, "Thin-Layer Chromatography: A Laboratory Handbook," p. 448. Springer-Verlag, New York (1969).
91. D. Waldi, K. Schnackerz, and F. Munter, J. Chromatog. 6, 61 (1961).
92. L. Manara, Eur. J. Pharmacol. 2, 136 (1967).
93. K. Teichert, E. Mutschler, and H. Rochelmeyer, Z. anal. Chem. 181, 325 (1961).
94. M. Ikram and M.K. Bakhsh, Anal. Chem. 36, 111 (1964).
95. W.E. Court, Can. J. Pharm. Sc. 1, 76 (1966).
96. W.E. Court and M.S. Habib, J. Chromatog. 80, 101 (1973).
97. M.S. Habib and W.E. Court, Can. J. Pharm. Sc. 8, 81 (1973).
98. J.J. Menn and J.B. Bain, Nature 209, 1351 (1966).
99. M.E. Getz and S.F. Friedman, J. Assoc. Offic. Agr. Chem. 46, 707 (1963).
100. N.A. Mirzazade, Azerb. Med. Zh. 47, 26 (1970): Chem.

Abstr. 73, 38588t (1970).
101. W.J. Weaver, J. Pharm. Sc. 55, 1111 (1966).
102. M.F. Bartlett, B.F. Lambert, H.M. Werblood, and W.I. Taylor, Jr., J. Am. Chem. Soc. 85, 475 (1963).
103. J. Baumler, and S. Rippstein, Pharm. Acta Helv. 36, 382 (1961).
104. Seyson, Mottet, Crispen, Hennau, Gloesener, Dvred, Vanhaelen, Louvet, Toth, Stainier, J. Pharm. Belg. 26, 292 (1971): Chem. Abstr. 75, 101330M (1971).
105. R.R. Paris, R. Rousselet, M. Paris, and J. Fries, Ann. Pharm. Franc 23, 473 (1965).
106. D. Giacopello, J. Chromatog. 19, 172 (1965).
107. M.S. Habib and W.E. Court, J. Pharm. Pharmacol. 23, 230S (1971)
108. M. Sahli, Arzneimittel-Forsch. 12, 55 (1962).
109. M. Sahli, Arzneimittel-Forsch. 12, 155 (1962).
110. F. Schlemmer and E. Link, Pharm. Ztg. 104, 1349 (1959).
111. E. Ullman and H. Kassalitzky, Arch. Pharm. 295, 37 (1962).
112. K. Macek and S. Vanecek, Coll. Czechoslov. Chem. Commun. 26, 2705 (1961)
113. D. Banes, J. Carol, and J. Wolff, J. Am. Pharm. Assoc. 44, 640 (1955).
114. J. Carol, D. Banes, J. Wolff, and H.O. Fallscheer, J. Am. Pharm. Assoc. 45, 200 (1956).
115. B.P. Korzun, A.F. St. Andre, and P.R. Ulshafer, J. Am. Pharm. Assoc. Sc. Ed. 46, 720 (1957).
116. H.J. Pazdera, W.H. McMullen, L.L. Ciaccio, S.R. Missan, and T.C. Grenfell, Anal. Chem. 29, 1649 (1957).
117. L. Reio, J. Chromatog. 4, 458 (1960).
118. R.J. Boscott and A.B. Kar, Nature 176, 1077 (1955).
119. J. Bayer, Ceskoslov. farm. 9, 396 (1960): Chem. Abstr. 55, 9791f (1961).
120. F. Kaiser and A. Popelak, Ber. 92, 278 (1959).
121. H. Kaneko, Yakugaku Zusshi 78, 512 (1958): Chem. Abstr. 52, 13190e (1958).
122. A.B. Kar and R.J. Boscott, Indian J. Pharm. 18, 204 (1956).
123. K. Macek, J. Hacoperkova, and B. Kakac, Pharmazie 11, 533 (1956).
124. F. Machovicova, Ceskoslov. farm. 6, 310 (1957).
125. N.Y. Tsarenko, V.P. Georgievskii, and M.S. Shraiber, Aptechn. Delo, 14, 49 (1965): Chem. Abstr. 64, 5257d (1966).
126. E. Vidic and J. Schuette, Arch. Pharm. 295, 342 (1962).
127. A. Becker, Z. anal. Chem. 174, 161 (1960).
128. W.F. Bartelt and E.E. Hamlow, J. Am. Pharm. Assoc. 44,

660 (1955).
129. S. Barkan, J. Ass. Offic. Anal. Chem. 52, 113 (1969).
130. F.M. Kunze and S. Barkan, J. Assoc. Offic. Anal. Chem. 51, 1324 (1968).
131. M.L. Dow and R.C. Grant, J. Assoc. Offic. Anal. Chem. 53, 1106 (1970).
132. A.L. Hayden, L.A. Ford, and A.E.H. Houk, J. Am. Pharm. Assoc. 47, 157 (1958).
133. D. Banes, A.E.H. Houk, and J. Wolff, J. Am. Pharm. Assoc. 47, 625 (1958).
134. R. Klaus and G. Weber, J. Chromatog. 48, 446 (1970).
135. E. Haberli and E. Beguin, Pharm. Acta Helv. 34, 65 (1959).
136. K. Yamaguchi, H. Shoji, and M. Ito, Eisei Shikenjo Hokoku No. 76, 87 (1958): Chem Abstr. 53, 17418e (1974).
137. E.E. van Tamelen and C.W. Taylor, J. Am. Chem. Soc. 79, 5256 (1957).
138. M. Sahli, Pharm. Acta Helv. 33, 1 (1958).
139. D.A.A. Kidd and P.G.W. Scott, Pharm. Pharmacol. 9, 176 (1957).
140. Banerjee and Hausler, Bull, Calcutta School Trop. Med. 3, 111 (1955).
141. C. Djerassi, M. Gorman, A.L. Nussbaum, and I. Reynoso, J. Am. Chem. Soc. 76, 4463 (1954).
142. M.W. Klohs, M.D. Draper, F. Keller, and W. Malesh, Chem. & Ind. 1264 (1954).
143. R. Vasiliev, V. Scintee, and M. Mangu, Faracia Bucharest) 9, 239 (1961).
144. R. Vasiliev, V. Scinteie, and M. Mangu, Rev. Chim. (Bucharest) 11, 180 (1960).
145. I.M. Roushdi and R.M. Shafik, J. Pharm. Sc. U.A.R. 9, 65 (1968).
146. F. Pellerin, J.A. Gautier, and D. Demay, Ann. Pharm. Franc 22, 495 (1964).
147. S. Cheng, Chemistry (Taiwan), 213 (1957).
148. H. Sakurai, Y. Yoneda, and H. Machida, Takamine Kenkyusho Nempo 10, 154 (1958).
149. C.W.R. Phaf, Pharm. Weekblad 9, 517 (1960).
150. E. Kahane and M. Kahane, Ann. pharm. franc. 16, 726 (1958).
151. M.J. Allen and V.J. Powell, J. Electrochm. Soc. 105, 541 (1958).
152. L. Hruban, A. Nemeckova, and F. Santavy, Acta Univ. Palackianae Olomucensis 16, 5 (1958): Chem. Abstr. 54, 12490f (1960).
153. A.E. Kalinowska and J. Bartnik-Kurzawinska, Acta Polon.

Pharm. 19, 45 (1962).
154. W.R. Maynard, Jr., J. Assoc. Offic. Anal. Chem. 41, 676 (1958).
155. E.H. Sakal and E.J. Merrill, J. Am. Pharm. Assoc. 43, 709 (1954).
156. D. Banes and J. Carol, J. Assoc. Offic. Agr. Chem. 38, 866 (1955).
157. B.C. Bose and R. Vijay vargirja, J. Pharm. Pharmacol. 11, 456 (1959).
158. W.A. Mannell and M.G. Allmark, Drug Standards 24, 6 (1956).
159. D. Banes, J. Assoc. Offic. Agr. Chem. 41, 488 (1958).
160. C.R. Szalkowski and W.J. Madar, J. Am. Pharm. Assoc. 45, 613 (1956).
161. N.Ya. Tsarenko and M.S. Shraiber, Farmatsevi. Zh. (Kiev) 19, 34 (1964): Chem. Abstr. 63, 6861c (1964).
162. Analyst 85, 755 (1960).
163. R.P. Haycock and W.J. Mader, J. Am. Pharm. Assoc. 46, 744 (1957).
164. R.P. Haycock, P.B. Sheth, R.J. Connolly and W.J. Mader, J. Assoc. Offic. Agr. Chemists 42, 613 (1959).
165. D. Banes, J. Assoc. Offic. Agr. Chemists 40, 798 (1957).
166. D. Banes, J. Wolff, H.O. Fallscheer and J. Carol, J. Am. Pharm. Assoc. 45, 708 (1956).
167. A.W.M. Indemans, I.M. Jakovljevic, and J.J.A.M. v.d. Langerijt, Pharm. Weekblad 94, 1 (1959).
168. Z. Nowakowska, I. Wilczynska and W. Zyzynski, Acta Polon. Pharm 21, 161 (1964).
169. A.W.M. Indemans, Pharm. Weekblad 108, 641 (1973).
170. V. Scarselli, Boll. soc. ital. biol. sper. 34, 1132 (1958).
171. D. Banes, J. Am. Pharm. Assoc. 44, 408 (1955).
172. D. Banes, J. Assoc. Offic. Agr. Chemists 39, 620 (1956).
173. H. Wachsmuth and L. van Koeckhoven, J. Pharm. Belg. 18, 378 (1963).
174. Z. Jung, Ceskoslov. farm. 6, 299 (1957).
175. N. Nakamura and T. Yoshida, Takeda Kenkyusho Nempo 18, 25 (1959).
176. J. Bartos, Ann. Pharm. Franc. 19, 610 (1961).
177. J.E. Gardner and S.J. Dean, Drug Standards 28, 50 (1960).
178. H.P. Rieder and M. Bohmer, Helv. Chim. Acta 42, 1793 (1959).
179. R.E. Booth, J. Am. Pharm. Assoc. 44, 568 (1955).
180. H.-S. Chou, C.-Y. Ku, C.-L. Chin, C.-Y. Ch'en, C.-L.

Hu, S.-L. Lo, and H.-S. Chao, Yao Hsueh Hsueh Pao 13, 280 (1966): Chem. Abstr. 65, 8673a (1966).
181. E.A. DeFelice, Experentia 14, 159 (1958).
182. D. Banes, J. Assoc. Offic. Agr. Chemists 40, 796 (1957).
183. D.P. Page, J. Assoc. Offic. Anal. Chemists 53, 815 (1970).
184. D. Banes, J. Am. Pharm. Assoc. 46, 601 (1957).
185. F. Tishler, P.B. Sheth, and M.B. Giaimo, J. Assoc. Offic. Agr. Chemists 46, 448 (1963).
186. W.J. Mader, R.P. Haycock, P.B. Sheth, and R.J. Connolly, J. Assoc. Offic. Agr. Chemists 43, 291 (1960).
187. W.J. Mader, R.P. Haycock, P.B. Sheth, R.J. Connolly, and P.M. Shapoe, J. Assoc. Offic. Agr. Chemists 44, 13 (1961).
188. W.J. Mader, G.J. Papariello, and P.B. Sheth, J. Assoc. Offic. Agr. Chemists 45, 589 (1962).
189. B.N. Kabadi, A.T. Warren, and C.H. Newman, J. Pharm. Sc. 58, 1127 (1969).
190. B.N. Kabadi, J. Pharm. Sc. 60, 1862 (1971).
191. E.B. Dechene, J. Am. Pharm. Assoc. 44, 657 (1955).
192. E.B. Dechene, J. Am. Pharm. Assoc. 47, 757 (1958).
193. R.B. Poet and J.M. Kelly, Abstr. 126th Meeting Amer. Chem. Soc., p. 83c (1954).
194. S.M. Hess, P.A. Shore, and B.B. Brodie, J. Pharmacol. Exptl. Therap. 118, 84 (1956).
195. I.M. Jakovljevic, J.M. Fose, and N.R. Kuzel, Anal. Chem. 34, 410 (1962).
196. R. Stainier, Farmaco 24, 167 (1968).
197. Pharmacopoeia Belgique IV.
198. R. Stainier, J. Pharm. Belg. 28, 115 (1973).
199. R. Stainier, H.P. Husson, and C.L. Lapiere, J. Pharm. Belg. 28, 307 (1973).
200. T. Urbanyi and A. O'Connell, Anal. Chem. 44, 565 (1972).
201. S. Barkan, J. Assoc. Offic. Anal. Chemists 55, 149 (1972).
202. T. Urbanyi and H. Stober, J. Assoc. Offic. Anal. Chemists 55, 180 (1972).
203. T. Urbanyi and H. Stober, J. Pharm. Sc. 59, 1824 (1970).
204. G. Kallistratos, J. Less-Common Metals 29, 226 (1973).
205. M.H. Malone, H.I. Hochman, and K.A. Nieforth, J. Pharm. Sc. 55, 972 (1966).
206. M. Gibaldi, S. Feldman, and T.R. Bates, J. Pharm. Sc. 57, 708 (1968).
207. T.R. Bates, J. Pharm. Pharmacol 21, 710 (1969).
208. R.G. Stoll, T.R. Bates, K.A. Nieforth, and J.

Swarbrick, J. Pharm. Sc. 58, 1457 (1969).
209. L. Decato, Jr., M.H. Malone, R.G. Stoll, and K.A. Nieforth, J. Pharm. Sc. 58, 273 (1969).
210. L. Decato, Jr., R.G. Stoll, T.R. Bates and M.H. Malone, J. Pharm. Sc. 60, 140 (1971).
211. E.I. Stupak and T.R. Bates, J. Pharm. Sc. 61, 400 (1972).
212. B. Rubin, M.H. Malone, M.H. Waugh, and J.C. Burke, J. Pharmacol. Exptl. Therap. 120, 125 (1957).

SPIRONOLACTONE

John L. Sutter and Edward P. K. Lau

Contents

1. Description

1.1 Name, Formula, Molecular Weight

Spironolactone is 17-hydroxy-7α-mercapto-3-oxo-17α-pregn-4-ene-21-carboxylic acid γ-lactone 7-acetate.

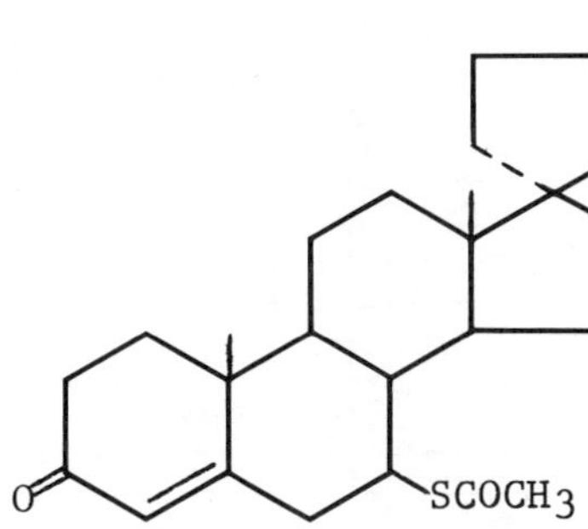

$C_{24}H_{32}O_4S$ Molecular Weight: 416.59

1.2 Appearance, Color, Odor

Spironolactone is a yellowish-white crystalline powder, with a faint mercaptan odor.

2. Physical Properties

2.1 Infrared Spectrum

The infrared absorption spectrum of a spironolactone reference standard compressed in a KBr disc is shown in Figure 1. Essentially the same infrared spectrum is observed in chloroform solution. The following assignments have been made for absorption bands in Figure 1:

Wavenumber, cm^{-1}	Assignment
1775	5-membered lactone carbonyl
1670 - 1690	3-ketone, 7-thioester carbonyl
1620	4,5 - double bond

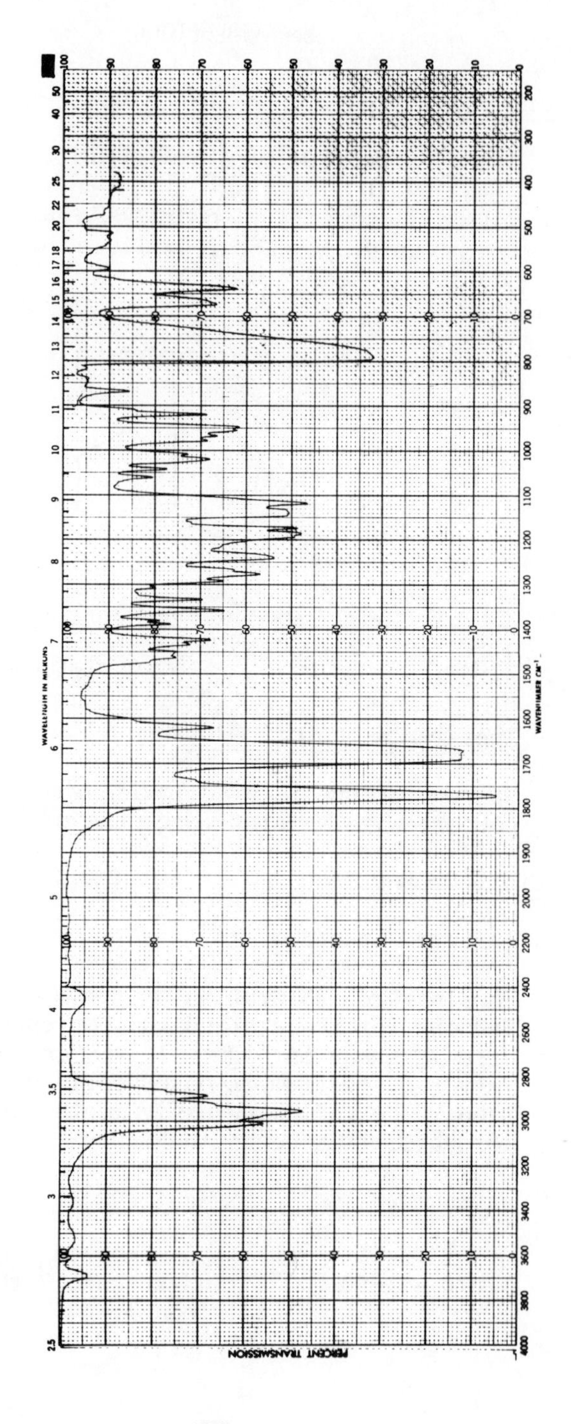

Fig1. Infrared Spectrum of Spironolactone

2.2 Nuclear Magnetic Resonance Spectrum

The NMR spectrum of a spironolactone reference standard in deuterated chloroform is shown in Figure 2. Spectral assignments are as follows:

Chemical Shift (ppm)	Type	Assignment
5.7	Broad quartet	4 - H
4.0	Broad singlet	7 - H
2.4	Singlet	7-thioacetate protons
1.2	Singlet	19-methyl protons
1.0	Singlet	18-methyl protons

2.3 Ultraviolet Spectrum

The ultraviolet absorption spectrum of a spironolactone reference standard in methanol is shown in Figure 3. The molar absorptivity of spironolactone in methanol is 19.6×10^3 at the absorbance maximum at about 238 nm.[1]

2.4 Mass Spectrum

The low resolution mass spectrum of spironolactone shown in Figure 4 was obtained with an AEI Model MS-30 mass spectrometer. A molecular ion was observed at m/e 416. The base peak in the spectrum was at m/e 341, corresponding to loss of $.SCOCH_3$. Structure assignments are as follows:

m/e	Assignment	% Relative Intensity
43	CH_3CO^+	89
267	340 - Lactone ring	26
325	340 - methyl	12
340	$M^{+\cdot}$ - $HSCOCH_3$	43
341	$M^{+\cdot}$ - $.SCOCH_3$	100
359	$M^{+\cdot}$ - (CH_2CO+CH_3)	20

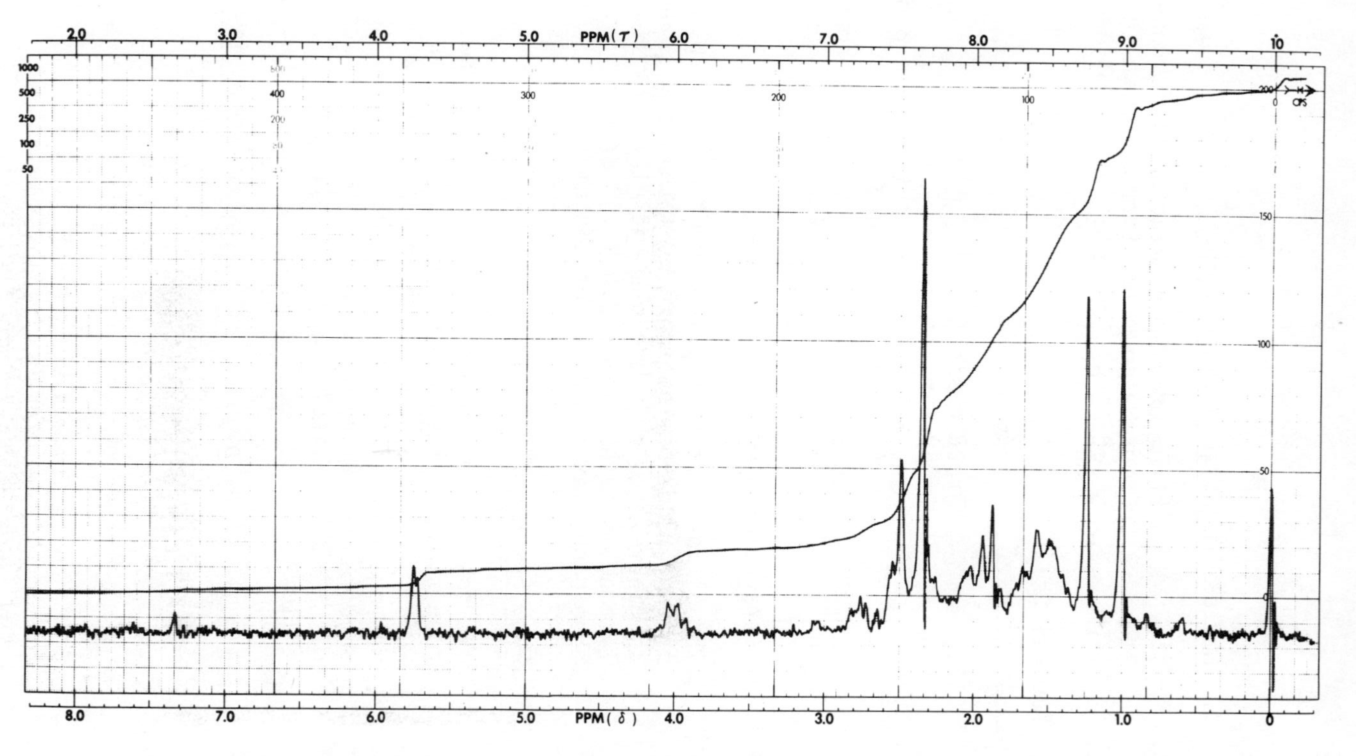

Fig. 2 Nuclear Magnetic Resonance Spectrum of Spironolactone

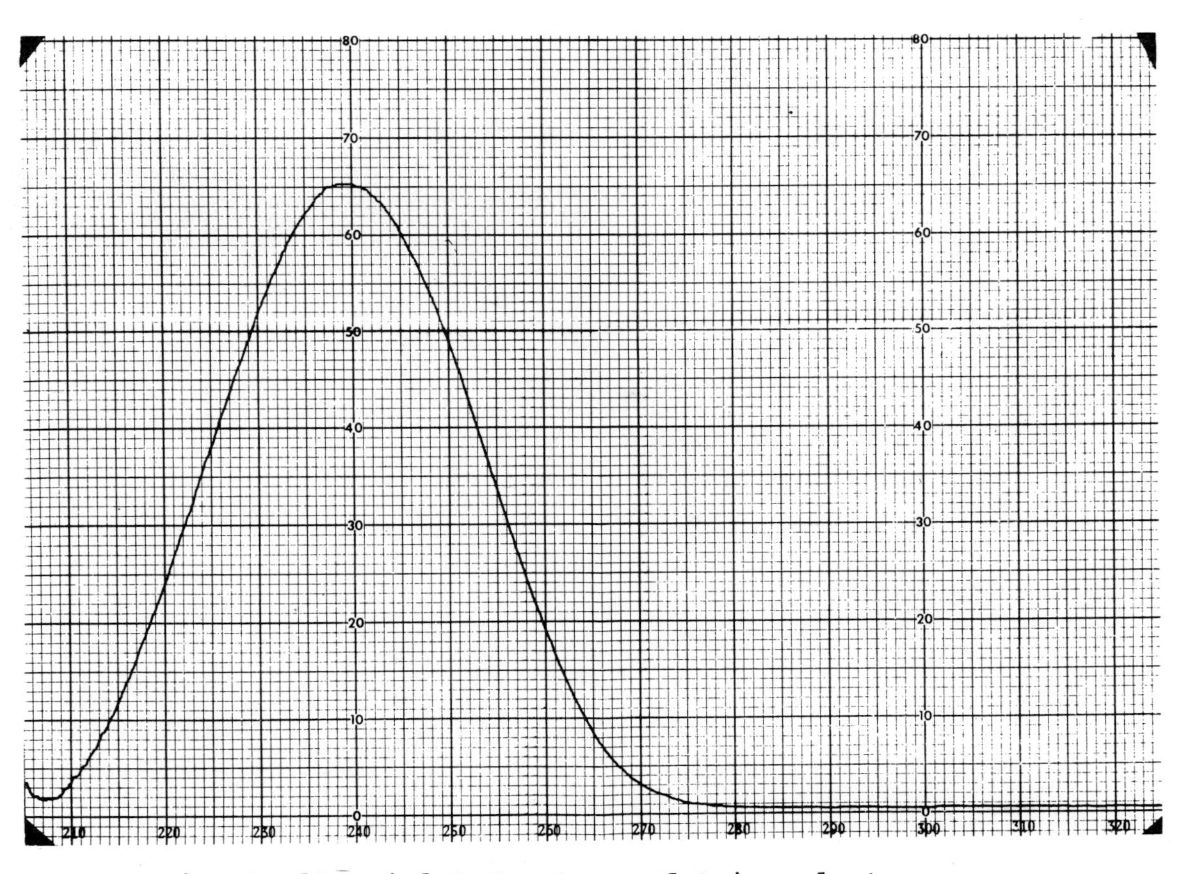

Fig. 3 Ultraviolet Spectrum of Spironolactone

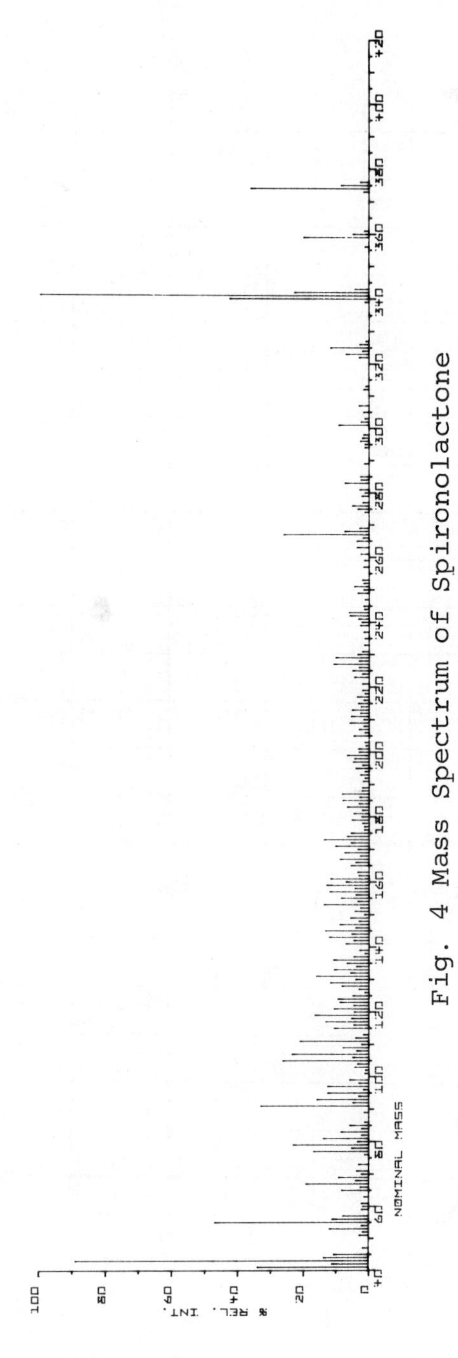

Fig. 4 Mass Spectrum of Spironolactone

374	$M^{+\cdot}$ - (CH_2CO)	36
383	$M^{+\cdot}$ - $(H_2O + CH_3)$	1
416	$M^{+\cdot}$	0.4

2.5 Optical Rotation

The following specific rotation values in chloroform[1] have been reported for spironolactone:

$[\alpha]_{356}^{27} = -403.8^{o}$

$[\alpha]_{436}^{27} = -110.0^{o}$

$[\alpha]_{546}^{27} = - 45.8^{o}$

$[\alpha]_{578}^{27} = - 38.8^{o}$

$[\alpha]_{589}^{27} = - 36.3^{o}$

2.6 Melting Range

The USP 1a melting range of spironolactone is 198° - 207°C.[2] A reference standard lot of the compound was found to melt from 205.4° to 206.9°C. Melting and resolidification at lower temperatures is sometimes noted.

2.7 Differential Scanning Calorimetry

The DSC thermogram of a spironolactone reference standard crystallized from ethanol and water shown in Figure 5 was obtained on a Perkin-Elmer DSC-1B differential scanning calorimeter at a heating rate of 20°C/minute in an atmosphere of nitrogen. A slight exothermic peak was observed beginning at about 160°C, and an endothermic melting peak was found at about 210°C.[3] Spironolactone chemical crystallized from different solvents can yield different thermograms. Samples crystallized from methanol have

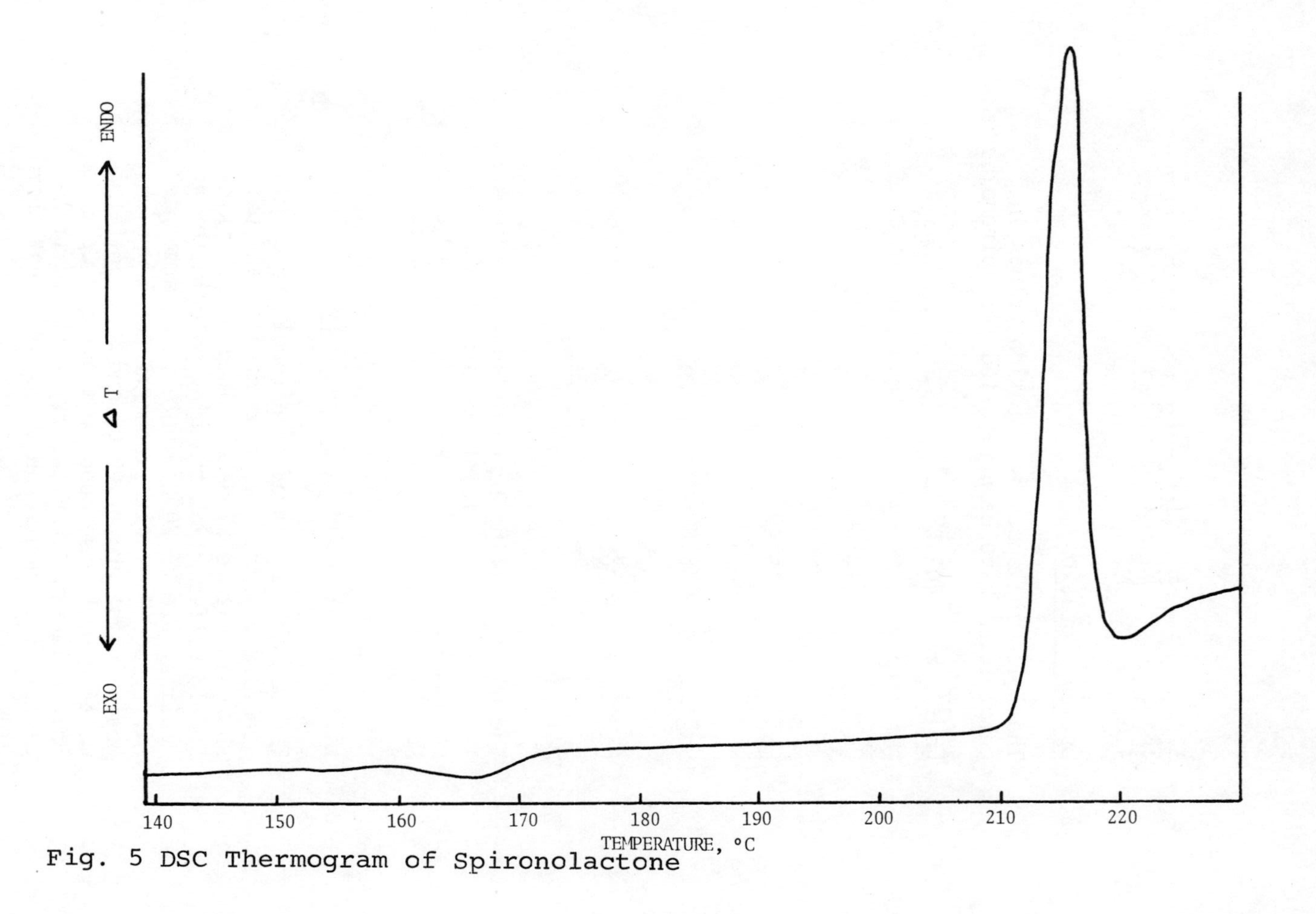

Fig. 5 DSC Thermogram of Spironolactone

been shown to give thermograms with two transition peaks: an endotherm at about 120°C and an exotherm at about 140°C. These transition peaks can be removed by heating the sample at 105°C or refluxing in water for four hours. A sample crystallized from n-propanol, however, gave only one endothermic peak at 210°C. Thermograms of spironolactone crystallized from ethyl acetate have shown two transition peaks: one endotherm at about 110°C and an another at about 207°C. This behavior suggests that spironolactone can exist in a number of polymorphic forms.

2.8 Solubility

Solubilities of spironolactone in various solvents at 25°C are given in the following table:[4]

Solvent	Solubility, mg./ml.
Water	2.8×10^{-2}
Methanol	6.9
Ethanol (USP)	27.9
Chloroform	50
Heptane	2.4×10^{-1}

3. Synthesis

The synthetic route to spironolactone shown in Figure 6 has been reported in papers by Cella, Brown and Burtner[5], and Cella and Tweit.[6]

Carbonation of the Grignard reagent of 17α-ethynyl-5-androstene-3β,17β-diol (I) yielded an acetylenic acid (II). Selective reduction of the acetylenic bond was accomplished by catalytic hydrogenation over palladium on calcium carbonate, using dioxane and pyridine as solvents. Treatment of the product with mineral acid yielded the unsaturated lactone, 3-(3β,17β-dihydroxy-5-androstene-17α-yl) propenoic acid lactone (III),

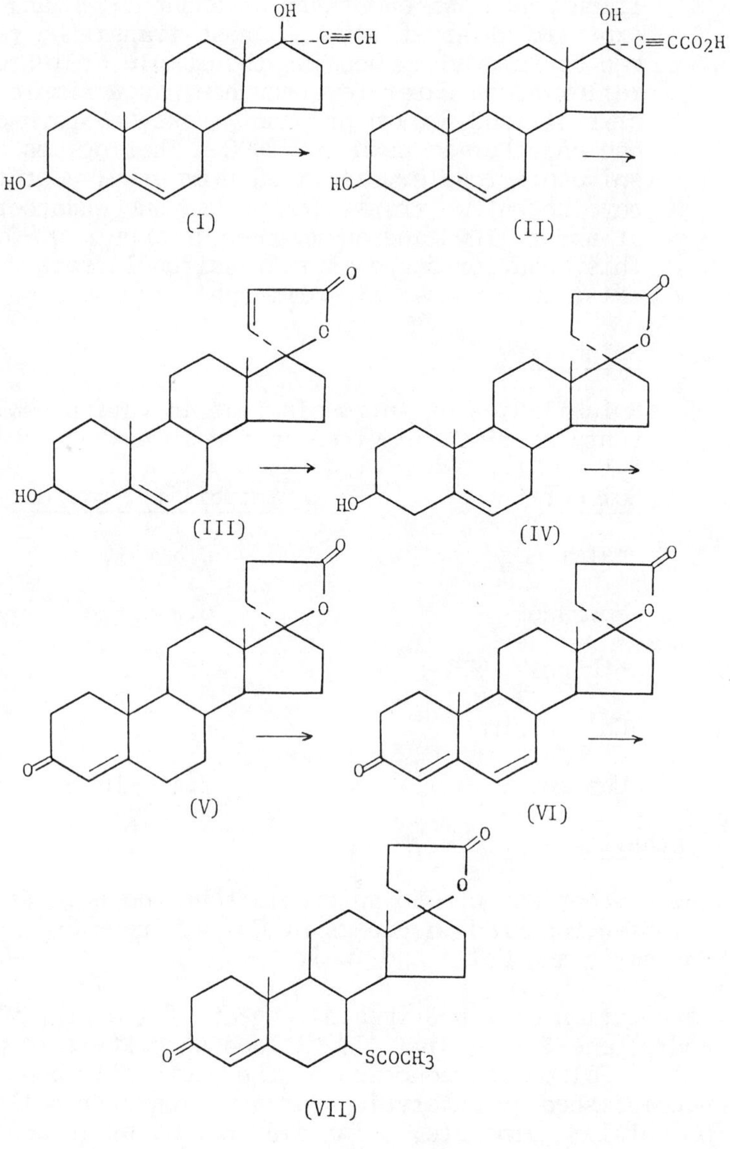

OH
$C\equiv CH$
HO
(I)

OH
$C\equiv CCO_2H$
HO
(II)

O
O
HO
(III)

O
O
HO
(IV)

O
O
O
(V)

O
O
O
(VI)

O
O
O
$SCOCH_3$
(VII)

Figure 6. Spironolactone Synthesis

which was easily reduced to the saturated lactone (IV) by hydrogen over palladium on charcoal. Oppenauer oxidation of the product afforded 3-(3-oxo-17β-hydroxy-4-androsten-17α-yl) propionic acid lactone (V). Unsaturation at C_6 was then introduced by treatment with chloranil. Finally, the resulting 17-hydroxy-3-oxo-17α-pregna-4,6-diene-21-carboxylic acid-γ-lactone (VI) was reacted with thioacetic acid, yielding 17-hydroxy-7α-mercapto-3-oxo-17α-pregn-4-ene-21-carboxylic acid γ-lactone 7-acetate, spironolactone (VII).

4. Stability and Degradation

Spironolactone has been found to decompose to the dienone, canrenone. The reaction is not a facile one in either the pure chemical or its dosage forms, since canrenone forms only to the extent of 1% or less over a period of five years at 40°C.[8]

5. Drug Metabolic Products and Pharmacokinetics

The structures of spironolactone metabolites in man identified so far are summarized in Figure 7.

Gochman and Gantt, using the fluorimetric method described in Section 6.5, demonstrated that spironolactone (VII) is readily converted to the dethioacetylated metabolite, canrenone (VI), in humans.[8] Subsequent studies by Zicha et al[9] and by Wagner et al[10] showed the presence of at least six metabolites in the urine of humans after ingestion of spironolactone. Tentative structures for the unknown metabolites were proposed by these investigators;[9,10] however, rigorous identification of the metabolites was not attempted.

Karim and Brown established that elimination of the thioacetate group of spironolactone *in vivo* is not quantitative.[11] Besides canrenone, these investigators identified a major sulfur-containing metabolite, 3-(3-oxo-7α-methylsulfinyl-6β,17β-dihydroxy-4-androsten-17α-yl) propionic acid γ-lactone (Metabolite C, (VIII), Figure 7), in the urine of humans. Tentative

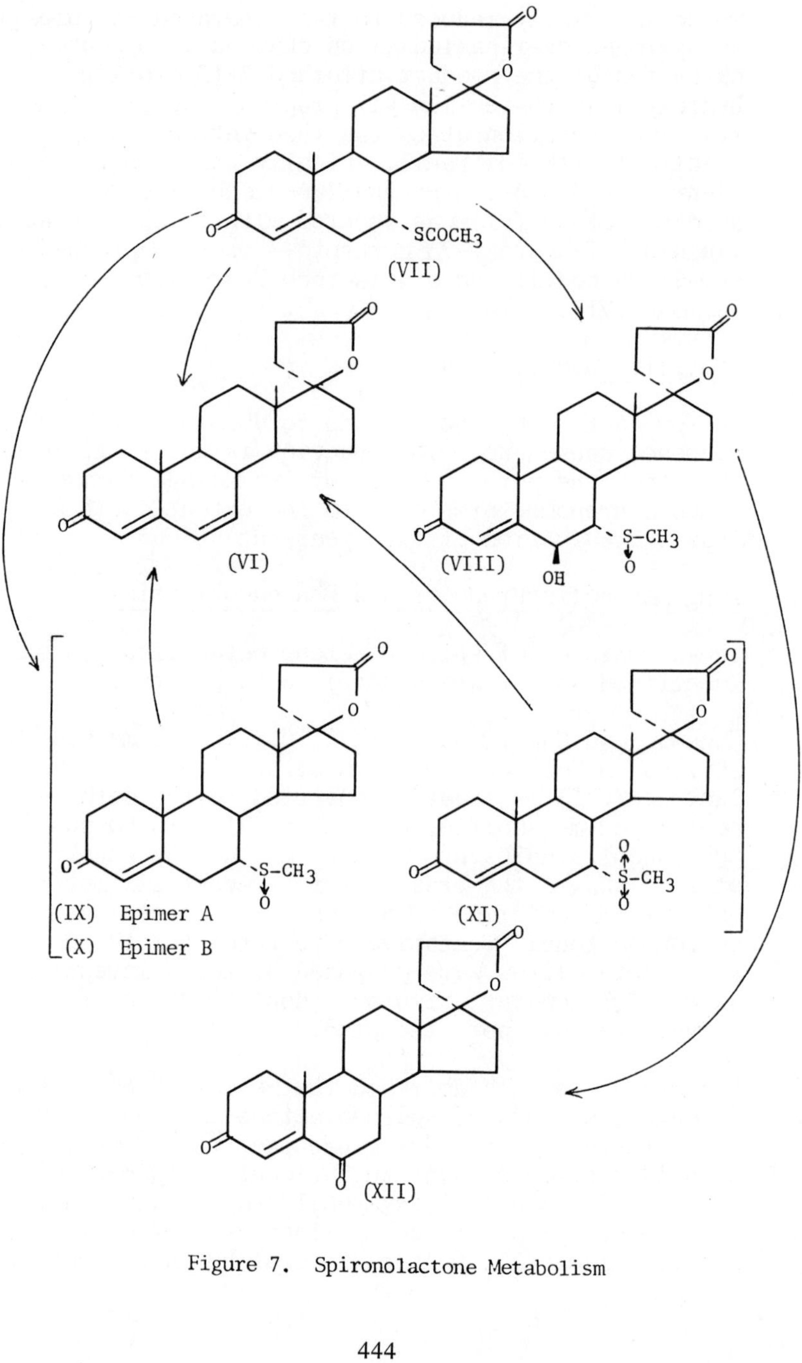

SCOCH3
(VII)

(VI)

S-CH3
O
OH
(VIII)

S-CH3
O
(IX) Epimer A
(X) Epimer B

O
S-CH3
O
(XI)

(XII)

Figure 7. Spironolactone Metabolism

structures were also proposed for three other minor sulfur-containing metabolites, (IX), (X), and (XI). Under the gas chromatographic conditions reported by Chamberlain for the determination of canrenone, it was found that Metabolite C (VIII) was converted to Δ^4-3,6-dioxo steroid (XII). Further, under these conditions, spironolactone (VII) as well as labile metabolites (IX), (X), and (XI), were converted to canrenone (VI).

Further unpublished observations[13] have shown that metabolite C (VIII) had the same retention time as an unknown urinary polar metabolite observed by Chamberlain.

The pharmacokinetics of spironolactone and canrenone were investigated by Sadee, Dagcioglu and Schroder[14]. Determination of plasma levels following equivalent oral administration of spironolactone or canrenone led to the conclusion that 79% of spironolactone is dethioacetylated to canrenone. The terminal log-linear phase half-life of canrenone in plasma following oral administration of spironolactone ranged from 17 to 22 hours. Fluorescence assay revealed that from 14 to 24% of oral doses of spironolactone were excreted in the urine as fluorogenic metabolites within five days.

6. Methods of Analysis

6.1 Partition Coefficient

Spironolactone is preferentially extracted into heptane from water. The partition coefficient is 3.5 at 25°C.

6.2 Phase Solubility

Phase solubility analysis of spironolactone can be carried out by equilibration in methanol at 25°C. Figure 8 shows the phase solubility diagram of a spironolactone reference standard.[4]

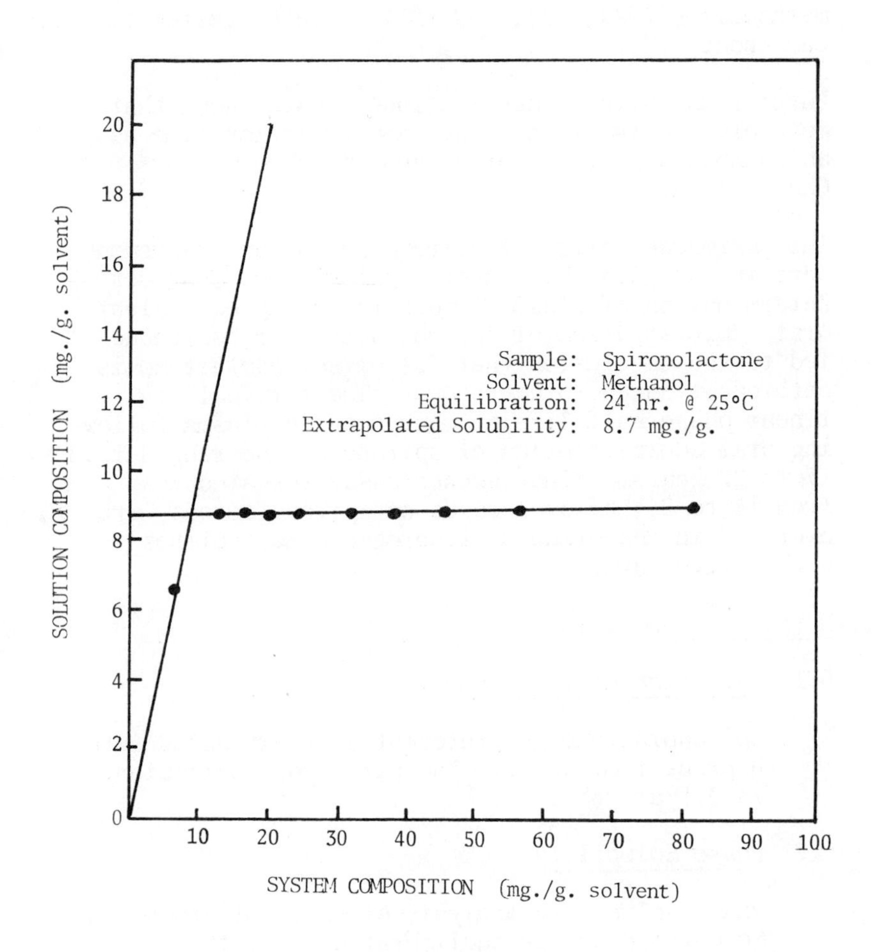

Figure 8. Phase Solubility Diagram of Spironolactone

6.3 Spectrophotometric Analysis

The ultraviolet absorption spectrum of spironolactone is the basis for the USP XVIII assay for the compound. The absorbance maximum at about 238 nm. in methanol is used for quantitation.[2]

6.4 Colorimetric Analysis

Reaction of spironolactone with methanolic hydroxylamine hydrochloride and ferric perchlorate yields a red ferric hydroxamate complex having an absorbance maximum at about 515 nm.[15] The absorbance at this wavelength is linear with concentration over a range of 5 mg./ml. to 29 mg./ml. The colored complex is stable for up to 2 hours. The method is used for analysis of the compound and a variety of its dosage forms. Canrenone, the principal degradation product of spironolactone, does not interfere.

Isonicotinic acid hydrazide may be reacted with spironolactone in methanolic solution, yielding a soluble yellow product whose absorbance rises to a plateau beginning at about 375 nm.[16] The molar absorptivity is 2.2×10^3 at about 380 nm. The method has not yet been adopted for quantitative analysis.

6.5 Fluorometric Analysis

Spironolactone may be dethioacetylated under mild acid or alkaline conditions, yielding the corresponding 4,6 -dienone,[17] canrenone, which, in 62% sulfuric acid, is converted to a fluorescent trienone.[18] This compound has an excitation maximum at 483 nm. and an emission maximum at 525 nm. The fluorescence is useful for quantitation of canrenone in plasma over the range 40-160 ng./ml.[8] The procedure may be used to determine spironolactone in the presence of its major dethioacetylated metabolites, by measuring fluorescence in 62% H_2SO_4 both with and without prior dethioacetylation.[17]

6.6 Chromatographic Analysis

6.61 High Pressure Liquid Chromatography

Spironolactone may be separated from canrenone on an octadecyl silane column using methanol - water mobile phase, and detected with the aid of a continuously variable wavelength UV detector. Spironolactone, which is eluted first, may be detected at about 238 nm. and canrenone may be observed at its absorption maximum at about 283 nm.[19]

6.62 Thin Layer Chromatography

Thin-layer chromatographic systems and corresponding R values for spironolactone are summarized in the following table:

Solvent System	Adsorbent	Detection	R
Ethyl acetate, 100%	Silica Gel GF (Woelm)	1,2,3	0.53 [4]
Benzene: Ethyl acetate: Methanol 73:25:2	Silica Gel G	4	0.67 [4]

Detection:

1. Observe under short wave UV.
2. Spray with 50% H SO , heat at 80 C for 10 minutes, observe under long wave UV.
3. Spray with Phosphomolybdic Acid.
4. Spray with Phosphomolybdic Acid, heat at 80 C. for 10 minutes.

7. Acknowledgements

The authors wish to express their appreciation to Mr. E. Brown, Dr. H. Dryden and Mr. R. Tweit for their helpful discussions on synthesis; to Dr. R. Bible, Mr. A. Damascus, Dr. J. Hribar and Miss Lydia Swenton for their assistance in the preparation of the sections on IR, NMR and Mass Spectra; to Dr. A. Karim for sharing with them his knowledge of spironolactone metabolism and pharmacokinetics; and to Mrs. Lorraine Wearley for searching the literature and assembling various sections of this manuscript. The expert secretarial assistance of Mrs. Corey Leone is also gratefully acknowledged.

8. References

1. Anthony, G., Searle Laboratories, personal communication.

2. "The United States Pharmacopeia" XVIII, p. 681.

3. Marshall, S., Searle Laboratories, personal communication.

4. Aranda, E., Searle Laboratories, personal communication.

5. Cella, J. A., Brown, E. A. and Burtner, R. R., J. Org. Chem. 24, 1109 (1959).

6. Cella, J. A. and Tweit, R. C., J. Org. Chem. 24, 1109 (1959).

7. Baier, M. E., Searle Laboratories, personal communication.

8. Gochman, N. and Gantt, C.L., J. Pharmacol. Exp. Ther. 135, 312 (1962).

9. Zicha, L., Weist, F., Scheiffarth, F. and Schmid, E., Arzneimittel-Forsch. 14, 699 (1964).

10. Wagner, H., Weist, F. and Zicha, L., Arzneimittel-Forsch. 17, 415 (1967).

11. Karim, A. and Brown, E. A., Steroids 20, 41 (1972).

12. Chamberlain, J., J. Chromatog. 55, 249 (1971).

13. Karim, A., Searle Laboratories, personal communication.

14. Sadee, W., Dagcioglu, M. and Schroder, R., J. Pharmacol. Exp. Ther. 185, 686 (1973).

15. Seul, J., Searle Laboratories, personal communication.

16. Wearley, L., Searle Laboratories, personal communication.

17 Sadee, W., Dagcioglu, M., and Riegelman, S., J. Pharm. Sci. 61, 1126 (1972).

18. Sadee, W., Riegelman, S. and Johnson, L. F., Steroids 17, 595 (1971).

19. Runser, D. J. and Fortier, C., Searle Laboratories, personal communication.

TESTOSTERONE ENANTHATE

Klaus Florey

CONTENTS

1. Description

1.1 Name, Formula, Molecular Weight

Testosterone enanthate (heptanoate) is 17-[(1-oxoheptyl)oxy]-4-androsten-3-one.

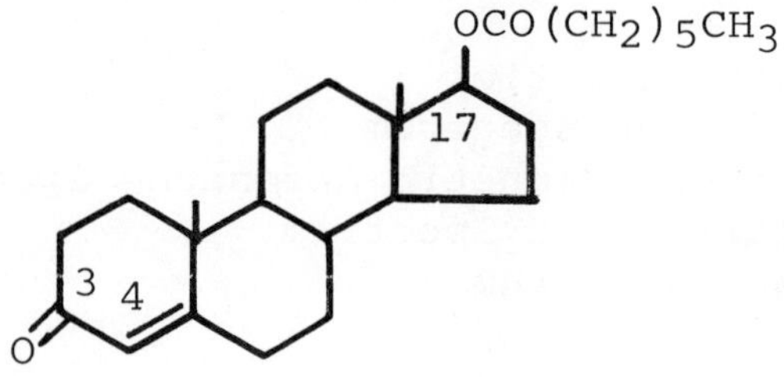

$OCO(CH_2)_5CH_3$

$C_{26}H_{40}O_3$ Mol. Wt. 400.61

1.2 Appearance, Color, Odor

White, crystalline, odorless powder.

2. Physical Properties

2.1 Infrared Spectrum

The infrared spectrum (KBr pellet of testosterone enanthate is presented in Figure 1[1].

2.2 Nuclear Magnetic Resonance Spectrum

The NMR spectrum is presented in Figure 2. It was obtained on a 60 MHz spectrometer in deuterochloroform containing tetramethylsilane as an internal reference. The following proton assignments were made[2]:

Protons at	Chemical Shift		Coupling Constants T(in H_2)
C-4H	4.20	singlet	-
C-17αH	5.38	triplet	16H,17H =8.5
C-18H	9.17	singlet	-
C-19H	8.82	singlet	-
ester methyl	9.13	multiplet	-

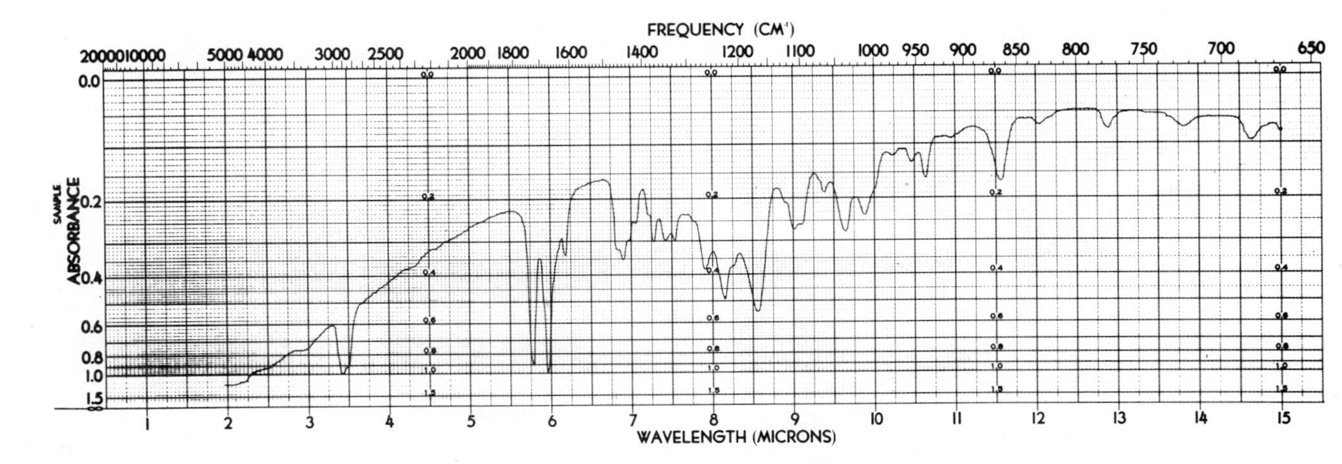

Figure 1. Infrared Spectrum of Testosterone Enanthate(KBr Pellet).
Instrument: Perkin-Elmer 621.

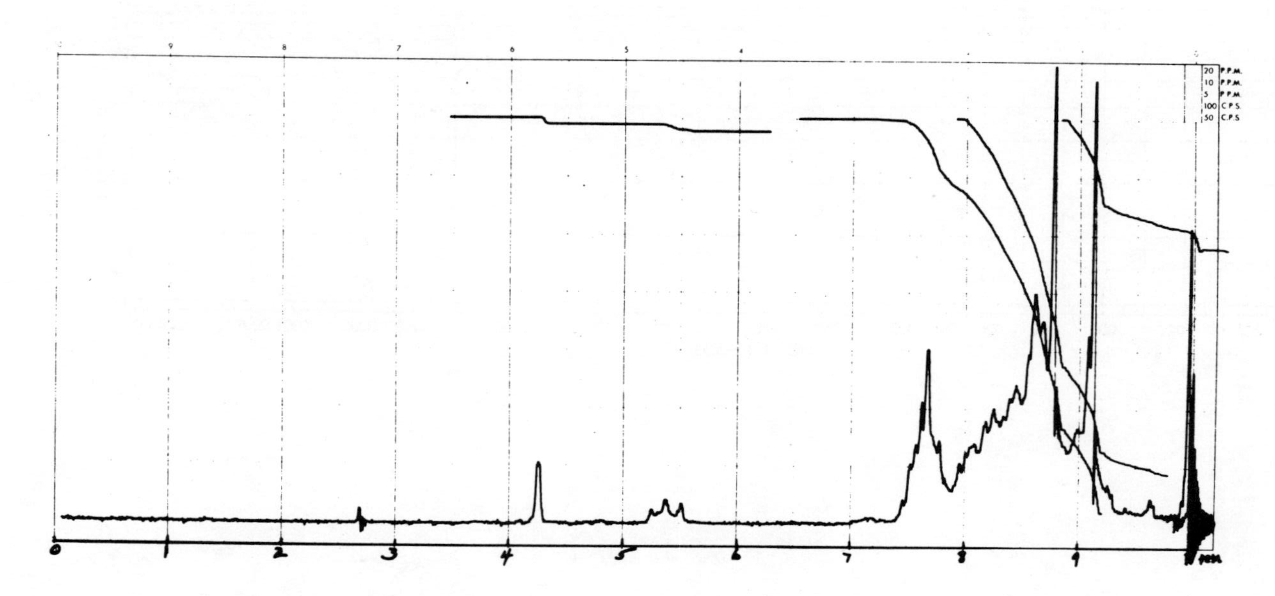

Figure 2. NMR Spectrum of Testerone Enanthate in Deuterated Chloroform. Instrument: Perkin-Elmer R12B.

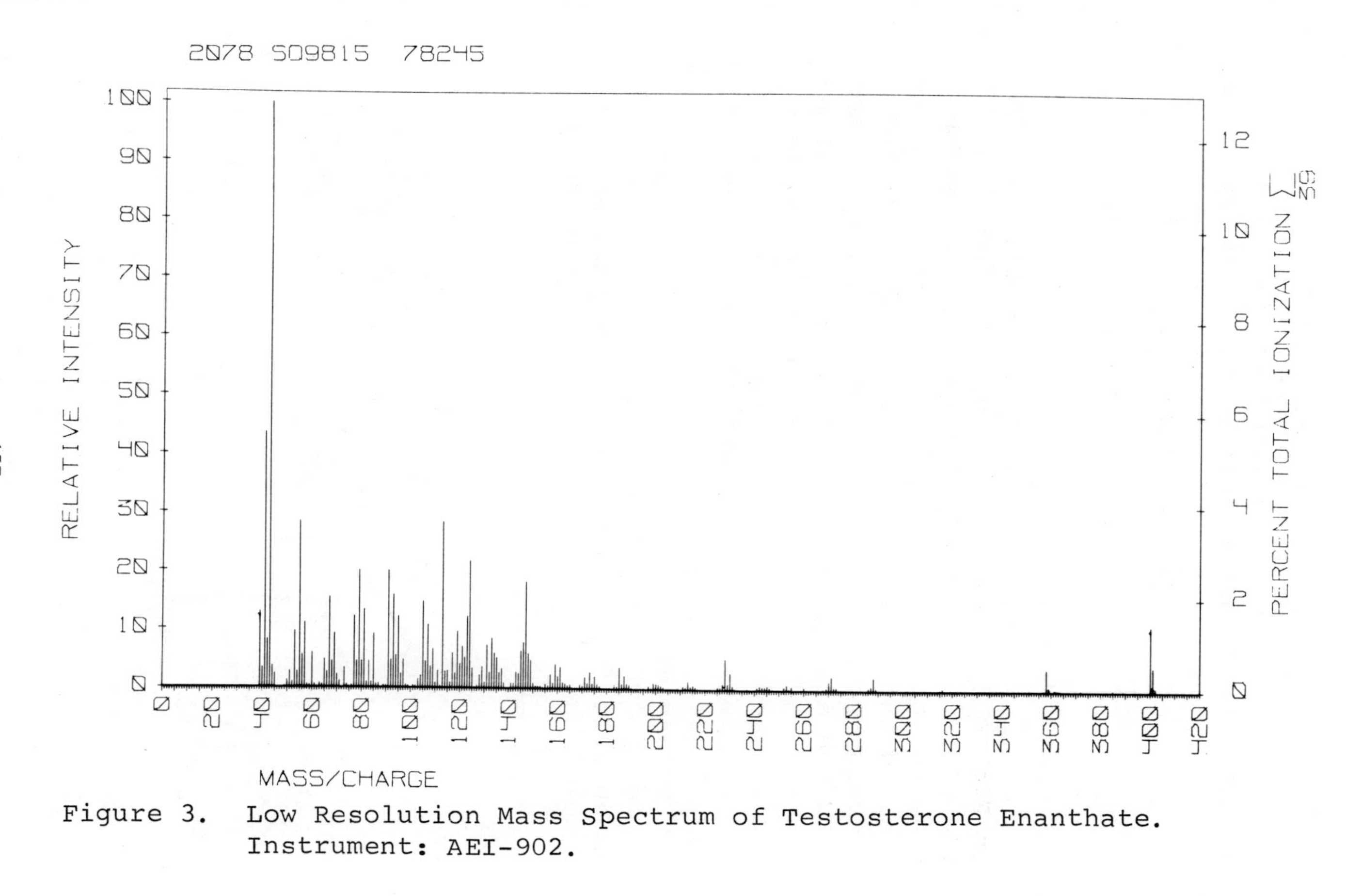

Figure 3. Low Resolution Mass Spectrum of Testosterone Enanthate. Instrument: AEI-902.

2.3 Ultraviolet Spectrum

Squibb House Standard (Lot 78245) gave a single band, due to the 3-keto-4 ene system[3]:

λ max 240 nm $E^{1\%}_{1cm}$ 429[3]

λ max 240 nm $E^{1\%}_{1cm}$ 410[7]

2.4 Mass Spectrum

The low-resolution mass spectrum, shown in Figure 3, demonstrates the expected M^+ of m/e 400. There is progressive loss of carbons from the alkyl portion of the acylate which gives rise to the ions at m/e 315, 316 from the hydrocarbon portion of the chain and the m/e 288 ion with the additional loss of the acylate carbonyl. The loss of the entire C-17 group yields the m/e 270 and 271 ions. The ions at m/e 85 and 113 represent the hydrocarbons, C_6H_{13}, and $C_7H_{13}O$, respectively. Fragment ions occur through a progressive loss of D-ring, C-ring and B-ring carbons. The m/e 147 and 124 ions are diagnostic for Δ^4-3-ketones. The assignment of some of the diagnostic ions is depicted below[2].

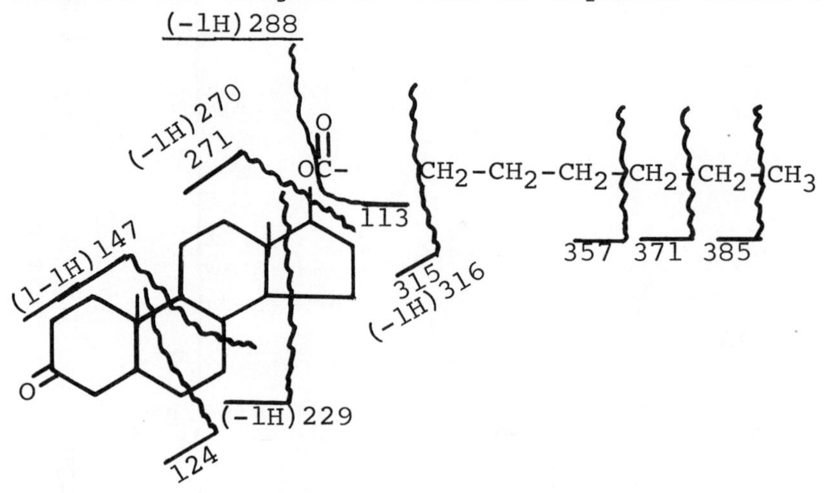

2.5 Rotation

The following rotation were reported:

+78° (dioxan)[4]

+78.0° (Squibb House Standard Lot 78245)[3]

+79°[7]

+75-76°[8]

2.6 Melting Range

The following melting range temperature temperatures (°C) were reported:

36-37.5[4]

37-38[3] (Squibb House Standard Lot 78245)

35 (from alcohol-ether)[5]

36[6]

36-37[7]

36-37[8]

2.7 Differential Thermal Analysis

Endotherm at 37°C.[3]

2.8 Solubility

Insoluble in water; very soluble in ether, soluble in vegetable oils[23].

2.9 Crystal Properties

The powder X-ray diffraction pattern is presented in Table I[9]. It is not too well resolved, since due to the low melting point of testosterone enanthate, the sample partially liquified during analysis.

Table I
Powder X-ray Diffraction
Pattern of Testosterone Enanthate[9]

d(Å)*	K/I_o**	d(Å)*	K/I_o**
14.5	0.39	4.80	1.00
9.20	0.15	4.69	0.25
8.00	0.12	4.60	0.30
7.30	0.15	4.49	0.27
6.40	0.25	4.20	0.45
6.30	0.25	4.11	0.27
6.18	0.25	4.03	0.27
6.00	0.25	3.72	0.22
5.80	0.30	3.52	0.20
5.46	0.28		
5.27	0.33		
5.10	0.77		

*d = interplanar distance $\frac{n\lambda}{2 \sin\theta}$

λ = 1.539 Å;

Radiation: $K\alpha_1$ and $K\alpha_2$ Copper

**Relative intensity based on highest intensity of 1.00

3. Synthesis

Testosterone enanthate is prepared by acylation of testosterone with enanthic anhydride in pyridine[4].

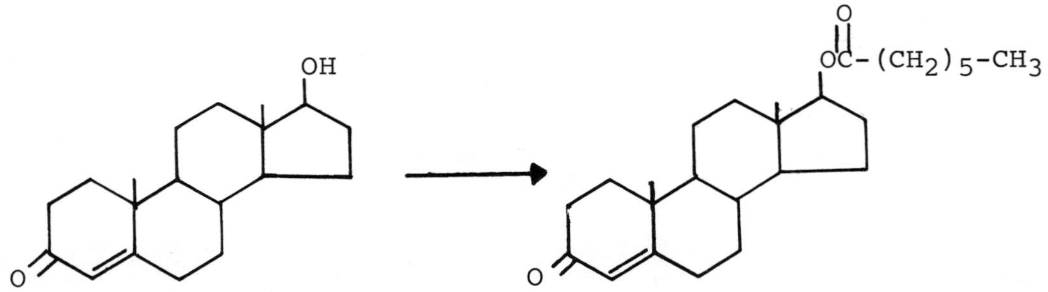

OH

O

$\longrightarrow$

$OC(=O)-(CH_2)_5-CH_3$

O

Enanthic acid and an ion exchanger[8] and enanthic acid chloride[6,7] have also been used instead of enanthic anhydride. It can also be prepared by transesterification of testosterone propionate and methyl enanthate[5].

4. Stability, Degradation

Testosterone enanthate is stable as a solid. In solution photolytic degradation of the A-ring is possible when exposed to ultraviolet light or ordinary fluorescent laboratory lightning (cf. 16). The kinetics of saponification have been studied[17,18]. Mucor fungi possess esterases able to saponify testosterone enanthate among other steroid esters[19].

5. Drug Metabolic Products

Generally, the metabolic fate of testosterone enanthate is deacylation to testosterone. It then follows testosterone metabolism, mostly to 17-keto steroids such as 4-androsten-3,20-dione and estrogenic substances. Data have been reported for human urine after intramuscular injection[10,11], rats[12], their prostates and livers[13]. The fate in maternal peripheral blood and cord venous blood has also been studies[14]. Tissue distribution and excretion of labeled testosterone enanthate was studied in steers[15].

6. Methods of Analysis

6.1 Elemental Analysis

	C	H	O
Calc.%	77.95	10.07	11.98

6.2 Spectrophotometric Analysis

The U.V. maximum at 240 nm can be used for quantitation determination in pharmaceutical preparations[20,21,22].

6.3 Colorimetric Analysis

Reaction with isoniazid and determination of the resulting hydrazone at 380 nm is the basis of the compendial assay[23].

6.4 Chromatographic Analysis

6.41 Paper

The quantitative determination and separation from testosterone by paper chromatography in oily vehicles has been described by Roberts and Florey[26]. After spotting, the paper strips were impregnated with 30% diethylene glycol monoethyl ether (Carbitol) in chloroform and developed for 2-1/2 hours with methyl cyclohexane saturated with diethylene glycol monoethyl ether. After locating of the steroid spots with a fluorescent paddle, the spots are eluted and reacted with isonicotinic acid hydrazide in methanol containing also hydrochloric acid. The absorbance of the yellow hydrazone is read at 415 nm.

A similar procedure, using impregnation with propylene glycol, phenylglycol ethyl ether and methanol (1:1:2) and propylene glycol, phenyl glycol ethyl ether and heptane (1:1:200) as developing solvent and 2,4-dinitrophenylhydrazone as spray reagent has also been reported[27].

6.42 Thin-Layer

Thin-layer chromatographic systems have been compiled in Table II.

Table II

Absorbent	Solvent System	Rf.	Ref.
Silica gel G impreg. with mineral oil	50% Acetic Acid	31	24
	50% Ethanol	26	24
Silica gel	Benzene-Acetone(4:1)	--	25
Silica gel	Benzene-Methanol(9:1)	--	25
Silica gel	Pet-ether, Benzene, Acetic Acid, Water (67:33:85:35)	--	25
Alumina	Benzene-Acetone(4:1)	--	25
Silica gel impreg. with corn oil	Methanol-Water(9:1)	--	23

Detection systems: U.V. light, sulfuric acid - alcohol.

6.43 Column

Column chromatography on silanized chromatographic siliceous earth using 95% ethanol-heptane solvent system followed by the isoniazid reaction is the basis of the compendial assay of testosterone enanthate in oil preparations[23].

7. References

1. B. Toeplitz, The Squibb Institute, Personal Communication.
2. A. I. Cohen, The Squibb Institute, Personal Communication.
3. G. A. Brewer, Jr., The Squibb Institute, Personal Communication.
4. K. Junkmann, J. Kathol, H. Richter, U. S. Patent 2,840,508(1958); K. Junkmann, Arch. Expt. Pathol. Pharmacol. 215,85(1952).
5. S. A. Alter, Spanish Patent 241,206(1958); C.A.54,3532a(1960).
6. T. Kasugai, Japan. Patent 14,089(1962); C.A.59,10182a(1963).
7. H. Cieslik, H. Koprowska, A. Walczak, Pol. Patent 56,358(1968)C.A.71,39286p(1969).
8. M. Ulrich, A. Novacek and F. Stejskal, Czech. Patent 95,825(1960)C.A. 55,15551d (1961).
9. Q. Ochs, The Squibb Institute, Personal Communication.
10. J. A. Schneider and A. Schuchter, Arztl. Wochschr. 9,392(1954)
11. G. Doerner, F. Stahl and R. Zabel, Endokrinologie 45,121(1963).
12. M. Wenzel, L.Pitzel and P. E. Schulze, Angew.Chem.Int.Ed.Engl. 7,211(1968).
13. J. Shimazaki, H. Kurihara, Y. Ito, K.Shida and K. Kirao, Gunma J. Med. Sci. 14,313

(1965); C.A. 66,263296(1967) ibid. 14,326(1965);C.A. 66,62263j(1967).

14. H. H. Simmer, M. V. Frankland and M. Greipel, Steroids 19,229(1972).
15. F.X. Gassner, R. P. Martin, W. Shimoda and J. W. Algeo, Fertility and Sterility 11,49(1960).
16. D.R. Barton and W. C. Taylor, J.Am.Chem. Soc. 80,244(1958); J.Chem.Soc.1958,2500.
17. M. Schenk and K. Junkmann, Arch.Exper. Path. and Pharmacol, 227,210(1955).
18. Z. Vesely, J. Pospicek and J.Trojanek, Collect.Czech. Chem. Commun. 34,685(1969).
19. K.A. Kohcheenko, G. K. Skryabin, B. K. Eroshin, L. M. Kogan and J.V.Torgor, Prikl. Biokhim.i.Microbiol. 1,181(1965); C.A.63,9018c(1965).
20. N. H. Coy, The Squibb Institute, Personal Communication.
21. A. Reginato, Anales fac. guim. y. farm., Univ. Chile 10,48(1958)C.A.54,6028f(1960).
22. S. Kanna, M. Takuma and S. Watanabe, Yakuzaigaku 25,145(1965)C.A. 64,15676d (1966).
23. U.S.P. 18th edition.
24. D. Sonanini and L. Anker, Pharm. Acta Helv. 42,54(1967).
25. S. Hara and K. Mibe, Chem. Pharm. Bull. 15,1036(1967).
26. H. R. Roberts and K. Florey, J.Pharm.Sci. 51,794(1962).
27. A. Nilufer, Turk. Ijiyen Tecrubi. Biyol. Dergisi 21,149(1961)C.A.59,6200f(1963).

Literature surveyed through December 1972.

The help of H. Gonda and A. Mohr in the preparation of this profile is gratefully acknowledged.

THEOPHYLLINE

Jordan L. Cohen

Table of Contents

1. Description

1.1 Name: Theophylline

Theophylline[1,2,3] is designated by Chemical Abstracts as 1, 3-dimethylxanthine. It is also known as theocin.

1.2 Formula and Molecular Weight

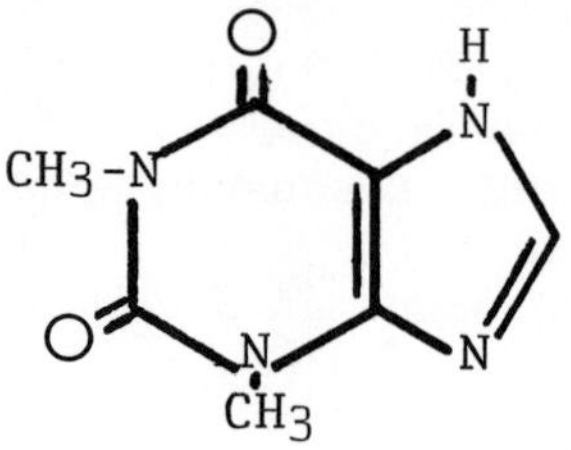

$C_7H_8N_4O_2$ 180.17

1.3 Hydrates

Theophylline has been reported to exist in both anhydrous and monohydrate forms. The anhydrous form is obtained by drying finely powdered drug at 150°C for three hours[4].

1.4 Salts

Sodium and potassium salts as well as a large number of less basic salts and/or complexes have been prepared to increase the water solubility of theophylline for parenteral administration. The ethylenediame salt (aminophylline)[5,6] is the most widely utilized form of this drug. Other theophylline salts which have been prepared include sodium and aluminum glycinates[7], hexamethylenediamine[5], monoethanolamine[8], triethanolamine[8], glucamine[9], methylglucamine[9], ethylglucamine[9], 2-diethylaminoethanol[10], choline[11] and sodium acetate[12]. Theophylline aminoisobutanol[13] and 2-carbamoylphenoxyacetic acid, sodium salt[14] have been prepared and studied clinically as has a niacinamide-theophylline complex [15,16]. Theophylline has also been shown to form stable, complexes with saccharin[17], phenobarbital[18], papaverine[19], caffeine[15], sulfosalicylic acid[15], benzyl alcohol[15], and α-, and β- naphthalene acetate[15].

1.5 Appearance, Color, Odor and Taste

Theophylline occurs as a white, odorless, crystalline powder with a bitter taste.

2. Physical Properties

2.1 Spectra

2.11 Infrared Spectrum

The IR spectrum of anhydrous theophylline is shown in Figure 1. This was recorded on a KBr pellet with a Perkin-Elmer model 21 spectrophotometer[20]. This spectrum is consistent with the literature interpretation which is presented in Table I[21].

Table I

Infrared Spectrum of Theophylline

IR Absorption Band (μ)	Interpretation
2.90	N-H (stretch)
5.86, 5.98	C=O (stretch)
6.20	C=C (stretch)
6.40	C=N (stretch)
6.92	C-H (bending)
7.7, 8.0	C-N, C-O (vibration)

The spectrum is generally consistent with those of other purine bases although the placement of the methyl groups prevents enolization and is reflected in differences in the 2.5-3.5μ region of the spectra. The region from 5.8-8.2 is nearly identical for the three methylated xanthines; i.e. theophylline, theobromine and caffeine, however, the fingerprint region between 8.5 and 13.0μ is quite dissimilar and could be used for differentiation.

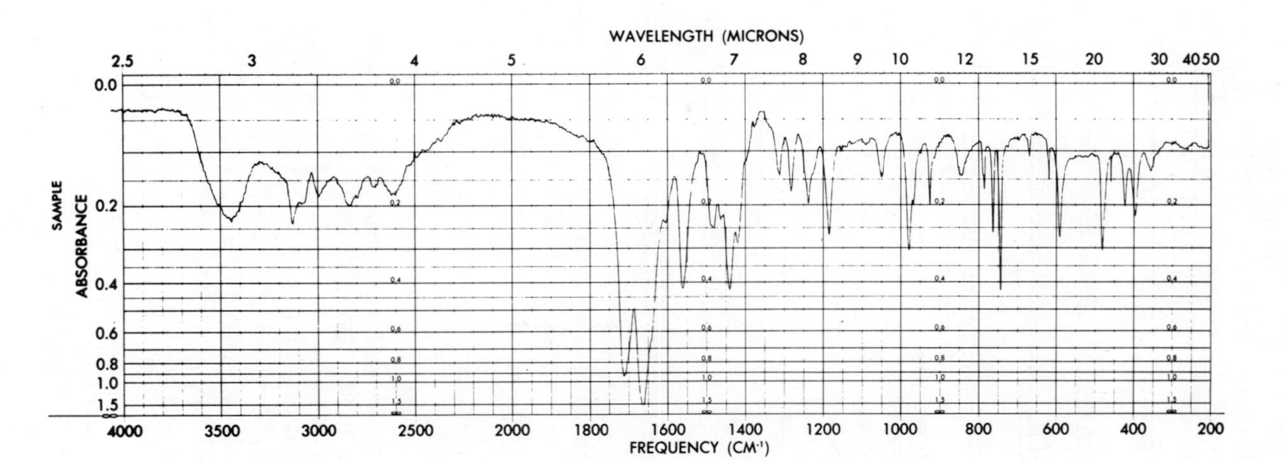

Figure 1. Infrared Spectrum of Theophylline Reference Standard.

2.12 Nuclear Magnetic Resonance Spectrum

The 60 MHZ proton magnetic resonance spectrum run in deuterodimethylsulfoxide on a Perkin-Elmer R-12 B spectrometer is shown in Figure 2[20]. The NMR of theophylline has been reported in the leterature[22] and the assignments are given in Table II.

Table II

NMR Spectral Assignments for Theophylline

Proton Assignment	No.	Chemical Shift (τ)
N-CH (position 3)	3	6.83 (s)
N-CH (position 1)	3	6.63 (s)
N-H	1	-3.4 (b)
C=C-H	1	2.01 (s)
Impurity	-	7.55

s= singlet, b= broad

2.13 Mass Spectrum

The low resolution, electron-impact mass spectrum of theophylline using a 70 e.v. ionizing energy is illustrated in Figure 3.[20] The resultant fragmentation represented by this spectrum is not readily explainable and is only in partial agreement with a literature mass-spectral report[23]. The peak at 180 m/e corresponds to the base peak, which is also the molecular ion and the peak at m/e 123, which represents loss of $CONCH_3$, are common to both spectra. Other peaks observable in Figure 3 occur at m/e 95, 68, 53 and 41 while the literature report indicates major peaks at 137, 109, 82, 67 and 55 m/e. Another, more recent, literature report[24] suggests that fragmentation from m/e 123→95 is due to loss of CO and that m/e 95→68 is due to loss of HCN from metastable ion calculations. Apparently ring opening occurs which results in a further complex fragmentation pattern.

2.14 Ultraviolet Absorption Spectrum

The ultraviolet spectrum of theophylline, K and K Laboratories, recrystallized from water, recorded on a Perkin-Elmer Hitachi, Model 124 Recording Spectrophoto-

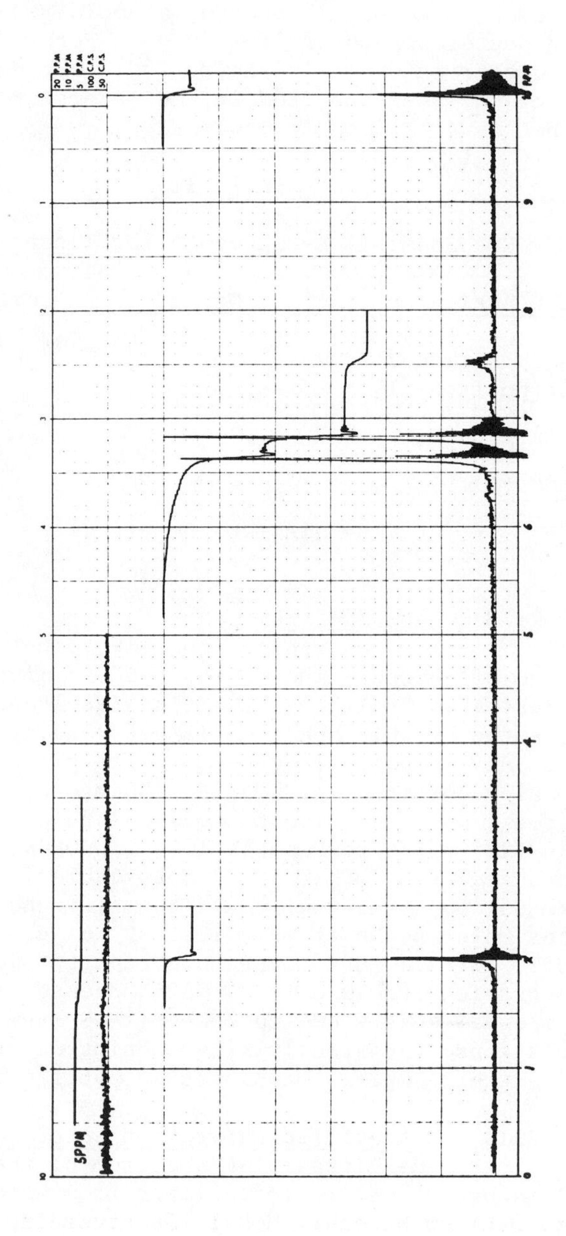

Figure 2. NMR Spectrum of Theophylline Reference Standard in DMSO-d6

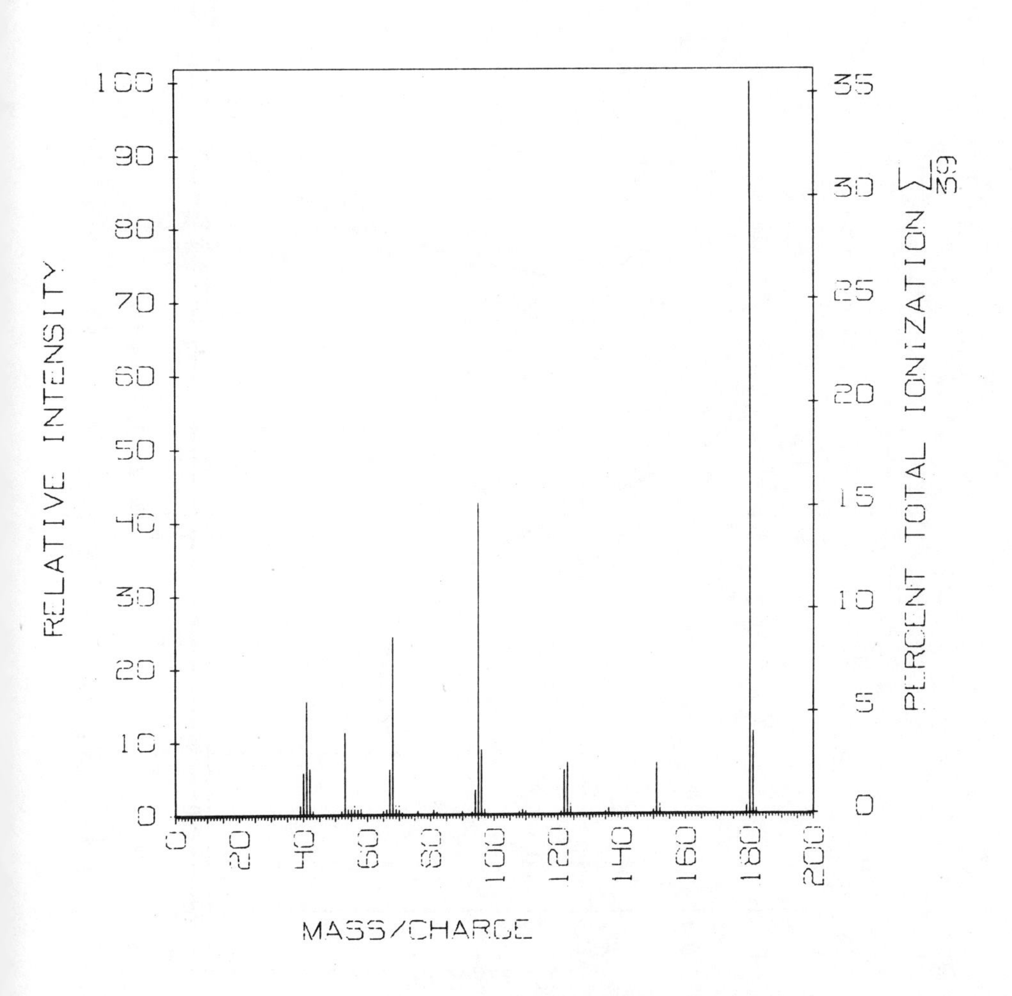

Figure 3. Low Resolution Mass Spectrum of Theophylline

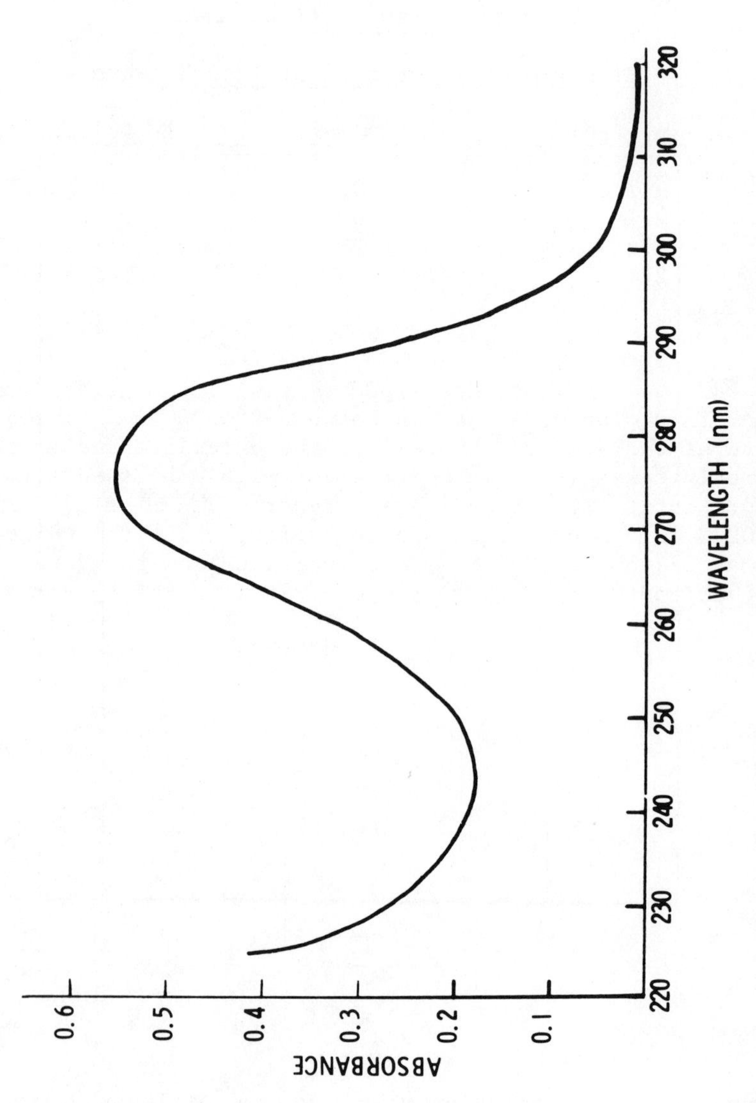

Figure 4. Ultraviolet Spectrum of Theophylline in 0.1 N NaOH.

meter in 0.1 N NaOH is illustrated in Figure 4. In addition the spectra was also recorded in 0.1 N HCl, pH 6.3 phosphate buffer and $CHCl_3$ with all of the results given in Table III.

Table III

Ultraviolet Absorption of Theophylline

λ max. (nm)	$10^4 \epsilon$ max.	Solvent
271	1.02	0.1 N HCl
271	1.04	pH 6.3 buffer
274	1.28	0.1 N NaOH
271	1.04	$CHCl_3$

All of these spectra exhibited a single maximum in the spectral region between 230-300 nm. The nature of the spectra, as well as the calculated molar absorptivities are in good agreement with published literature data. Turner and Osol[25] reported an ϵ at 271 nm. of 1.02×10^4 in pH 6.3 phosphate buffer, while the value of ϵ at 274 nm. in 0.1 N NaOH calculated from the data of Schach and Waxler[26] compares closely with the observed value above.

2.2 Optical Rotation

Theophylline exhibits no optical activity.

2.3 Melting Range

The original synthetic literature[27] reported a melting point of 264° C for theophylline subsequently the melting range has been reported to be between 271° and 274° C.[28]

2.4 Solubility

The reported solubility of theophylline is 8.3 mg/ml in water; 12.5 mg/ml in ethanol; 11.6 mg/ml in chloroform; and sparingly soluble in ether[1,2]. Several basic salts have been prepared which enhance the water solubility of theophylline and have been utilized therapeutically intended for oral, rectal and parenteral administration. (See Section 1.4) In addition, there are numerous literature reports of altered theophylline solubility due to specific interactions with a variety of chemical substances. Leuallen and Osol[29] have demonstrated marked increases in theophylline solubility in the presence of a wide variety of primary, secondary andd tertiary aliphatic amines. They attributed this increased solubility to salt formation as well as hydrophobic interactions. Paruta and Irani[30] reported a three-fold increase in theophylline solubility in alcohol-water and dioxane-water mixtures with maximal solubility occurring at a dielectric constant of 40 in these mixed solvent systems. A British patent[31] reports a six-fold increase in theophylline water solubility in the presence of guiacol. Several other papers[32-36] have reported specific interactions of theophylline with a multitude of planar organic molecules in aqueous solution usually resulting in an increase in the solubility of of both interacting species. For a thorough discussion of the nature of these specific molecular interactions and the magnitude of this solubility effect see Cohen and Connors.[37] Not surprisingly there has been some NMR evidence suggesting that theophylline also forms 1:1 complexes with itself (dimerizes).[38,39]

Theophylline also demonstrates altered solubility in the presence of several inorganic salts as illustrated in Table IV.

Table IV

Effect of Inorganic Salts on Theophylline Solubility

Salt	Effect
NaI, KI, NaCNS, KCNS	Increased theophylline solubility
NaCl, KCl, Na_2SO_4	Decreased theophylline solubility
NaBr, KBr	No effect on theophylline solubility

2.5 Dissociation Constant

Theophylline is a weakly acidic compound with the proton on the nitrogen in position 7 being dissociable. The pKa of 8.6 determined both potentiometrically and spectrophotometrically[4] is in good agreement with other literature data[25,41,42] reporting a pKa~8.6 in aqueous solution. Theophylline is also very weakly basic and pKb's of 13.5[25,43] and 11.5[44] have been reported in the literature from aqueous potentiometric measurements.

2.6 Dipole Moment

The dipole moment of theophylline has been determined in three different solvents with the following values being reported (all in DeBye units): μ = 3.94 in dioxane[45]; μ = 4.6 in 90% phenol[46]; and μ = 3.94 in benzene[47].

3. Synthesis

Although present in coffee, tea and cocoa as well as other natural sources, theophylline is made available commercially by total synthesis. Theophylline was first synthesized by Traube[27] as an intermediate in the total synthesis of caffeine starting from urea. This classic procedure, which is still utilized, is illustrated in Figure 5. Several subsequent syntheses which are modifications of this procedure and are claimed to either increase the reaction yield or simplify the process, have also been reported[48,49].

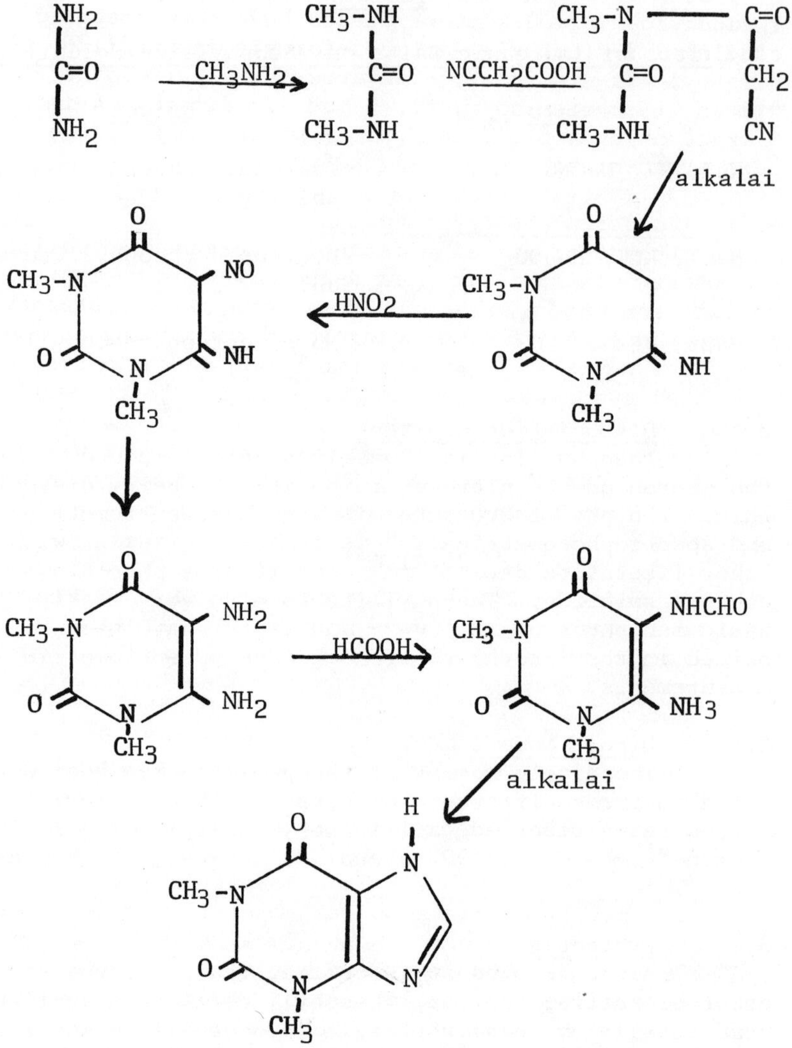

Figure 5. Chemical Synthesis of Theophylline

Uric acid has been used as the starting compound.[50,51] In addition various intermediates have been prepared, including 1,3-dimethyl-4-amino-5-formylaminouracil[52], 1,3-dimethyl-4,5-diamino-2,6-dihydroxyformylpyrimidine[53], 1,3-dimethyl-5,6-diaminouracil,[54] and 1,3-dimethyl-4-imino-2.6-dioxyhexahydropyrimidine[55], from which theophylline has been synthesized.

4. Isolation and Purification

Because of its significantly increased solubility in hot water[2], theophylline can be purified by recrystallization from water. The anhydrous form can be obtained by finely powdering the resulting solid and drying at 150° C. for three hours[4]. Adjusting the pH of an aqueous solution to 4-6 allows optimal extraction with chloroform while pH 5-7 is required using 1,2-dichloroethane.[56] Saturation of the aqueous phase with sodium or amonium sulfate and addition of 5-15% isopropyl alcohol to the organic phase has been shown to markedly increase the extraction efficiency[26,57,58].

Sephadex G-10 adsorption chromatography has been used to isolate theophylline from biological samples also containing several closely-related xanthines[59]. Thin layer chromatographic separation of theophylline, caffeine and theobromine on silica gel-G using an ethylacetate, acetone, butanol, 10% NH_4OH (5:4:3:1) mixture has also been reported[60]. Cohen and Garrettson have demonstrated base line separation of these same three compounds using reverse-phase high performance liquid chromatography[58].

5. Theophylline Stability and Compatibility

Theophylline solutions are generally quite stable over the entire pH range. Strongly alkaline solutions (pH > 12) show decomposition and apparent ring opening after several weeks[4]. Solutions of theophylline have been shown to be susceptible to oxidation at position 8 forming 1,3-dimethyluric acid in the presence of methylene blue which acts as a photosensitizing dye[61]. Because of its low solubility and relatively high pKa theophylline will precipitate from aqueous solutions if the pH drops below 9 unless present in concentrations less than the water solubility. All acidic salts, therfore, are potentially incompatible with theophylline in solution. As discussed earlier theophylline forms generally soluble complexes with a variety of other compounds. In the solid state however, these complexes may lead to formation of eutechtic mixtures as in the cases of theophylline, phenobarbital and papaverine hydrochloride[62] and theophylline riboflavin interactions[63] which have been reparted.

Theophylline ethylenediame (aminophylline), which is commonly employed in alkaline parenteral solutions in concentrations considerably greater than the solubility of theophylline, has many incompatibilities associated with it. Since the solutions are alkaline, but essentially unbuffered, all acidic substances produce precipitation of the free theophylline when mixed together with aminophylline solutions. Limited exposure to air results in carbon dioxide absorption and also leads eventually to precipitation of the free drug. For a more complete discussion and listing of the incompatibilities of aminophylline Remmington[3], Martindale[11] or a series of papers in the hospital pharmacy literature[64-68] may be consulted.

6. Methods of Analysis

6.1 Identification Tests

Theophylline has been qualitatively identified by infrared[21], NMR[22] and mass spectroscopy[23], thin-layer[60] and paper chromatography[69] using color localization reactions as well as by a multitude of specific color formation tests designed to distinguish it from caffeine, theobromine and other xanthines and purines. Brown needles are produced from the aqueous reaction of theophylline and ammoniacal $AgNO_3$ [70] while reaction with ammoniacal silver and subsequent treatment with thallium acetate yields light brown ball shaped masses[71]. A combination of 2,6-dichloroquinone chlorimide and theophylline in sodium borate solution gives a blue color and subsequently forms a red-violet precipitate[72]. Sanchez[73] has reported theophylline to produce a semi-quantitative red color when reacted with diazo-p-nitroaniline. More recently Feigl, et. al.[74], have shown that microgram quantities of theophylline can be detected in the presence of other xanthines by heating at 150°-180° C. with excess $Hg(CN)_2$.

6.2 Quantitative Analytical Methods

6.21 Ultraviolet Spectrophotometry

Shack and Waxler described a remarkable ultraviolet spectrophotometric method for the analysis of theophylline in aqueous solutions as well as biological fluids in 1949 [26]. This procedure involves extraction from the sample adjusted to pH 7.4 using 5% isopropyl alcohol-chloroform followed by back-extraction of the drug into 0.1 N NaOH and measurement of absorbance at 274 nm. They found the method to be sensitive to about 1 mcg/ml and demonstrated its applicability to theophylline analysis in water, plasma, urine and even tissues. Although it lacked specificity, it was rapid and sensitive and is still widely utilized for routine blood level determinations of theophylline. Gupta and Lundberg[75] reported a modified procedure using differential spectrophotometry and estimating the theophylline concentration from absorbance readings at 285 nm and a standard curve. Although their method was somewhat less sensitive, 10 mcg/ml in plasma, it was specific for theophylline in the presence of phenobarbital; a drug frequently taken with theophylline and a serious interference in the Shack and Waxler procedure.

A similar ultraviolet procedure was developed for the analysis of theophylline in dosage forms which avoided the back-extraction into NaOH. The absorbance of the organic extract was measured at 271 nm and the theophylline concentration determined from a beers law plot of standards which was linear from 2-15 mcg/ml.[76] A method utilizing extraction and combination of two successive volumes of $CHCl_3$-isopropyl alcohol (20:1) from an aqueous sample to which sodium chloride had been added reports improved sensitivity of 0.5 mcg/ml from aqueous samples[77]. A method involving oxidation of theophylline with $K_2Cr_2O_7$ in acid and steam distillation of the product which is measured at 257 nm has also been reported applicable to plasma samples[78]. It is, however, not specific for theophylline and marginally sensitive for therapeutic plasma levels of 10-20 mcg/ml.

6.22 Visible Spectrophotometry

Several methods involving derivitization of theophylline and subsequent colorimetric analysis have been reported. Coupling reactions have been described with p-nitroaniline[79], and p-aminobenzenesulfonic acid[80] exhibiting colorimetric absorption maxima at 510 nm and 482 nm respectively. While the sensitivity was less than ultraviolet methods, acceptable standard curves enabled quantitation of 50-5,000 mg. Chemical reaction of theophylline with murexide and thiophene enabled quanitation of 300-2,500 mg. amounts by measuring absorbance at 660nm.[81]

6.23 Gas-Liquid Chromatography

A gas chromatographic method for theophylline in dosage forms using a 6 foot, 1.5%, SE-30 chromosorb W column has been reported[82]. Helium was used as the carrier gas and temperature program was required to separate the individual xanthines. Elefant, et al[83] reported a GLC procedure for tablets containing ephedrine, phenobarbital and thophylline. Using a 6', ¼" o.d., 3% HI-EFF 8 BP on 100/120 mesh Gas Chrom Q, theophylline eluted as a moderately sharp peak after 14 minutes at a column temperature of 250° C.

Recently, Shah and Riegelman[57] reported a GLC procedure applicable to therapeutic plasma and saliva determinations of theophylline with a sensitivity of 1 mcg/ml. The dipropylated analog was formed by derivitization using tetrapropylammonium hydroxide and chromatographed on a 3% OV-17 column at 190°. The retention time was 3.6 minutes with a sharp peak resulting.

6.24 High Pressure Liquid Chromatography

The author has developed a high performance liquid chromatographic method for the analysis of theophylline in biological fluids[58]. A reverse-phase 2', Phenyl Corasil column with a 5% CH_3CN in pH 8.5 phosphate buffer mobile phase permits determination of as little as 0.5 mcg/ml using an ultraviolet detector. Baseline separation of theophylline, theobromine and caffeine is also achieved.

6.25 Thin Layer Chromatography

Several thin layer chromatographic systems for the separation of theophylline, theobromine and caffeine and subsequent quantitation have been developed and are summarized in Table V.

Table V[84]

Summary of Thin Layer Chromatographic Literature Data

Solid	Solvent	Rf Theophylline	Rf Caffeine	Rf Theobromine
Silica Gel G	$CHCl_3$-Ether (85:15)	.35	.20	.40
"	$CHCl_3$-CH_3OH (95:5)	.50	.23	.26
"	CCl_4-$CHCl_3$-CH_3OH (8:5:1)	.54	.46	.31
"	$CHCl_3$-acetone-CH_3OH (1:1:1)	.50	.49	.44
Aluminum Oxide	$CHCl_3$-BuOH (98:2)	.54	.30	.15
"	$CHCl_3$-C_2H_5OH (99:1)	1.00	.10	.10
"	C_6H_6-C_2H_5OH (98:2)	.15	.06	.02
"	C_6H_6-C_2H_5OH (95:5)	.47	.08	.04
"	C_6H_6-C_2H_5OH (90:10)	.58	.11	.06
"	C_6H_6-C_2H_5OH (80:20)	.75	.27	.19

Quantitation was most often achieved by densitometry at 254nm. Fluorescence inhibition scanning at 254 nm. enabled quantitation of 50 mcg/ml of theophylline on a silica gel plate and a 95:5 $CHCl_3$-C_2H_5OH developing solvent[85]. Two papers report the use of Draggendorfs Reagent for the quantitative estimation of theophylline in amounts as low as 0.5 mcg. Schunack, et al[86] utilized two different solvent systems, benzene-acetone (30:70), R_f= 0.06 and chloroform-ethanol-formic acid (88:10:2), R_f= 0.45 while Senanayake, et al[87] utilized an 8:5:1 mixture of chloroform, carbontetrachloride and methanol and found an R_f for theophylline of 0.31.

6.26 Paper Chromatography

There are several references in the literature to paper chromatographic systems for the separation and quantitation of theophylline.[69,88] A typical system is illustrated by Berger and Hedrick[89] who developed two different percentage mixtures of acetonitrile and aqueous buffer to achieve separations of theophylline from theobromine and caffeine. Quantitation by diazotization and fluorescence quenching densitometry following a paper chromatographic separation has been described by Kala, et al.[90]

6.27 Column Chromatography

Theophylline has been separated and quantitatively determined by column adsorption chromatography.[91] A silicic acid column with 5% butanol in chloroform was utilized and the eluent analyzed by ultraviolet spectrophotometry.

6.28 Polarography

Dusinsky and Cavanak[92] report on a non-specific polarographic determination of theophylline involving the bromine oxidation to methyl parabanic acid. The half-wave potential was reported to be -0.72 v. and accuracy for dosage form analysis was of the order of 3%.

6.29 Titrimetry

There are numerous literature reports describing titrimetric quantitative analytical methods for theophylline due to its acidic properties and reactivity with a variety of compounds.

6.291 Complexometric Titration

Theophylline has been determined by reaction with excess copper (II) ions, usually as amine salts and then back-titration with Complexon III (Murexide) to a visual indicator endpoint.[93,94] Theophylline also forms molecular complexes with silver ions and has been determined by both direct titration with silver nitrate using visual[95] or potentiometric detection[96] and by back-titration after the addition of excess ammoniacal silver nitrate using potassium thiocyanate[97].

6.292 Photometric Titration

Fleischer[98] has reported a photometric titration determination of theophylline by coupling with Fast Blue. Sensitivity from aqueous solutions is claimed to be 2 mcg/ml.

6.293 Non-Aqueous Titration

Theophylline can be determined by titration in a nonaqueous solvent using perchloric acid in acetic acid as a titrant.[99] Endpoint determination can be done either with a visual indicator such as Nile Blue[100] or Crystal Violet[101] or potentiometrically[102]. Theophylline has also been titrated as an acid in dimethylformamide using sodium methoxide and thymol blue as an indicator.[103]

6.294 Miscellaneous Titrations

Other titrimetric procedures reported for theophylline determination include iodometric[104,105], bromometric[106], amperometric[107], coulometric[108] and radiometric[109].

6.3 Bioassay Methods

The diuretic activity of theophylline was determined on a relative scale by Lipshitz, et al[110] in dogs by plotting the log dose versus the log of the diuretic effect. With urea normalized to 1.0, thophylline was found to be 115 times as effective.

7. Pharmacokinetics

Because of the availability of methodology for the analysis of theophylline in plasma as early as 1949[26] there are numerous literature reports detailing the disposition of this drug in humans following administration by several routes. Although theophylline is generally well absorbed following oral administration, considerable variation does occur depending upon dosage form, formulation and which salt form of theophylline being administered. Hydroalcoholic solutions produce peak plasma levels in 60 minutes which decline below acceptable therapeutic levels four hours after single dose administration.[111,112] Tablets demonstrate peak levels from 2-4 hours with availability dependent upon formulation, while sustained release tablets provide peak levels at 6 hours and provide therapeutic levels up to 10 hours[113]. A non-alcoholic liquid suspension administered as a soft gelatin capsule showed absorption characteristics between those for a hydroalcoholic solution and a sustained release tablet[114]. This same study suggests that there is significantly slower absorption of theophylline ethylenediamine (aminophylline) than plain theophylline in comparable oral doses in the same subjects on a cross-over study design. This contradicts earlier findings by Waxler and Moy[115] and additional studies are being performed.

Theophylline is commonly administered rectally, particularly in asthmatic children, but has been shown to be poorly and erratically absorbed from normal suppository base preparations showing peak plasma levels considerable lower than from comparable oral doses, and occurring 1-3 hours after administration[116]. Theophylline retention enemas, however, have been shown to produce plasma consideration--time profiles similar to those following an intravenous dose[114]. The drug appears to be fully bioavailable from this route if allowance is made for a normal 30-60 minute absorption delay and subsequent effect of partial first-pass metabolism.[117] Correlation of plasma levels and clinical response is readily achievable and this route has been proposed as the optimal method of theophylline administration in children[118].

Theophylline disposition following intravenous, intramuscular and oral administration can be described by an open two-compartment pharmacokinetic model[119] with an average plasma half-life for β-phase dissappearance of

4.4 hours. Wide patient-to-patient variation with half-lives ranging from 2.5-9.5 hours presents possible serious management problems for asthmatics on chronic theophylline therapy[120]. The drug is relatively widely and rapidly distributed into all tissues with high rates of blood flow and has a volume of distribution (V_d) of 0.31/kg. and a half-life for distribution (α-phase) of 0.12 hours. Erythrocytes apparently also take up significant amounts of theophylline rapidly. While there was found to be no statistically significant differences in the pharmacokinetic parameters between normal volunteers and asthmatic subjects, Maselli, et al[121] showed that asthmatic children evidenced significantly shorter β-phase half-lives than adults with a mean of 2.65 hours compared to the 4.4 hour half-life for the adult population.

A recent study[122] has shown theophylline to be bound about 60% to plasma proteins at therapeutic plasma concentrations of 12-15 mcg/ml. This same study, as well as that of Shah and Riegelman[57], indicates that saliva levels of theophylline are approximately 50% of those in plasma and the disappearance of theophylline from saliva parallels that from plasma over the normal therapeutic concentration range. This may lead to a simplified method for monitoring theophylline concentrations in clinical pharmacokinetic studies for patients undergoing chronic therapy.

Theophylline has been shown to be extensively metabolized in vivo with 13% of the administered dose appearing in the urine as 3-methylxanthine, 35% as 1,3-dimethyl-uric acid and 19% as 1-methyluric acid[123]. Although considerably more work needs to be performed in this area, the variations in plasma half-life in adults and the marked decrease in plasma half-life in children is probably related to differences in metabolic activity in individual subjects.

References

1. The United States Pharmacopoiea, XVIII, p.724 (1970)
2. Merck Index, 8th Ed., p. 1034 (1968).
3. Remmington Pharmaceutical Sciences, 14th Ed., p. 1157, (1970).
4. J. L. Cohen, Ph.D. Dissertation, University of Wisconsin, (1969).
5. German Patent 223,695 (Oct. 18, 1970).
6. U.S. Patent 919,161 (April 20, 1909).
7. Brit. Patent 624,696 (June 15, 1949).
8. U.S. Patent 1,867,332 (July 12, 1932).
9. U.S. Patent 2,161,114 (June 6, 1932).
10. German Patent 739,334 (Aug. 12, 1943).
11. Martindale, The Extra Pharmacopoiea, 26th Ed., p. 357, (1972).
12. Ibid, p. 362, 1972.
13. F. Steinberg and J. Jensen, J. Lab. Clin. Med., 30, 769 (1945).
14. German Patent 1,106,923 (May 18, 1961).
15. R. Reiss, Arzneimittel-Forsch, 11, 669 (1961); C.A. 55, 25976 b (1961).
16. B. M. Cohen, J. of Asthma Res., 4, 75 (1966).
17. Brit. Patent 1,109,964 (April 18, 1968).
18. U.S. Patent 2,017,279 (Oct. 15, 1935).
19. A. Mossini, Boll. Chim. Farm., 78, 261 (1939).
20. This spectrum was kindly provided by Dr. K. Florey, Squibb, Medical Research, New Brunswick, N.J.
21. E. R. Blout and M. Fields, J. Am. Chem. Soc., 72, 479, (1950).
22. R. Otinger, G. Boulin, J. Reisse and G. Chiurdogiu, Tetrahedron, 21, 3435 (1965).
23. G. Spiteller and M. Friedmann, Monatsh, 93, 632 (1962) C.A. 58, 1018 g (1964).
24. G. S. Rao, K. L. Khanna and H. H. Cornish, J. Pharm. Sci., 61, 1822 (1972).
25. A. Turner and A. Osol, J. Am. Pharm. Assoc., Sci. Ed., 38, 158 (1949).
26. J. A. Schack and S. H. Waxler, J. Pharmacol. Exp. Ther., 97, 283 (1949).
27. W. Traube, Ber. 33, 3052 (1900).
28. H. Doser, Arch. Pharm., 281, 251 (1943).
29. E. E. Leuallen and A. Osol, J. Amer. Pharm. Assoc., Sci. Ed., 38, 92 (1949).

30. A. N. Paruta and S. A. Irani, J. Pharm. Sci., 55,1055 (1966).
31. British Patent 932,874 (July 31, 1963).
32. J. L. Cohen and K. A. Connors, Amer. J. Pharm. Ed., 31, 476 (1967).
33. T. Higuchi, J. Pharm. Sci., 53, 644 (1964).
34. R. Huettenrauch, W. Suess and U. Schmeiss, Pharmazie, 24, 646 (1969).
35. British Patent 1,109,964 (April 18, 1968).
36. T. Higuchi and K. A. Connors, Advances in Anal. Chem. and Instrumentation, Vol. 4, Interscience, New York, 1965, p.117-212.
37. J. L. Cohen and K. A. Connors, J. Pharm. Sci., 59, 1271 (1970).
38. A. L. Thakkar, L. G. Tensmeyer and W. L. Wilham, J. Pharm. Sci., 60, 1267 (1971).
39. D. Guttman and T. Higuchi, J. Pharm. Sci., 60, 1269 (1971).
40. V. P. Gusyakov, Aptechnoe Delo., 8, 30 (1959); C.A. 54: 3033 f (1960).
41. H. V. Maulding and M. A. Zoglio, J. Pharm. Sci., 60, 309 (1971).
42. H. Miyamoto, Sci. Rep. Niigata Univ., 80c, 23 (1969); C.A. 72: 83560 t (1969).
43. K. Linek and C. Peciai, Chem. Zvesti., 16, 692 (1962); C.A. 58: 8409 h (1963).
44. K. J. Evstratova and A. I. Ivanova, Farmatsiya, 17, 41 (1968); C.A. 69: 46128 a (1968).
45. H. Weiler-Feichenfeld and E. Bergmann, Israel J. Chem., 6, 833 (1968); C.A. 70: 77174 n (1969).
46. H. Miyazaki, Osaka Dalgaku Igaku Zasshi, 11, 4306 (1959); C.A. 54: 14339 a (1961).
47. H. Weiler-Feilchenfeld and Z. Neiman, J. Chem. Soc., B, 596 (1970).
48. B. Gepner and L. Kreps, J. Gen. Chem. (U.S.S.R.); C.A. 41: 96[d] (1947).
49. E. I. Abromova and V. I. Khmelevskii, Med. Prom. S.S.S.R., 18, 35 (1964); C.A. 61: 1865 c (1964).
50. V. I. Khmelevskii, Med. Prom. S.S.S.R. 12, 11(1958); C.A. 53: 11761 g (1958).
51. E. I. Abramova, V. I. Khmelevskii and Y. L. Shneidermann, Med. Prom. S.S.S.R., 15, 31 (1961); C.A. 56: 6097 g (1962).
52. German Patent (East) 64722 (November 20, 1968); C.A.

71: 22136 s (1969).
53. F. L. Grinberg, J. Appl. Chem. U.S.S.R., 13, 1461 (1940); C.A. 35: 3974[2] (1941).
54. German Patent 2,058,912 (June 16, 1971); C.A. 75: 63836 w (1971).
55. B. Bobranski and Z. Synowiedski, J. Amer. Pharm. Assoc., Sci. Ed., 37, 62 (1948).
56. A. I. Shkadova, 19, 56 (1970); C.A. 72: 136462 w (1970).
57. V. P. Shah and S. Riegelman, J. Pharm. Sci., 63, 1283 (1974).
58. J. L. Cohen and L. K. Garrettson, Acad. of Pharm. Sci., Annual Meeting, Nov. 1974 (abstract).
59. L. Sweetman and W. L. Nyhan, J. Chromatog., 32, 662 (1968).
60. V. M. Pechenikov, A. Z. Knizhnik and P. L. Senov, Farm. Zh., 23, 41 (1968); C.A. 70: 22932 c (1969).
61. P. A. Friedman, Biochem. Biophys. Acta, 166, 1 (1968).
62. E. G. Glass, Pharm. Praxis Beil Pharmazie, 8, 197 (1964); C.A. 61: 14474 a (1964).
63. A. Koshiro, Yakuzaigaku, 27, 349 (1967); C.A. 69: 99331 t (1968).
64. R. Misgen, Amer. J. Hosp. Pharm., 22, 92 (1965).
65. J. A. Patel and G. L. Phillips, Amer. J. Hosp. Pharm., 23, 409 (1966).
66. Amer. J. Hosp. Pharm., 24, 440 (1967).
67. Amer. J. Hosp. Pharm., 27, 67 (1970).
68. J. Hosp. Pharm., 28, 228 (1970).
69. H. K. Hegi, Pharm. Acta Helv., 34, 105 (1959); C.A. 53: 16466 e (1959).
70. C. K. Glycart, J. Assoc. Off. Agr. Chem., 19, 512 (1936).
71. L. Rosenthaler, Pharm. Acta Helv., 14, 218 (1939); C.A. 34: 7782[1] (1940).
72. H. W. Raybin, J. Amer. Chem. Soc., 67, 1621 (1945).
73. J. A. Sanchez, Boll. Soc. Quim., Peru, 9, 197 (1943); C.A. 38: 2289[3] (1944).
74. F. Feigl, D. Goldstein, E.K. Libergott, Anal. Chim. Acta, 49, 573 (1970).
75. R. C. Gupta and G. D. Lundberg, Anal. Chem., 45, 2403 (1973).
76. L. O. Kircheuko, F. Kagan and F. Yu, Farm. Zh., 25, 42 (1970); C.A. 73: 28951 s (1970)

77. B. K. Kim, Rachelle Laboratories, Personal Communication, Jan. 1973.
78. A. Villa and A. Pistis, Farmaco. Ed. Pract., 25, 717 (1970); C.A. 74: 30565 h (1971).
79. H. Raber, Sci. Pharm., 32, 122 (1964); C.A. 61: 11853 e (1964).
80. M. P. Castro and R. R. Mendoza, Inform. Quim. Anal., 17, 158 (1963); C.A. 61: 537 d (1964).
81. H. Wachsmuth and L. Van Koechoven, J. Pharm. Belg., 18, 366 (1963); C.A. 61: 8131 b (1964).
82. J. Reisch and H. Walker, Pharmazie, 21, 476 (1966); C.A. 66: 22234 u (1967).
83. M. Elefant, L. Chafetz and J. M. Talmadge, J. Pharm. Sci., 56, 1181 (1967).
84. L. J. Kraus and E. Dumont, J. of Chromatog., 48, 96 (1970).
85. V. Massa, F. Gal, P. Suspulgas and G. Maestre, Trav. Soc. Montpellier, 31, 167 (1971); C.A. 75: 83949 p (1971).
86. W. Schunack, E. Mutschler and H. Rochelmeyer, Deut. Apotheker-Ztg., 105, 1551 (1965); C.A. 65: 16791 f (1966).
87. U. M. Senanayake and R. O. B. Wijesekrea, J. Chromatog., 32, 75 (1968).
88. E. G. C. Clarke and S. Kalayci, Nature, 198, 783 (1963).
89. B. A. Berger and C. E. Hedrick, Anal. Biochem., 16, 260 (1966).
90. H. Kala, H. Moldenhauer and K. Wolff, Pharmazie, 14, 519 (1959); C.A. 54: 10239 i (1960).
91. A. J. Shingler and J. F. Carlton, Anal. Chem., 31, 1679 (1959).
92. C. Dusinsky and T. Cavanak, Ceskosiov. Farm., 7, 511 (1958); C.A. 53: 8539 g (1959).
93. A. G. Mindermann, K. Leupin and R. Baum, Pharm. Acta Helv., 39, 390 (1964); C.A. 61: 4154 d (1959).
94. G. H. Morait, M. Coada, V. Turailet and G. H. Clogolea, Farmacia, 12, 521 (1964); C.A. 62: 403 d (1965).
95. M. E. Shub and E. N. Tsitrin, Med. Prom. S.S.S.R., 11, 43 (1957); C.A. 52: 10500 a (1958).

96. F. Pellerin and G. Leroux-Mamo, Ann. Pharm. Fr., 29, 153 (1971); C.A. 75: 52853 y (1971).
97. A. N. Stevens and D. J. Wilson, J. Amer. Pharm. Assoc., 26, 314 (1937).
98. G. Fleischer, Pharmazie, 11, 208 (1956); C.A. 51: 8365 a (1957).
99. M. B. Devani, C. J. Shishoo, and D. J. Bhut, J. Pharm. Sci., 57, 1051 (1968).
100. H. Ellert, T. Jasinski and F. Pawelczak, Acta Polon. Pharm., 16, 235 (1959); C.A. 54: 1182 d (1960).
101. R. Reiss, Z. Anal. Chem., 167, 16 (1959); C.A. 53: 18729 e (1959).
102. C. Peciar and K. Linek, Chem. Zvesti., 15, 895 (1961); C.A. 58: 3271 g (1963).
103. M. A. McEniry, J. Assoc. Off. Agr. Chem., 40, 926 (1957); C.A. 52: 5750 e (1958).
104. C. Runti, Farmaco Ed. Prat., 11, 218 (1956); C.A. 53: 6537 f (1959).
105. A. P. Plummer, J. Pharmacol. Exp. Ther., 93, 142 (1948).
106. H. Raber, Sci. Pharm., 34, 202 (1966); C.A. 66: 22249 c (1967).
107. G. Bozsai and L. Mosonyi, Pharm. Zentrahalle, 103, 250 (1964); C.A. 61: 8131 e (1964).
108. Z. Kalinowska, Acta Polon. Pharm., 20, 193 (1963); C.A. 62: 1516 e (1965).
109. J. Kormicki and E. Mikolajek, Dissertations Pharm., 16, 565 (1964); C.A. 62: 11628 c (1965).
110. W. L. Lipshitz, Z. Hadidian and A. Kerpcsar, J. Pharmacol., 79, 97 (1943); C.A. 38: 1615 (1944).
111. A. R. Flora, Curr. Ther. Res. Clin. Exp., 12, 611 (1970).
112. C. Bruda, R. B. Miller, S. T. Leslie, E. G. Nicol and I. Thomson, J. Clin. Pharmacol., 14, 385 (1973).
113. E. B. Truitt, V. A. McKusick and J. C. Krantz, J. Pharmacol. Exp. Ther., 10, 309 (1950).
114. R. H. Jackson, R. Garrido, H. I. Silverman and H. Saleno, Ann. Allergy, 31, 413 (1973).
115. S. H. Waxler and H. B. Moy, J. Amer. Pharm. Assoc., Sci. Ed., 49, 619 (1960).
116. A. S. Ridalfo and K. G. Kohlstaedt, Am. J. Med. Sci.

116. 237, 585 (1959).
117. M. S. Segal and E. B. Weiss, Annals of Allergy, 29, 136 (1971).
118. J. W. Yunginger, M. Shigeta, I. Smith, M. Gwen and H. Keitel, Annals of Allergy, 24, 469 (1966).
119. P. A. Mitenko and R. I. Ogilvie, Clin. Pharmacol. Ther., 14, 509 (1973).
120. J. W. Jenne, E. Wyze, F. S. Roodand and F. M. MacDonald, Clin. Pharmacol. Ther., 13, 349 (1972).
121. R. Maselli, G. L. Casal and E. F. Ellis, J. Pediatr. 76, 777 (1970).
122. R. Koyoosko, E. F. Ellis and G. Levy, Clin. Pharmacol. Ther., 15, 454 (1974).
123. H. A. Cornish and A. A. Christman, J. Biol. Chem., 228, 315 (1957).

TYBAMATE

Philip Reisberg, John Kress, and Jerome I. Bodin

CONTENTS

1. Description

1.1 Name, Formula, Molecular Weight

Tybamate is 2-(hydroxymethyl)-2-methylpentyl butylcarbamate carbamate. It is also known as 2-methyl-2-propyltrimethylene butylcarbamate carbamate, and as N-butyl-2-methyl-2-propyl-1,3-propanediol dicarbamate.

```
      O         CH3      O
      ||         |       ||
H2N-C-O-CH2-C-CH2-O-C-NH-CH2-CH2-CH2-CH3
                 |
                CH2-CH2-CH3
```

$C_{13}H_{26}N_2O_4$, Molecular Weight: 274.36

1.2 Appearance, Color, Odor, Mode of Occurrence

Tybamate occurs as a white, crystalline powder or as a clear, colorless, viscous liquid, which may congeal to a solid form on standing. It has a mild characteristic odor and a bitter taste.

2. Physical Properties

2.1 Infrared Spectrum

The infrared spectrum of N. F. Tybamate Reference Standard (1) is shown in Figure 1. The spectrum was obtained on a 0.5% dispersion of tybamate, previously dried under vacuum at 30° to 35°C for 4 hours, in a potassium bromide disc.

The following band assignments have been made (2):

Wave Number, $cm.^{-1}$	Characteristic of
3350	NH stretching
1720	Amide I band
1540	Amide II band
1255	Amide III band
620	Amide IV band

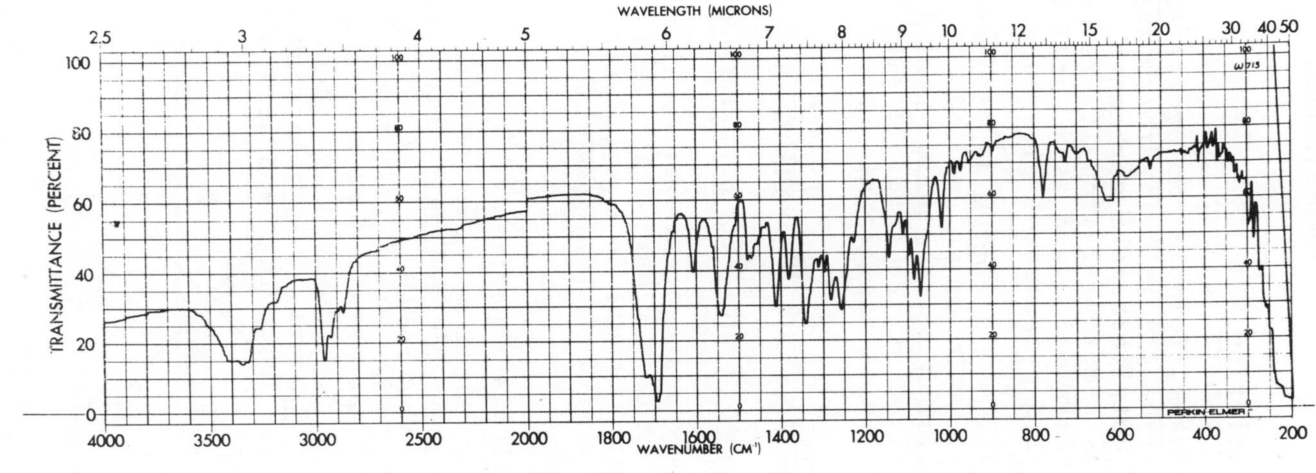

Figure 1 - IR spectrum of N.F. Tybamate Reference Standard in KBr dispersion.

2.1 Infrared Spectrum Cont'd.

A mineral oil (Nujol) mull of crystalline tybamate can be prepared much more easily than a satisfactory KBr disc, mainly because of the low melting range of tybamate.

The infrared spectrum of a 1% mineral oil mull of tybamate is shown in Figure 2.

2.2 Nuclear Magnetic Resonance Spectrum

The NMR spectrum as presented in Figure 3 was obtained by using a 12.5% solution of tybamate in carbon tetrachloride containing tetramethylsilane as the internal standard. The following peak assignments have been made:

Chemical Shift, ppm.	Proton	No. of Protons
(a) 0.9	CH_3-C	9
(b) 1.3	C-CH_2-CH_2-C	8
(c) 3.05	N-CH_2	2
(d) 3.8	C-CH_2-O	4
(e) 5.2 - 5.7	N-H N-H_2	3
		26

The peak assignments indicated in Figure 3 are shown on the following structural representation of tybamate:

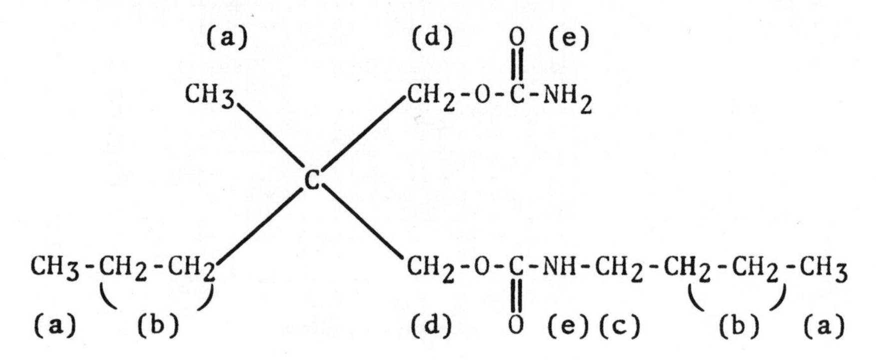

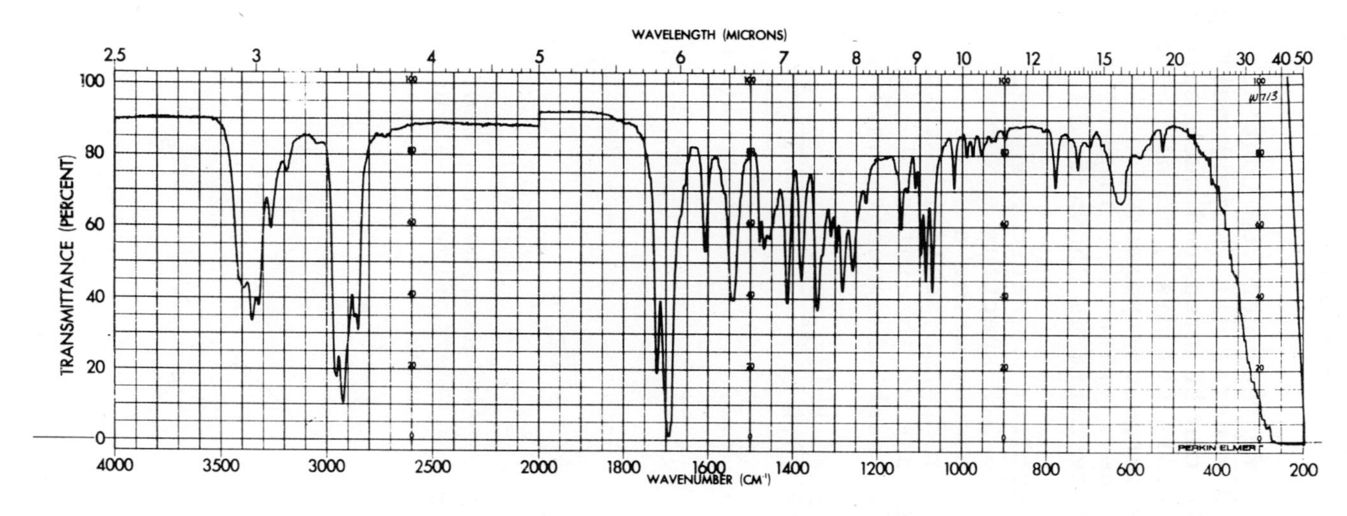

Figure 2 - IR spectrum of N.F. Tybamate Reference Standard in Nujol mull.

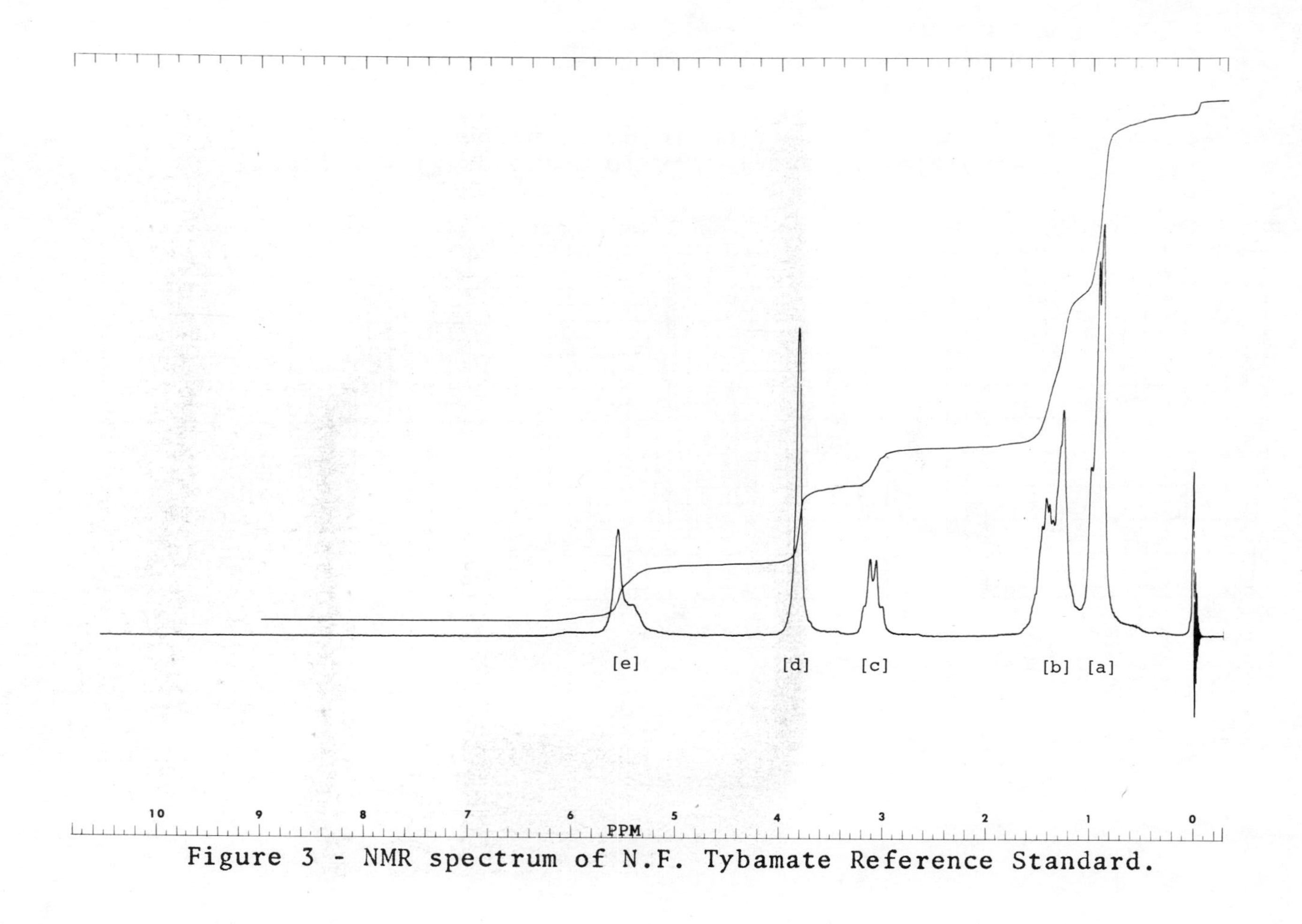

Figure 3 - NMR spectrum of N.F. Tybamate Reference Standard.

2.3 Mass Spectrum

The low resolution mass spectrum of tybamate (3), shown in Figure 4, exhibits the expected M+ of m/e 274. Long chain alkyl groups, particularly those attached to nitrogen, fragment on the carbon-carbon bond β to the nitrogen function with retention of the charge on the alkyl moiety. Cleavage of the N-butyl group results in an m/e 43 ion. The m/e 41 ion, C_3H_5, arises in the cleavages of the branched hydrocarbon portion of the molecule. Likewise, an m/e 43 ion arises in the fragmentation of the C_3H_7 moiety of the branched hydrocarbon. The m/e 55-57 ions have the composition C_4H_7 to C_4H_9. The fragment ions resulting from direct bond cleavage are: m/e 29, 43, 44, 72, 74, 158, 200, 214, 231, and 245.

The diagnostic peak for carbamates, m/e 62, was present and was shown by Coutts (4) to be $NH_2COOH_2^+$. Through rearrangement, $HOCONH_2$ is lost from the M+ to form the m/e 213 ion. The same loss can occur from the m/e 213 ion to form the m/e 170 ion. In a parallel fragmentation mode, the elements of $OCONH_2$ are lost from the m/e 231 ion to yield the m/e 171 ion. The very intense high mass m/e 158 ion can be formed from either the m/e 200 ion through loss of CH_3-CH=CH_2 (42 amu) or from the M+ by the loss of $OCONHC_4H_9$. Finally, the m/e 184 ion can arise through the loss of C_2H_5 from the m/e 213 ion.

The mass spectrum is in agreement with the structure proposed for tybamate in 1.1.

2.4 Thermal Characteristics

2.41 Melting Range

Crystalline tybamate in fine powder form, dried under vacuum for 4 hours at 30°-35°C, melts within a range of 2° between 49° and 54°C.

2.42 Boiling Range

Tybamate boils at approximately 150°-152°C at a pressure of 0.06 mm. of mercury.

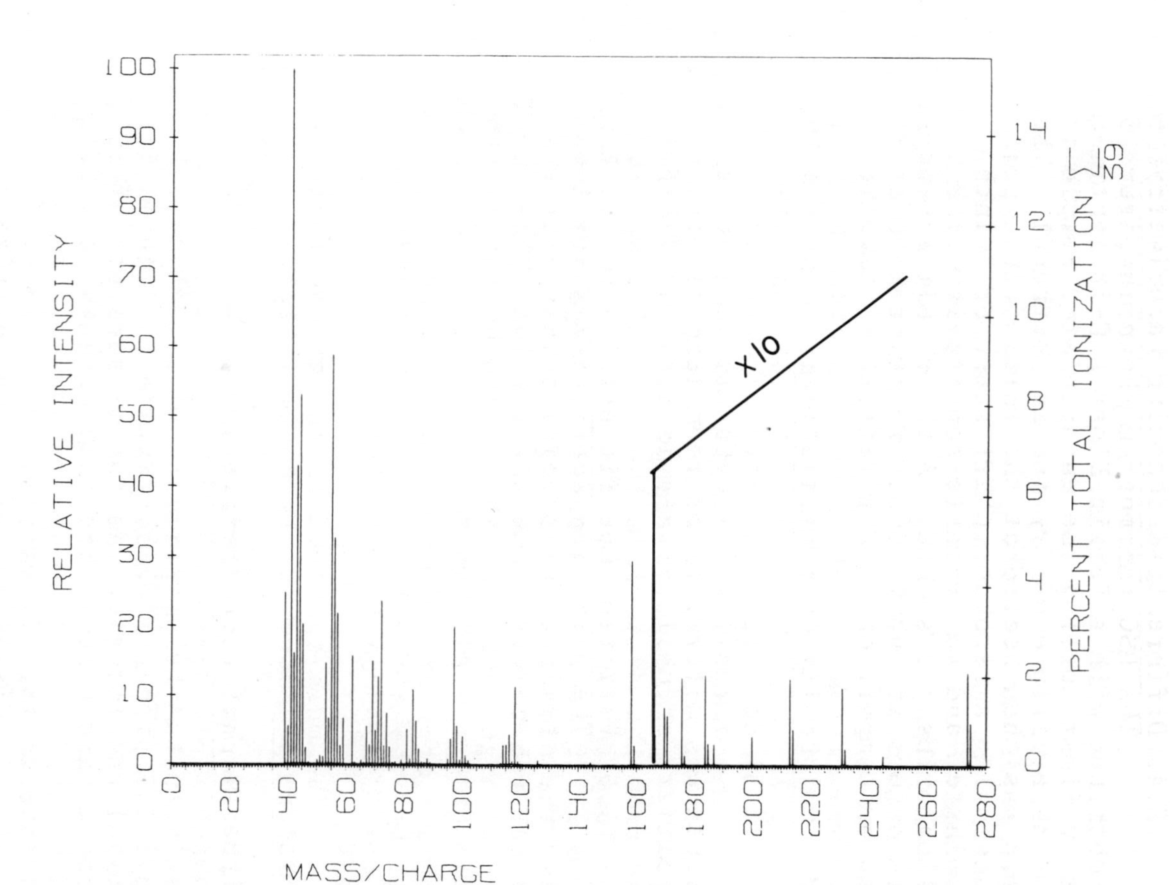

Figure 4 - Mass Spectrum of N.F. Tybamate Reference Standard.

2.4 Thermal Characteristics Cont'd.

2.43 Differential Scanning Calorimetry

The DSC thermogram shown in Figure 5 was obtained with a Perkin-Elmer DSC-1b instrument by first cooling the encapsulated sample to 0°C with a mixture of dry ice and methanol. The sample was then heated at the rate of 10°C per minute using a recorder chart speed of 1 inch per minute, and an ordinate sensitivity of 8 millicalories per second. A sharp melting endotherm occurs at about 46°C. Purity can be estimated by comparison with a reference standard.

2.5 Solubility Characteristics

2.51 Solubility in Several Solvents

Solubility of tybamate in several solvents was determined at room temperature. Approximate solubility data is shown in Table I.

2.52 Partition Coefficients

The partition coefficients for tybamate were determined for several solvent systems at room temperature. These are shown in Table II.

2.6 Optical Rotation

A 5% alcoholic solution of tybamate is optically inactive when tested by the USP procedure.

3. Synthesis

Tybamate has been synthesized (5,6) by the sequence of reactions shown in Figure 6. The first step is the preparation of 2-methyl-2-propyl-1,3-propanediol by reacting 2-methylpentanal and formaldehyde in the presence of potassium hydroxide. The diol is then reacted with diethyl carbonate in the presence of sodium methylate to form 5-methyl-5-propyl-2-m-dioxanone. This is distilled and reacted with 28% aqueous ammonia solution forming 2-methyl-2-

Figure 5 - DSC thermogram of N.F. Tybamate Reference Standard.

TABLE I

Approximate Solubility of Tybamate

Solvent	mls. Solvent Required per g. of Tybamate
Water	2000
Ethyl Alcohol	0.6
Chloroform	0.5
Ethyl Ether	1
Propylene Glycol	2

TABLE II

Partition Coefficients for Tybamate

Solvent System	Partition Coefficient
Cottonseed Oil/Water	28
Chloroform/Water	>100
Carbon Tetrachloride/Water	10
Benzene/Water	43
Heptane/Water	0.2

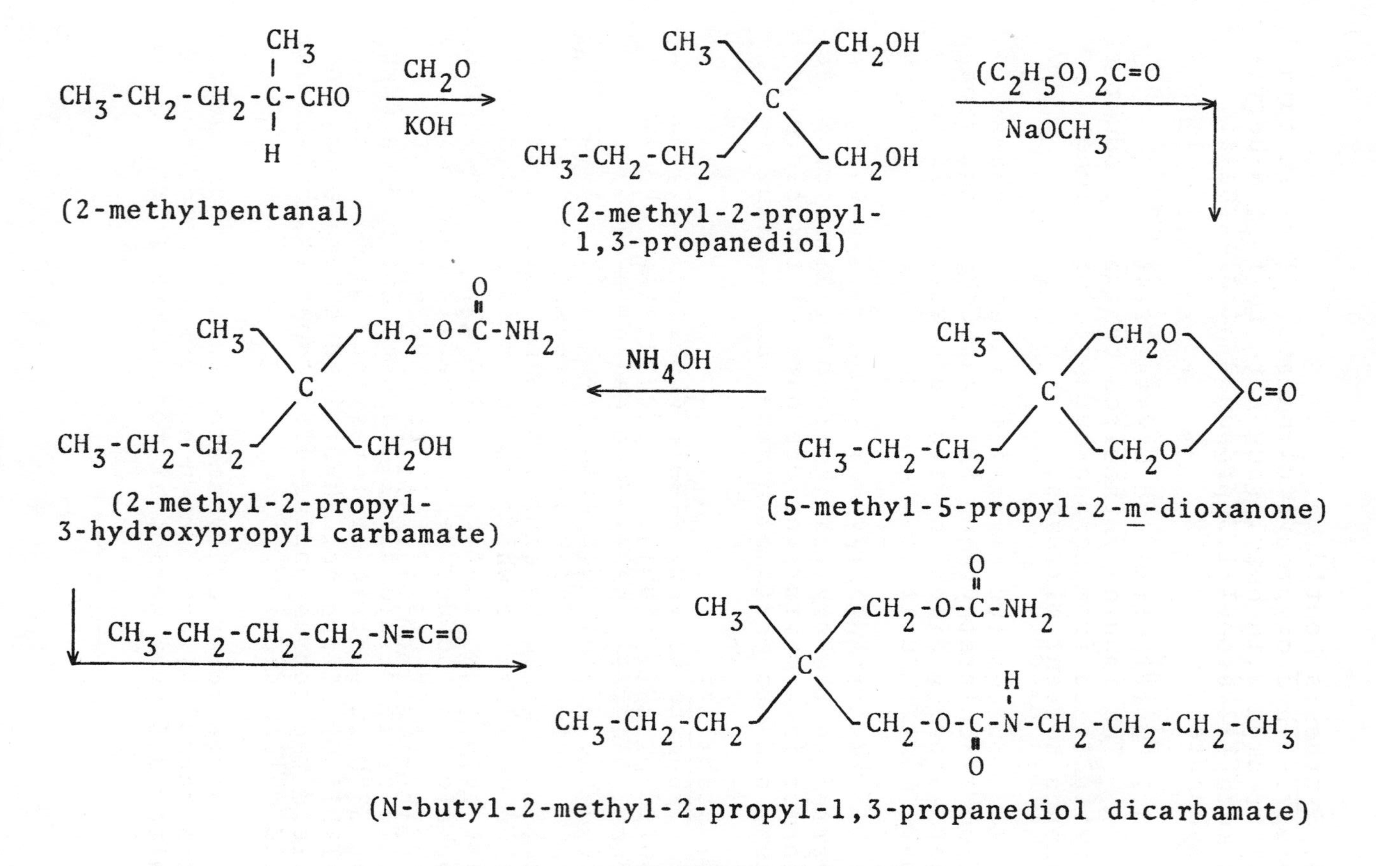

FIG. 6: SYNTHESIS OF TYBAMATE

3. Synthesis Cont'd.
propyl-3-hydroxypropyl carbamate. This in turn is treated with butylisocyanate to form N-butyl-2-methyl-2-propyl-1,3-propanediol dicarbamate (tybamate).

3.1 Purification and Crystallization

The impure synthesized tybamate product is crystallized from hexane-butanol, or xylene-naptha solvent mixtures.

Successive crystallizations of tybamate from trichlorethylene, SDA-3A alcohol, and from hot water by dilution with cold water will remove most impurities.

The liquid form of tybamate may be converted to the crystalline form by dissolving the former in a mixture of one part of trichlorethylene and two parts of hexane, and cooling.

4. Stability

Tybamate is very stable chemically (7) as a solid or as a liquid. If degradation should occur, it could be via thermal or hydrolytic cleavage of the carbamate group. As N-substitution increases, stability increases and thus tybamate would be expected to be more stable than meprobamate which has no N-substitution. It has been shown that even carbamates of the meprobamate type do not degrade readily under the above conditions.(7) It is stable in dilute acid or in dilute alkali, and is not broken down at 37°C in gastric or intestinal fluid. Hot alkali or strong acid hydrolyse tybamate to yield the corresponding diol, ammonia, butylamine and carbon dioxide.

Dosage forms of tybamate (tablets, soft gelatin capsules) have shown no loss of potency after a 5 year period at room temperature.

5. Drug Metabolic Products

Douglas, Ludwig, Schlosser, and Edelson (8) have identified 4 compounds in the urine of both dogs and cats after administration of tybamate. These are: the unchanged drug, meprobamate (I), hydroxymeprobamate (II), and the major metabolite, hydroxytybamate (III). The metabolites are shown in Figure 7.

6. Methods of Identification and Analysis

6.1 Elemental Analysis

A typical result obtained on medicinal grade tybamate using the F & M Model 185 CH & N Analyser (9) is shown below:

Element	% Theoretical	% Found
Carbon	56.90	56.51
Hydrogen	9.55	9.45
Nitrogen	10.21	10.30

6.2 Identification Tests

6.21 Derivative Formation

A crystalline derivative of tybamate with xanthydrol in glacial acetic acid can readily be prepared. This reaction is characteristic of carbamates.(10)

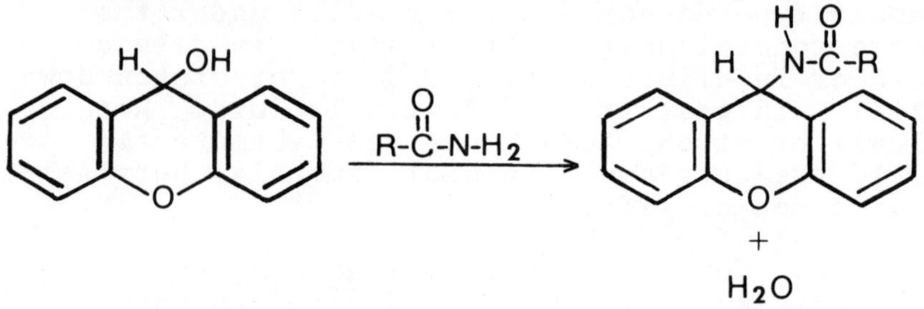

H OH
O
R-C-N-H2
H O
H N-C-R
O O
+
H_2O

R = balance of Tybamate molecule

(I) Meprobamate

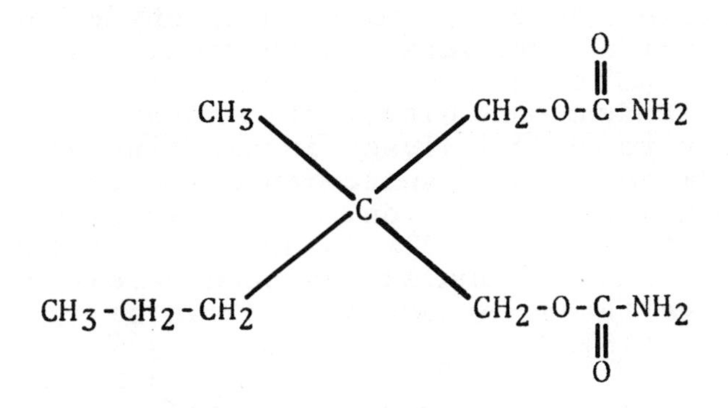

(II) Hydroxymeprobamate

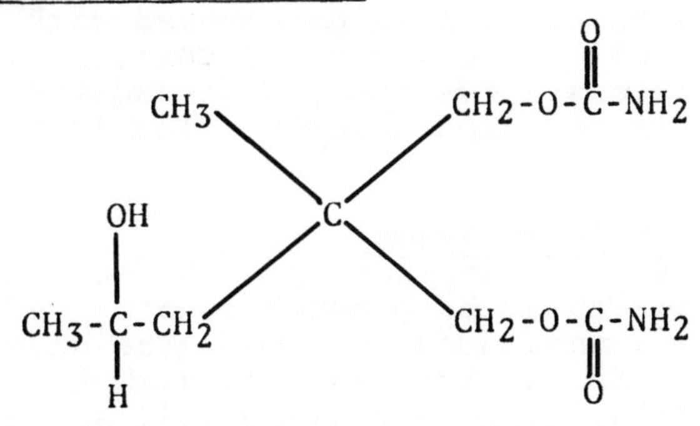

(III) Hydroxytybamate

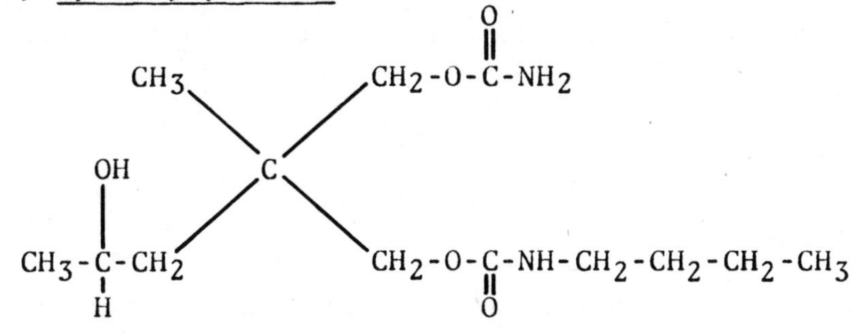

FIG. 7: METABOLIC PRODUCTS OF TYBAMATE

6.22 Colorimetric Identification Test

A red color is formed when tybamate is heated with p-dimethylaminobenzaldehyde, antimony trichloride, and acetic anhydride. (11)

6.3 Cobalt Cyanate Colorimetric Assay

Tybamate is hydrolysed in alkaline alcoholic medium to yield cyanate ion which forms a blue complex with cobalt ion. The absorption is determined at 590 nm. (12) An automated method based on the cobalt complex has been developed for N-unsubstituted carbamates and can be applied to tybamate. (13)

6.4 Hypochlorite Colorimetric Assay

At pH 10.5 tybamate reacts with hypochlorite to form an "active" chlorine derivative. The excess hypochlorite is decomposed with phenol in dilute acid. The chlorinated compound is reacted with excess potassium iodide and the liberated iodine is measured colorimetrically at 357 nm. (14)

6.5 Volumetric Analysis

The analysis depends upon the solvolysis of the unsubstituted carbamate group by sodium methoxide in a nonaqueous medium. The excess sodium methoxide is titrated with 0.1N hydrochloric acid using phenolphthalein T.S. as the indicator. (11) Cerri, et al (15) have shown that this method is specific for the non-N-substituted carbamates. They have postulated the following mechanism for the reaction:

(1)
$$R\text{-}O\text{-}\overset{\overset{\displaystyle O}{\|}}{C}\text{-}NH_2 \rightleftharpoons R\text{-}O\text{-}\overset{\overset{\displaystyle OH}{|}}{C}{=}NH$$

(2)
$$R\text{-}O\text{-}\overset{\overset{\displaystyle OH}{|}}{C}{=}NH + CH_3ONa \longrightarrow R\text{-}O\text{-}\overset{\overset{\displaystyle ONa}{|}}{C}{=}NH + CH_3OH$$

6.5 Volumetric Analysis Cont'd.

(3)

$$R\text{-}O\text{-}\underset{}{\overset{ONa}{\overset{|}{C}}}\text{=}NH \longrightarrow R\text{-}OH + NaO\text{-}C\equiv N$$

In anhydrous pyridine, the equilibrium shown in the first equation is strongly shifted to the right. The enolic species reacts with the sodium methoxide. In aqueous medium, the keto form is favored and the reaction with alkali is no longer quantitative.

6.6 Spectrophotometric Analysis

6.61 Infrared Identification

The infrared absorption spectrum of a 10% solution of tybamate in chloroform is compared with the spectrum obtained with N.F. Tybamate Reference Standard. No extraneous bands should be found. Mineral oil (Nujol) mulls have also been used for obtaining tybamate IR spectra.

6.62 Nuclear Magnetic Resonance

An NMR method was developed for the analysis of meprobamate tablets using malonic acid as an internal standard. (16) A simple adaption of this method can be used for tybamate. (17) Possible degradation of tybamate to N-butyl-2-methyl-2-propyl-3-hydroxypropyl carbamate can be detected by observing the NMR signals of the protons of the methylene groups attached to oxygen. In tybamate, there are 2 such groups and they appear as a singlet. If hydrolysis occurs, the singlet will decrease and a new signal will start appearing down field from the methylene singlet.

6.7 Gas Chromatographic Methods

6.71 Tybamate in Dosage Forms

A method for the determination of tybamate in dosage forms is made available by adaptation of the procedure of Rabinowitz, et

6.71 Tybamate in Dosage Forms Cont'd.
al.(18) This involves a simple extraction procedure, followed by chromatography on a 3.8% OV-17 column with meprobamate as the internal standard. Other methods using different internal standards have been reported.(19,20)

6.72 Tybamate in Biological Fluids

A method for the determination of tybamate in biological fluids has been reported by Douglas, et al,(21) which employs glass columns packed with 3.8% UC-W98 methyl silicone on 80-100 mesh Diataport S. A linear relationship has been established using dibutyl phthalate as the internal standard. Other carbamates are determined similarly.

6.8 Thin-Layer Chromatographic Methods

6.81 Identity and Purity of Commercial Tybamate

Ten μl of a 10% solution of tybamate in chloroform is spotted on a silica gel plate. The chromatogram is developed with a mixture of chloroform and acetone (4:1), the plate is air-dried for several minutes, sprayed with a saturated solution of antimony trichloride in chloroform and then with a 3% solution of redistilled furfural in chloroform. After 10 to 15 minutes dark spots are visible on a white to grey background which slowly darkens. Pure tybamate shows only one spot. Commercial material may have trace quantities of meprobamate and 2-methyl-2-propyl-3-hydroxypropyl carbamate with R_f values of 0.36 and 0.51 relative to tybamate. (22)

6.82 Identification in Presence of Other Drugs

Several other TLC methods are available for the identification of tybamate in the presence of other psychotropic drugs in blood or urine. Positive identification has been obtained by the use of 5 different chromatographic systems. (19,20)

7. Pharmacokinetics

Peak serum concentrations of tybamate were obtained about 1 hour after oral administration. The major metabolite found in the urine was hydroxytybamate. Serum levels of tybamate were determined in male mongrel dogs by the colorimetric method of Hoffman and Ludwig (25) after oral administration of 15 mg./Kg. Metabolic studies were carried out after tybamate, 100 mg./Kg., was given daily for 5 days. The distribution of tybamate in rat tissue was observed after oral administration of 3 to 7 mg. of carbon-14 labeled drug. Tybamate was rapidly absorbed; a maximum blood serum level in dogs of about 10 mcg./ml. occurred within 1 hour after administration. Serum half-life was about 3 hours and no drug was detectable after 24 hours. Neither tybamate nor its metabolites were retained by rat tissue, but were excreted almost entirely in the urine during a 24 hour period. (5)

8. References

1. A. Z. Hayden, W. L. Brannen, and C. A. Yaciuw, J. Ass. Offic. Agr. Chem., 49, 1109 (1966).
2. L. J. Bellamy, "The Infrared Spectra of Complex Molecules", 2nd ed., John Wiley and Sons, Inc., New York, N.Y., 1964, chap. 12.
3. K. Florey and P. T. Funke, Squibb Institute for Medical Research, personal communication.
4. R. T. Coutts, J. Pharm. Sci., 62, 769(1973).
5. F. M. Berger and B. J. Ludwig, U. S.Patent 2,724,720 (1955).
6. B. J. Ludwig and E. C. Piech, J. Amer. Chem. Soc., 73, 5779 (1951).
7. P. Adams and F. A. Baron, Chem. Revs., 65, 567 (1965).
8. J. F. Douglas, B. J. Ludwig, A. Schlosser, and J. Edelson, Biochemical Pharmacology, 15, 2087, (1966).
9. I. A. Brenner, Carter-Wallace, Inc., unpublished data.
10. E. F. Salim, J. I. Bodin, H. B. Zimmerman, and P. Reisberg, J. Pharm. Sci., 55, 1439 (1966).
11. "The National Formulary", 13th ed., Mack Publishing Co., Easton, Pa., 1970, p.749.
12. G. Devaux, P. Mesnard, and J. Cren, Prod. Pharm., 18, 221 (1963).
13. L. F. Cullen, L. J. Heckman, and G. J. Papariello, J. Pharm. Sci., 58, 1537 (1969).
14. G. H. Ellis and C. A. Hetzel, Anal. Chem., 31, 1090 (1959).
15. O. Cerri, A. Spialtini, and U. Gallo, Pharm. Acta Helv., 34, 13 (1959)
16. J. W. Turczan and T. C. Kram, J. Pharm.Sci. 56, 1643 (1967).
17. I. A. Brenner and P. Reisberg, Carter-Wallace, Inc., unpublished data.
18. M. P. Rabinowitz, P. Reisberg, and J. I. Bodin, J. Pharm. Sci., 61, 1974 (1972).
19. R. C. Grant, F.D.A. By-Lines, 2 (1): 10-14, July (1971).

20. C. Cardini, V. Quercia, and A. Calo, J. Chromatogr., 37, 190 (1968).
21. J. F. Douglas, N. B. Smith, and J. A. Stockage, J. Pharm. Sci., 58, 145 (1969).
22. A. E. Martin, A. H. Robbins Co., personal communication.
23. I. Zingales, J. Chromatogr., 31, 405 (1967).
24. T. W. McConnell, J. Chromatogr., 29, 283 (1967).
25. A. J. Hoffman and B. J. Ludwig, J. Amer. Pharm. Ass., Sci Ed., 48, 740 (1959).

General References

1. B.J. Ludwig, L.S. Powell, and F.M. Berger, J. Med. Chem., 12, 462 (1969).
2. B.J. Ludwig and J.R. Potterfield, Adv. Pharmacology and Chemotherapy, 9, 173, (1971).

9. Acknowledgements

The authors wish to express thanks to Dr. K. Florey and Dr. P.T. Funke of the Squibb Institute for Medical Research for providing the mass spectral data, to Drs. B.J. Ludwig and D. Reisner of Wallace Laboratories for suggesting improvements in the manuscript, and to Mr. I.A. Brenner, Miss R. Rader, and Mrs. M. Belason for assistance in its preparation.

ADDENDA

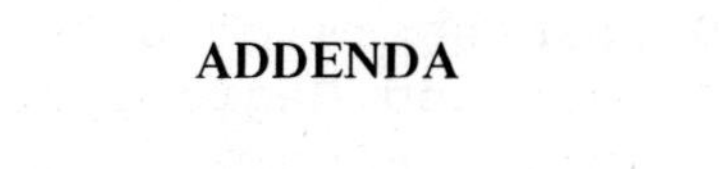

Ampicillin

6.21 Spectrophotometric determination of ampicillin sodium in the presence of its degradation and polymerization products. H. Bundgood, J. Pharm. Pharmac. 26, 385(1974).

Chloramphenicol

3. Biosynthesis of Chloramphenicol. Origin and Degradation of the Aromatic Ring. W. P. O'Neill, R. F. Nystrom, K. L. Rinehart Jr. and D. Gottlieb, Biochemistry 12, 4775(1973).

Chlordiazepoxide Hydrochloride

6.6 Determination of Chlordiazepoxide Hydrochloride and its Major Metabolites in Plasma by Differential Pulse Polarography. M. R. Hackman, M. A. Brooks, J. A. F. de Silva and T. S. Ma, Anal. Chem. 46,1075(1974).

Diazepam

6.33 Quantitative Determination of Medazepam, Diazepam and Nitiazepam in Whole Blood by Flame-Ionization Gas-Liquid Chromatography. M. S. Graeves, Clin. Chem. 20, 141(1974).

Dexamethasone

5. Radioimmunoassay for Dexamethasone in Plasma.
M. Hichens and A. F. Hogans, Clin. Chem. 20,266(1974).

Fluphenazine Hydrochloride

5. Metabolism in rats; in vitro and urinary metabolites.
H. J. Gaertner, U. Breyer and G. Liomin, Biochem.Pharmac. 23, 303(1974).

Formation of identical metabolites from piperazine and dimethylamino-substituted phenothiazine drugs in man, rat and dog.
U. Breyer, H. J. Gaertner and A. Prox, Biochem. Pharmac. 23,313 (1974).

6.72 Thinlayer chromatography of phenothiazine derivative and analogues.
A. deLeenheer, J. Chromatog. 75, 79(1973).

6.74 Gas-liquid chromatographic analysis of fluphenazine and fluphenazine sulfoxide in the urine of chronic schizophrenic patients.
M. I. Kelsey, A. Keskiner, E. A. Moscatelli, J. Chromatog. 75,294(1973).

Meprobamate

6.63 GLC Determination of Meprobamate in Water, Plasma and Urine. L. Mortis and R. H. Levy, J. Pharm. Sci. 63,834(1974).

Methadone Hydrochloride

5.4 Biliary excretion of methadone by the rat; identification of a para-hydroxylated major metabolite. R. C. Baselt and M. H. Bickel, Biochem. Pharmac. 22,3117(1973).

Methaqualone

5. Identification of Free and Conjugated Metabolites of Methaqualone by Gas Chromatography-Mass Spectrometry R. Bonnichsem, Y. Marde and R. Ryhage, Clin. Chem. 20,230(1974).

Qualitative and Quantitative Determination of Methaqualone in Serum by Gas Chromatography. M. A. Evenson and G. L. Lensmeyer, Clin. Chem. 20,249(1974).

Propoxyphene Hydrochloride

4. Fluorometric Determination of Propoxyphene. J. C. Valentour, J. R. Monforte and I. Sunshine, Clin. Chem. 20, 275(1974).

Sulfamethoxazole

5. Modification of an Automated Method for Measurement of Sulfamethoxazole and its Major Metabolite in Biological Fluids.
A. Bye and A. F. J. Fox, Clin.Chem. 20,288(1974).

Triamcinolone

2.4 Mass spectrometry of some corticosteroids and related compounds of pharmaceutical interest. P. Taft, B. A. Lodge and M. B. Simard, Can. J. Pharm.Sci. 7,53(1972).

Triamcinolone Acetonide

5. 6β-Hydroxylation of trimcinolone acetonide by a hepatic enzyme system.
D. Kupfer and D. Partridge, Arch. Biochem. Biophys., 140, 23 (1970).

Triflupromazine Hydrochloride

5. Formation of identical metabolites from piperazine and dimethylamino-substituted phenothaizine drugs in man, rat and dog.
U. Breyer, H. J. Gaertner and A. Prox, Biochem. Pharmac. 23,313 (1974).

6.72 Thinlayer Chromatography of phenothiazine derivatives and analogues.
A. deLeenheer, J. Chromatog. 75 (1973).

ERRATA

Fluphenazine Enanthate

6.2 Vol. II p. 258 Neutralization equivalent: calc. 275. not 285.

Triamcinolone

1.1 Vol. I p. 369 Molecular weight: 394.45 not 434.49

CUMULATIVE INDEX

Italic numerals refer to Volume numbers.

A 5
B 6
C 7
D 8
E 9
F 0
G 1
H 2
I 3
J 4